TECHNOLOGY

Engineering Our World
Seventh Edition

John B. Gradwell
McGill University

Malcolm Welch
Queen's University

Publisher
The Goodheart-Willcox Company
Tinley Park, Illinois
www.g-w.com

The Goodheart-Willcox Company, Inc. Brand Disclaimer: Brand names, company names, and illustrations for products and services included in this text are provided for educational purposes only and do not represent or imply endorsement or recommendation by the author or the publisher.

The Goodheart-Willcox Company, Inc. Safety Notice: The reader is expressly advised to carefully read, understand, and apply all safety precautions and warnings described in this book or that might also be indicated in undertaking the activities and exercises described herein to minimize risk of personal injury or injury to others. Common sense and good judgment should also be exercised and applied to help avoid all potential hazards. The reader should always refer to the appropriate manufacturer's technical information, directions, and recommendations; then proceed with care to follow specific equipment operating instructions. The reader should understand these notices and cautions are not exhaustive.

The publisher makes no warranty or representation whatsoever, either expressed or implied, including but not limited to equipment, procedures, and applications described or referred to herein, their quality, performance, merchantability, or fitness for a particular purpose. The publisher assumes no responsibility for any changes, errors, or omissions in this book. The publisher specifically disclaims any liability whatsoever, including any direct, indirect, incidental, consequential, special, or exemplary damages resulting, in whole or in part, from the reader's use or reliance upon the information, instructions, procedures, warnings, cautions, applications, or other matter contained in this book. The publisher assumes no responsibility for the activities of the reader.

Cover Source: Shutterstock (rf)

Library of Congress Cataloging-in-Publication Data

Gradwell, John B.

   Technology : engineering our world / by John B. Gradwell, Malcolm Welch. -- 7th ed.

      p. cm.

      Rev. ed. of Technology : shaping our world / John B. Gradwell, Malcolm Welch,
      Eugene Martin. 2004.

      Includes index.

   ISBN 978-1-60525-428-9

   1. Technology. 2. Engineering. I. Welch, Malcolm. II. Title.

   T47.G69 2012

   600—dc22

Preface

Technology: Engineering Our World is written to help you understand and contribute to the technological world around you. This book introduces you to the various technologies and shows how they have used basic scientific principles.

First you will study various methods of solving problems using a design process. Various chapters will introduce the major technologies: communication, manufacturing, construction, energy and power, transportation, medical, and agricultural technologies, as well as biotechnology. The importance of materials, structures, systems, energy, people, machines, and information to all technological activity will become clear as you study. You will also learn about job opportunities in technology.

The technology of our earliest ancestors was very basic. Often, new developments affected only a small group of people. Because travel was limited, a new plow might be developed and used in only one village. Even so, life was improved for that village through greater crop yields.

Technology today is usually more complicated. A technological decision might affect a whole city, a continent, or even the entire planet. Millions of people have experienced great changes in their lives as a result of new technology. Consider the technological products and services that affect your life. These may include computers, medicines, composites, microelectronics, biotechnology, and telecommunications, among others. In the near future, you are sure to see new inventions unheard of today.

It is very important that you begin to understand technology so that you can make intelligent decisions about its use. While a new product or service may improve your life, you must consider its side effects. With the development of new technology comes the responsibility to control that technology so that no harm is done to humans, animals, or the environment. Your understanding of technology will help you care for our Earth and its people. In this way you will help engineer our world. Your educated decisions will help provide a healthier environment.

Technology is an exciting subject. It affects what you wear, how you travel, what you eat, how you communicate with others, the comfort of your home, and much more. Technology is improving at an incredibly fast pace and expanding to reach into new areas every day. This book is written to help you become more informed about and involved in technology and the world that is daily being shaped by it.

John Gradwell

Malcolm Welch

4

Bamboo is very suitable for casual clothing. What other clothing, recreational, and home products could be made using bamboo?

Bonnie Siefers

Bonnie Siefers is an American fashion designer and eco-designer who believes that clothing must look chic and feel luxurious. But Bonnie also believes that fashion designers have a responsibility toward the environment and must begin to use certified organic or eco-friendly fabrics (ecotextiles). So Bonnie designs apparel made from ecoKashmere®, a soft, silky fabric that incorporates the pulp of bamboo grass. The result is clothes that are soft and flowing. Also, because bamboo absorbs water rapidly, it produces fabrics that "wick" moisture at twice the rate of cotton. Bamboo-based fabrics are also highly breathable, anti-static, and naturally antibacterial, making them helpful to people with sensitive or allergy-prone skin.

Ecotextiles such as bamboo, flax, cotton, and jute are renewable, biodegradable, and sustainable.

"A pure, natural environment is vital to children of all ages. Organic [textiles] are not only gentle on the skin, but also safer for the people who make the clothes, for the farmers who grow the crops, and for the environment." —Bonnie Siefers

Better by Design feature highlights designers who have made a difference in the area of technology explored in each chapter.

Reading Target, along with the **Reading Target organizer** at the end of each chapter, identifies strategies for reading the chapter material.

Key Terms list identifies new vocabulary in each chapter, enhancing student recognition of important concepts.

Objectives identify goals to be achieved by students.

Summarizing Information

A *summary* is a short paragraph that describes the main idea of a selection of text. Making a summary can help you remember what you read. As you read each section of this chapter, think about the main ideas presented. Then use the Reading Target graphic organizer at the end of the chapter to summarize the chapter content.

adhesion	heat joining
bending	jig
casting	laminating
chemical joining	marking out
chiseling	mechanical joining
coatings	molding
cohesion	planing
development	sawing
drilling	shearing
filing	volatile organic compounds
finishing	(VOCs)
forming	

After reading this chapter, you will be able to:
- Demonstrate responsible and safe work attitudes and habits.
- Design and shape a product using the correct tools and processes.
- Select the correct materials and methods for joining materials.
- Recall methods for applying a finish to a material.

Useful Web sites:
www.jonano.com
www.bambooclothing.co.uk/why_is_bamboo_better.html

Think Green features help students become aware of environmental concerns and encourage them to make environmentally friendly decisions.

Think Green
Sustainable Design

As you design a new product, be sure to consider its environmental impact. A *sustainable design* is one that has little or no negative impact on the environment and society. To be sustainable, a design should be:

- Constructed of renewable natural resources.
- Free of toxic chemicals and other items that cause damage to the environment or to human, plant, or animal life.
- Easily recyclable or reusable indefinitely, or be biodegradable in common landfill conditions.

Sustainable design can be applied to small items we use every day, as well as large projects such as community planning. However, for best results, sustainability should be considered at the design stage.

Developing Alternative Solutions

After you have gathered information, you are ready to develop your own designs. It is very important that you write or draw every idea on paper as it occurs to you. Making these design sketches is an excellent way to remember, explore, develop, and communicate ideas. **Figure 2-8** shows some solutions one designer identified for the pencil holder design problem.

You may find that you like several of the solutions. Eventually, you must choose one. Usually, careful comparison with the original design brief will help you select the best. You must also consider:

- Your own skills
- The materials available
- Time needed to build each solution
- Cost of each solution

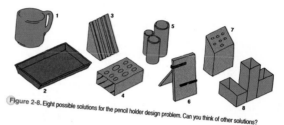

Figure 2-8. Eight possible solutions for the pencil holder design problem. Can you think of other solutions?

Concurrent Engineering

Concurrent engineering is a team effort that involves continuous communication among the entire design and production team from the very beginning of the design stage. The customer, project manager, and marketing staff also join in the process. Changes by any of these people are immediately passed on to others. Because of increased communication, concurrent engineering saves time. The product is produced quickly and meets the customer's needs.

3D Printing

Today, many industries use some form of *3D printing*. This term originally referred to small machines that could create plastic parts from a CAD file. Now, however, it can mean any system that uses *additive fabrication* to create a part from a CAD file. *Additive fabrication* is a long term that means the part is built by adding layer upon layer of plastic or metal powder. Design information is sent directly to the 3D printer. Here the data is numerically sliced into thin layers. The 3D printer then creates each two-dimensional cross section using a liquid, powder, or sheet material and bonds it to the previous layer. A complete part is built by stacking layer upon layer until the part is completed.

Specific 3D printing systems are often known by their output. For example, 3D printers that are used to create prototypes quickly during the design stage of product development are called *rapid prototyping* (RP) *systems*. See **Figure 3-37**. By using prototypes in real-life field tests, designers can better evaluate a product's strengths and weaknesses and avoid costly mistakes.

New Terms appear in bold italic green type where they are defined.

Figure 3-37. A student with her CAD file and 3D model of a three-dimensional puzzle she designed for the blind.

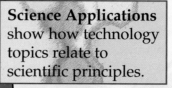

Science Applications show how technology topics relate to scientific principles.

Science Application

Chemical Symbols

There are just over 100 known substances, or elements, in the world. An *element* is the simplest form of matter and is made up of only one kind of atom or molecule. Each element is given a chemical symbol of one or two letters. Chemists, scientists, and technologists throughout the world use the same symbols, so there is never any need for translation.

For example, the letter H is the symbol for hydrogen, and the letter O is used for oxygen. These substances cannot be separated

Technology Applications expand on the use of technology related to chapter content.

Technology Application

Art, Nature, and Technology

The shapes and patterns of technological designs are often inspired by nature. Many patterns are based on shapes that we know quite well: leaves, flowers, birds, and animals. If you look around, you may find examples of sunflower tiles on a floor, leaf patterns on a carpet, sofas and fabrics decorated with wild grasses, or a necklace of ivory pieces, each carved in the shape of a bird. Usually, the patterns are not exact copies, but abstract versions that were inspired by the objects. For example, the design for the faucet shown below may have been inspired

Engineering Applications help students understand how engineering principles relate to technology.

Engineering Application

Testing Truss Strength

The various designs of trussed bridges are named after the engineers who first designed them. James Warren, an engineer, patented the first bridge that used equilateral triangles in 1848. An *equilateral triangle* is one in which all sides are the same length and all angles are the same. Today we use trusses more often in roofs of buildings than for bridges, but the reasons for using them are the same. They span distances of 10 to 100 feet (3 to 30 meters) using a minimum amount of material while providing the maximum amount of strength.

Math Applications explain the math needed to work with various designs and technologies.

Math Application

Using Mixed Fractions to Calculate Material Needed

Before manufacturers begin shaping materials, they have to decide how much material will be needed for a given production run. For example, a manufacturer of metal nameplates must determine how many nameplates can be made per linear foot of metal. This often involves working with mixed fractions. A *mixed fraction* is a number that includes both a whole number and a fraction, such as 1½.

To multiply a whole number by a mixed fraction, first change the mixed number to an improper fraction. To create an *improper fraction*, multiply the whole number part of a mixed fraction by the denominator of the fraction and add the result to the numerator. Place this number over the denominator to complete the fraction. For example, to convert 3½ to an improper fraction, multiply 3 by 2 and add the result to the 1: 3 × 2 = 6; 6 + 1 = 7. The improper fraction is written as ⁷⁄₂.

After you have created the improper fractions, multiply the numerators. Then multiply the denominators. Reduce the result to lowest terms. Finally, divide the denominator into the numerator to reduce the fraction to a mixed fraction.

For example, suppose the manufacturer has an order for 50 copper nameplates. Each nameplate will be 7½″ × 1¾″. If the nameplates are blanked (stamped out) of a roll of copper that is exactly 7½″ wide, what length of copper will be needed? (Note: The improper fraction for any whole number is that number over a denominator of 1.)

Problem statement: 50 × 1¾

Convert to improper fractions: 50⁄1 × 7⁄4

Multiply: 50⁄1 × 7⁄4 = 50 × 7⁄1 × 4 = 350⁄4

Divide: 350 ÷ 4 = 87, with a remainder of 2

Reduce: 87 2⁄4 = 87 1⁄2

Math Activity

A door manufacturer needs 150 door strikes (Figure A) measuring 2¼″ × 1¾″. These will be blanked from a roll of brass, using the most economical cutting pattern. A space of ⅛″ is needed between each piece. See Figure B for examples of cutting patterns. Determine the length of brass needed if the roll of brass is:

A. 2″ wide
B. 2½″ wide

All **STEM** application features include activities to strengthen student skill in these areas.

End Note takes a quick look at the past, present, and future of technology related to the chapter.

Reading Target organizer helps students apply the reading technique suggested at the beginning of the chapter.

Chapter 5 Processing Materials 167

Tools are used to shape and form materials into finished products. When designing and making a product, you must know how to select and use both hand and machine tools safely. In addition to shaping and ... need to join materials either temporarily or ... joined using mechanical devices, chemicals, ... cessing or coating the surface of the product. ... hnologies are developed, new processes are ... m. Manufacturers are looking for replacements ... the past and present, such as chromium and ... ay lie in nanotechnology. For example, nano- ... ied to finishing processes. A nanocrystalline cobalt and phosphorus compound has been developed to replace chromium coatings in some cases. Other products and processes are also under development to reduce carbon emissions and the use of hazardous chemicals.

○ Working safely with hand and power tools requires careful planning and attention to detail.
○ To create a product, materials are first marked out. Then they are ... ing one of several processes. Examples include sawing, ... ing, shearing, chiseling, drilling, bending, forming, ... d molding.
... an be joined using mechanical fasteners or other ... methods, chemicals, or heat.
... re finished to protect their surfaces or to improve their

Summarizing Information
 Copy the following graphic organizer onto a separate sheet of paper. For each chapter section (topic) listed in the left column, write a short, one-paragraph summary of the topic in the right column.

Chapter Section (Topic)	Summary
Learning to Work Safely	
Shaping Materials	
Joining Materials	
Finishing Materials	

90 Technology: Engineering Our World

Test Your Knowledge

Write your answers to these review questions on a separate sheet of paper.
1. Name the three basic types of communication.
2. List at least three different forms of communication.
3. What is freehand sketching?
... sides of a rectangular block

9. What is the alphabet of line...
10. You have been asked to cre... of a printer that measures... If you create the drawing... how long will the overall... and height lines be on yo...
11. List the advantages of c... design.
12. What is virtual reality?
...plain the concept of... engineering.
...hat is a 3D printer?
...hat is the differen... rototyping and rap... ystems?

Test Your Knowledge questions help students gauge their understanding of chapter topics.

Critical Thinking

1. Listen to a recording of whale or dolphin calls. Write an essay comparing and ... human speech.

3. Analyze the re... drafting techn... years. Prepare... following qu... trends helped... factors may... Given the tr... innovative... expect to b... future?
4. Write a p... how the... technolog... math an...

Critical Thinking questions encourage students to think about issues related to chapter content.

Chapter 3 Communicating Design Ideas 91

Apply Your Knowledge

1. Create a symbol that can be used at a zoo to communicate the message: "Do not feed the animals!"
2. Design a logo (name or symbol) that you could use on your own personalized worksheets. Letters, geometric shapes, natural shapes, and simplified pictures are most appropriate for a logo.
3. Draw an isometric sketch, a perspective sketch, and an orthographic projection of a toothbrush.
4. Use a CAD system to make an isometric sketch of a tool you have used in the technology lab.
5. International Morse Code is a means of communication invented before telephone and e-mail. Check the symbols used. Then design a series of symbols that could be used to send a message by e-mail without using letters or words.
6. Research one career related to the information you have studied in this chapter. Create a report that states the following:
 • The occupation you selected
 • The education requirements to enter this occupation
 • The possibilities for promotion to a higher level

Apply Your Knowledge activities allow students to apply skills and knowledge they have learned in the chapter.

STEM Applications

1. **TECHNOLOGY** In a group of four students, create a message to be transmitted to the rest of your class...... for ways to commu... Develop as many of... Use both electronic a... methods. Present yo... class in at least five c...
2. **ENGINEERING** In a gr... students, brainstorm... and futuristic commu... that may help solve c... human needs. Select... present it to the class...

3. **MATH** Select an object you use...

STEM Applications activities encourage students to integrate science, technology, engineering, and math principles related to chapter content.

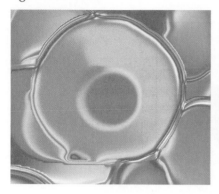

Contents in Brief

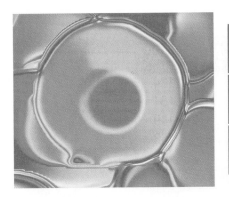

Contents

10

Think **Green**

STEM Applications

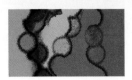

Technology and You—An Introduction

Better by Design

Designers make life better

The Better by Design page in each chapter features designers who have developed products to make people's lives better. Study the products presented in each chapter. Have these designers succeeded?

Compare and Contrast

As you read this chapter, make a list of tasks you perform every day that require the use of technology. Think about how your life would be different without this technology. Then use the Reading Target graphic organizer at the end of the chapter to organize your thoughts about what you have read.

designing
engineer
engineering
green technology
hypothesis
science
scientific method
scientist
technologist
technology

After reading this chapter, you will be able to:

- Describe how technology affects daily life.
- Recognize the broad range of technological products.
- Recall examples of technology in various historical periods.
- Explain the relationship between technology and science.
- Describe the benefits and disadvantages of technology.

Useful Web sites:
www.inventors.about.com/od/timelines/Timelines_of_Invention_and_Technology.htm
www.historyoftechnology.org/
www.scientificmethod.com/

Designing and making products has long been one of the most important human activities. See **Figure 1-1** and **Figure 1-2.** In the last 100 years, people have invented air conditioning, the electric guitar, television, jet planes, the microwave oven, cell phones, digital cameras, e-mail, and the Internet. Technology has completely changed the way we live. As new products are invented, they change society and create new needs and wants. The products people design and make are often good solutions to problems. Sometimes, however, the solutions are barely adequate, and other times they fail entirely.

Understanding Technology

Designing is the creative process for generating and developing ideas for new and improved products and services that satisfy people's needs. *Technology* includes the tools, materials, and processes used to make the designed products. These products allow us to do things we could not do without technology. They help us lead more comfortable and productive lives. When new products are developed to solve specific problems, the process is often called *engineering*.

Figure 1-1. Well-designed products focus on the user. Explain why one bottle is a better design than the other.

Figure 1-2. Function is another important design consideration. Look closely at the bottoms of these cups. Why is one cup a better design than the other?

One Day without Technology

Imagine that you are living thousands of years ago. You wake up when the sun rises. There are no alarm clocks to ring. You would crawl out of your bed—a pile of animal skins spread over branches cut from trees. Your animal skin clothing hangs loosely around your body. There are no zippers, Velcro™, or even strings to fasten them.

Leaving your cave, you find that a pile of fallen rocks has partially blocked the entrance. You move the rocks by hand. There are no carts, and the wheel has not yet been invented.

What about breakfast? Want to make waffles or pour a bowl of cereal? No chance! There might be a leftover bone from yesterday's kill, but no microwave oven to reheat it.

If your mouth still tastes of breakfast, too bad. Toothbrushes and toothpaste do not exist. You scrape congealed sap off the bark of a tree and chew it to freshen your breath. The day is warm and sunny, and you feel like taking the day off. Unfortunately, you can't if you want to eat. Your entire family spends most of each day collecting enough food to survive.

Are you starting to appreciate the comforts and conveniences provided by modern technology? Technology plays a crucial role in our lives. Almost everything we do depends in some way on the products and services that form our technological society. These products and services are frequently the result of very complex technologies. Often we take them for granted. It is only when we stop to consider what we do each day that we realize how important technology is to us.

One Day with Technology

How important is technology to you? How much do you depend on it? Think carefully about some of the items you encounter during a typical day. See **Figure 1-3**.

Your day might begin when your digital alarm clock/radio comes on. The music you hear is transmitted (sent) from a station a distance away. You stumble out of bed into a hot shower. The flow and temperature of the water can be adjusted. You style your hair and clean your teeth with plastic brushes. Your clothes are made from a mixture of natural and synthetic fibers.

In the kitchen, **Figure 1-4**, you prepare your breakfast using a variety of appliances. Toast is ejected from the toaster. The electric kettle shuts off when the water boils. A microwave oven cooks your bacon for a preselected period of time.

Before leaving the house, you dress in clothing made from waterproof synthetic fibers. As you leave, the timer built into the thermostat automatically lowers your home's temperature. If you live in an apartment building, you press a button to summon an elevator. Pressing a second button takes you to the ground floor. Once there, the doors open automatically. You leave the building through a front door controlled by an electromagnetic lock.

You may use a bicycle, bus or subway train to travel to school. See **Figure 1-5**. Your classroom uses electric lighting and may be heated by natural gas. The tables, desks, chairs, and cabinets have been made in a factory and transported to the school. The room is probably equipped with a computer, data projector, DVD player/recorder, and public address system.

Normally, we give little thought to these technological products and services on which we rely. However, think about them for a moment. How do they influence your life and the environment in which you live?

Figure 1-3. Even within the last 10 years, technology has changed the way we live and work. What products do you use for communication throughout your day?

Figure 1-4. Kitchen appliances help us prepare meals faster and with less effort. What products do you use to prepare food throughout your day?

Figure 1-5. Transportation technology makes it possible to live, go to school, and work farther away from home. What products and services do you use for transportation throughout your day?

Are you starting to recognize your dependency on the products of technology? Today, most people lead very "technological" lives. You have seen that, even during the first few hours of your day, technology is basic to your comfort and way of life. Are there any activities that do not depend on technology?

Technology's Effect on Health

Your health depends on medical technology. For example, dentists use a variety of tools and equipment to repair, replace, straighten, and keep your teeth in the best possible condition. See **Figure 1-6.** Surgeons replace damaged heart valves with valves made of metal and plastic. Diabetics can wear a tiny, computer-controlled infusion pump. The pump delivers a timed supply of insulin to the wearer. See **Figure 1-7.** Such advances in medical care help us lead healthier and happier lives.

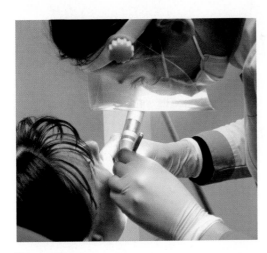

Figure 1-6. Most people pay an occasional visit to the dentist. Technology also provides other health care products you use every day. What products and services do you use to maintain your health throughout your day?

Figure 1-7. Some advances in technology allow people to manage diseases more easily. What does the designer need to think about when designing new medical equipment?

Technology for Leisure

Even in your leisure time, the products of technology surround you. Using a cell phone with wireless Internet access, you may decide to check movie listings, verify bank funds, or buy gifts, all while traveling home on a bus. Computer-designed tennis racquets made of fiberglass, graphite, or ceramic have replaced the older laminated wood. Recumbent (lying back) road-racing bicycle frames like the one in **Figure 1-8** are made of epoxy-glued aluminum or carbon fiber. They have a mass of less than half the average steel bicycle frame. Jogging and ski suits are made of fabrics that keep out wind, rain, and snow, yet allow perspiration vapor to escape.

Figure 1-8. This recumbent bicycle is designed for speed. What factors reduce its wind resistance?

A Brief History of Technology

Before humans designed and made technological products, their world must have appeared to be very small. For example, their view was limited to the height of the trees they could climb. Travel was limited by the distance their legs could carry them. Communication was limited by the distance their voices could carry.

Then, about two million years ago, humans made the first tools. They discovered that when a large pebble is struck with great force against another stone, pieces flake off. The sharp cutting edges formed in this manner could be used in axes, clubs, spears, and scrapers. See **Figure 1-9**.

Early hunters followed bison and mammoth across the frozen Bering Strait into North America, where they settled the Great Plains nearly 15,000 years ago. They were nomadic, obtaining their food by hunting and gathering. In 6000 B.C., they developed agriculture and domesticated animals. Stone tools became highly refined. About 2000 B.C., stone tools gradually gave way to tools cast in copper and bronze. See **Figure 1-10**. Later still, about 1000 B.C., iron began to replace copper and bronze. Iron was repeatedly heated and hammered to make ornaments, tools, and weapons.

Simple pulleys and levers were first used in Egypt and Mesopotamia. Later, Greek and Roman engineers developed these machines further.

In the period AD 1–1399, many of the technological products with which we are familiar were first developed, including clocks, pants, porcelain drinking cups, skates, and pens.

Beginning around 1450, a great explosion of scientific and technological creativity took place in Europe. This came about partly due to Gutenberg's use of movable type in printing that made books more readily available. At around the same time, it became acceptable to use everyday languages, rather than Latin, for writing and communicating ideas.

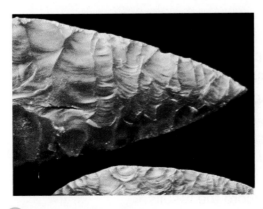

Figure 1-9. A spear point made from flint. Chipping away stone from both sides shaped the edges.

Figure 1-10. About 4000 years ago, humans learned how to cast tools from metal. These are examples of their craft.

Science Application

The Scientific Method

Sir Francis Bacon developed a method of inquiry in 1620 that is now known as the *scientific method*. This method has influenced every scientist since its publication. In the scientific method, scientists collect masses of facts by observing and experimenting. Then they analyze the facts and state a *hypothesis*—a guess or suggestion to account for the facts. Finally, they try to verify the hypothesis by further investigation, eventually forming a theory.

The scientific method can be applied to just about any type of technology. For example, you could use the scientific method to develop and attempt to verify a theory about something related to cell phone use. Have you observed how some people seem to act differently when they acquire a cell phone? Cell phones can cause people to change their habits in a number of ways:

- Freedom for the user (a person is not tied to a landline)
- Time and space (contact can be made at any time from almost any place)
- Sense of community (groups of "cyber friends" can extend connections)
- The way people interact with friends (there is less face-to-face communication)
- Sense of privacy (phone conversations happen in public and private spaces)

Science Activity

Follow these steps to create and test a theory using the scientific method:

1. Decide on a topic that is specific to cell phone use. Choose something that you have seen, but you are not sure if the actions of a few people are typical of the general population. What have you observed related to cell phone use?

2. Make a hypothesis that could explain what you think is generally happening.

3. Decide how you will organize a controlled experiment so as to reach a conclusion. Where will you find the information you want? How will you record it?

4. Carry out your experiment and record the results.

5. Analyze your results. Were you able to verify your hypothesis? If not, don't be concerned. It sometimes takes scientists many rounds of hypotheses and testing to prove or disprove a theory.

In the 1700s, the number of inventions increased dramatically. Many of these inventions were machines run by a new power source, the steam engine. This led to the Industrial Revolution. Applied to railways and ships, steam power completely changed transportation.

The start of the twentieth century also marked the first time most people saw or used automobiles, airplanes, telephones, radios, and electricity. The rate at which inventions appeared continued to increase. By the end of the twentieth century, inventions included nuclear reactors, synthetic fibers and industrial robots, suspension bridges and hydroelectric dams, telephones with storage memory, compact discs, microwave ovens, digital sound systems, portable computers, DVD players, and space shuttles.

Some of the most important inventions are summarized in **Appendix A**. What conclusion can you draw about the rate at which inventions have occurred? Remember that human evolution has developed over about 4 million years or 50,000 generations, but television and computers have been around for only 2 generations. What might happen in the next 10 generations?

How Science and Technology Are Related

Science and technology are often confused with one another, but they are very different. *Science* is the study of the laws of nature. *Scientists* (such as biologists, chemists, physicists, and oceanographers) try to discover and understand the natural world. See **Figure 1-11**. While some scientists gaze into microscopes, others record the songs of dolphins, look for undiscovered stars or planets, or piece together dinosaur skeletons. Scientists ask the questions: Why is the natural world as we see it? How did it come to be?

Technology, on the other hand, is concerned with the human-designed world. *Technologists* (designers, inventors, *engineers*, and craftspeople) design and make products and services that improve the designed world and fulfill human needs. See **Figure 1-12**. The products include everything from kitchen knives to jumbo jets. They range from the products we all use, such as toothbrushes, to products such as a dentist's drill that few people use. Services also range from small to large. A service may be as small as giving instructions for assembling furniture or as large as maintaining an airport system.

People often assume that scientific knowledge came before technology. This is not true. Many technological advances came before the understanding of the principles behind them. For example, the wheel and axle were used long before humans understood the physical principles governing levers.

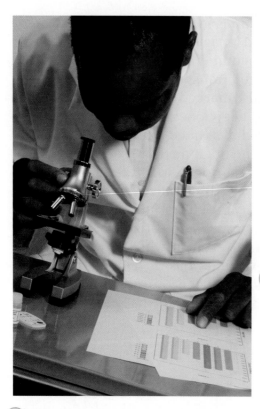

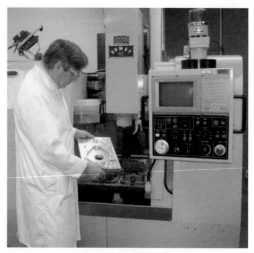

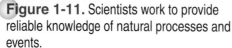

Figure 1-12. Technologists use tools and materials to make products that improve people's lives.

Figure 1-11. Scientists work to provide reliable knowledge of natural processes and events.

These examples will help you to understand the relationship between technology and science:

- Technologists invented and built the early telescopes. Scientists used these telescopes to observe and calculate the distance from Earth to the planets. In turn, these scientific observations were used by technologists in the design of space vehicles.

- Scientists study the flow and formation of rivers. Technologists design and build dams across rivers.

- Technologists built the first steam engines. Scientists studied these engines to develop the laws of thermodynamics.

- Scientists study the causes and control of diabetes. Technologists design and build portable, computer-controlled insulin pumps for diabetics.

- Technologists shape glass into tubes, bottles, and flasks. Scientists use these objects in experiments to analyze the chemical composition of substances.

- Scientists study atomic theory. Technologists use atomic theory to build nuclear power stations.

Think **Green**
Green Technology

In recent years, people have become more aware of the impact of technology on our environment. *Green technology* is a general term for technology that is environmentally friendly or that helps conserve resources. It includes the development of both processes and materials. Throughout this book, we will be looking at several examples of green technology.

As you read the section "Technology: For Better or Worse?," think about environmentally friendly alternatives. Can you come up with ideas that might allow us to keep the benefits of technology while removing or reducing the negative impacts?

Technology: For Better or Worse?

For the most part, technology has improved the quality of our lives. We live longer, healthier lives, and we have more goods and services. However, no matter what the benefits of any new technology, history shows that technology can also cause problems. See **Figure 1-13**.

Technology is neither good nor bad. However, the ways in which humans use the product can have good or bad effects. For example, some of the most important technologies of the 20th century were electricity, the telephone, and the internal combustion engine. But not all of the changes they brought about were beneficial.

Figure 1-13. As world demand for oil increases, wells are being dug in controversial places.

Some benefits we can probably agree on. Almost all of us like the comforts these technologies bring: electricity to warm or cool homes according to the season, a phone to call our friends, and quick transportation to stores or movie theaters. But these comforts have a cost. Pollution is caused by burning coal to generate electricity, the constant ringing of telephones can disrupt our quiet times, and car accidents can result in injury or death. In the end, people have to decide whether the positive effects of a specific technology outweigh the negative effects.

Take a moment to think about some of the technologies you use each day. For example, you may use a cell phone, television, video games, computers, and the Internet. These devices often replace real-world experiences. Does this mean that our modern world is becoming a society detached from reality? Are we living in a virtual world? What are the advantages of meeting a friend face-to-face, sharing a favorite drink , or watching a campfire, rather than accessing Facebook or Twitter? Over the last few decades, obesity has become more common in all age groups. Might this be because people spend long hours connected to computers or televisions, with snacks at hand? See **Figure 1-14**.

Figure 1-14. Technology influences the way we spend our leisure time. What are the positive and negative effects of the technologies we use for fun?

End Note

Technology has been around as long as humans. Even the very earliest humans tried to make their world better by designing tools and processes to meet their needs. Over hundreds of years, people have continued to build on earlier ideas to create new or improved products.

Tomorrow, who knows? Technology holds the keys to many world challenges. Throughout your life, you will share the exciting opportunities for designing new products. Remember, though, that you also share the responsibility to create and use those products wisely.

- The products and tools of technology greatly enhance our daily lives.
- The broad range of technological products affects every aspect of human activity.
- Over the last few centuries, the number of technological changes has increased dramatically.
- Science and technology are very different, but they often depend on each other.
- Technology itself is neither good nor bad. Human use of technology, however, can have good or bad effects.

Compare and Contrast

Copy the following chart to a separate piece of paper. In the Task column, list tasks you perform every day that require the use of technology. In the remaining columns, explain how you use the technology and what you would do if you did not have that technology. Add rows to the chart as necessary to list the tasks.

Task	With Technology	Without Technology
Example: Wake up in time for school	Use alarm clock	Let the sun wake me up

Test Your Knowledge

Write your answers to these review questions on a separate sheet of paper.

1. What is technology?
2. What is the relationship between technology and the "designed world"?
3. What areas of human behavior and activities have been changed by technology?
4. How might engineering and technology affect social and cultural values?
5. What do you consider to be the most important product designed in each of the following time periods? Explain why you think each product is important.
   A. 500,000 to 10,000 B.C.
   B. 10,000 B.C. to 1 B.C.
   C. AD 1 to 1400.
   D. 1401 to 1700.
   E. 1701 to 1850.
   F. 1851 to 1900.
   G. 1901 to 1945.
   H. 1946 to 1999.
   I. 2000 to present.

Test Your Knowledge (Continued)

6. What statement can be made about the number of objects invented during different periods throughout history?

7. Briefly describe the relationship between science and technology.

8. List three examples of how technology has impacted society, individuals, and the environment.

9. What are the major advantages of technology?

10. What are the biggest disadvantages of technology?

Critical Thinking

1. Choose a product you use every day. Think about the effects of this product on society in general. Consider social, cultural, political, economic, and environmental impacts. Write a one-page report on these effects. In your report, explain why you think the product's value to society does or does not outweigh its disadvantages. Be sure to use technical terms correctly.

2. Think about inventors you have learned about in other courses. How many were scientists? How many were technologists? Could any of them be considered both scientists and technologists? Explain your answer and give specific examples.

Apply Your Knowledge

1. Make a model of a tool that might have been used in the first century B.C.

2. Select what you consider to be the most important product ever invented. Make a scrapbook to show its impacts, both positive and negative, on society.

3. Make a list of five products that have been designed and made since you were born. Is the number of products increasing or decreasing each year? Explain your answer.

STEM Applications

1. **MATH**: Create a timeline to show a pictorial history of technological advances in one of the following categories:

   - Food
   - Shelter
   - Clothing
   - Transportation
   - Communication
   - Health

   Use one or more poster boards or a long sheet of roll paper to allow space to place pictures of the items at their proper locations on the timeline. For each item you include, find out the year or century (for older items) in which the item was invented. Then decide on a scale for the timeline. For example, you may decide to allow 10 inches (25.4 centimeters) for every 10,000 years. Label the timeline accurately according to your scale. Place the items at their correct locations on the timeline.

2. **TECHNOLOGY**: Name five technical objects you believe will be invented 20 years from now. For each object, describe:

   A. How it will work.
   B. Who will benefit from its use.
   C. What potential problems may occur as a result of using it.

Generating and Developing Design Ideas

The Aquaduct could help people in developing countries transport clean drinking water to their homes.

Better by Design

The IDEO team uses brainstorming to generate ideas quickly

In developing countries, people often carry heavy vessels over long distances every day to collect water. The water then has to be purified before it can be used. So the team at IDEO designed the Aquaduct, a cycle that allows a person to transport and sanitize water simultaneously. As the rider pedals, a pump attached to the pedal crank draws water from a large tank, through a carbon filter, and into a smaller, clean tank. The clean tank is removable and is closed for contamination-free home storage and use.

Where did the team get this good idea? The best way to have a good idea is to have a lot of ideas. The team used brainstorming to generate dozens of ideas for possible solutions very quickly. *Brainstorming* is a group problem-solving method of generating new ideas in which everyone's ideas are welcome—no idea is too crazy.

Designers brainstorm at the start of a project.

The team that created the Aquaduct design.

"Designers can get good ideas by looking at existing products, books, and magazines; and by sitting quietly thinking, doodling, and sketching." —Tom Kelley, IDEO

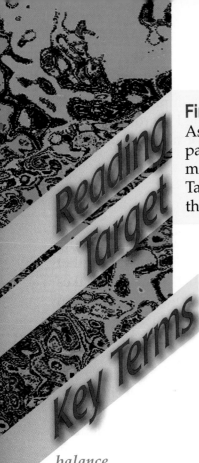

Reading Target

Finding the Main Idea
As you read this chapter, look for the key points, or main ideas, in each part of the chapter. Then look for important details that support each main idea. After you have read the entire chapter, use the Reading Target graphic organizer at the end of the chapter to organize your thoughts about what you have read.

Key Terms

balance
brainstorming
complementary colors
contrast
design brief

designer
design problem
design process
elements of design
engineering design process
ergonomics
experimentation
form
function
harmony
innovation
invention
lines
model

pattern
primary colors
principles of design
proportion
prototype
rhythm
secondary color
shape
style
sustainable design
tertiary color
texture
troubleshoot
unity

Objectives

After reading this chapter, you will be able to:
- Explain the role of the designer.
- Identify items you design in your daily life.
- Use design process skills to solve a design problem.
- Summarize other problem-solving techniques and explain when they should be used.
- Recall the elements of design.
- Identify the principles of design.
- Summarize the five basic types of design decisions.

Useful Web sites:
www.ideo.com/
www.pbs.org/wnet/innovation/

Every product we use has to be designed by someone, somewhere. These products may be as simple as a paper clip or as complex as a hospital operating theater. In most cases, a *designer* identified a group of users with a need that could be satisfied by a redesigned or new product.

The Role of the Designer

There is no such thing as "the perfect design." However, by keeping the needs of the user in mind, a designer increases the probability that the product or service will be a success. If the product or service makes people safer, more comfortable, more efficient, or just plain happier, then the designer has succeeded. Design determines the shape and height of a shoe heel, the special effects in films, and the curving sweep of the support of a bridge.

Design plays a key role in careers such as architecture, engineering, fashion design, graphic design, industrial design, interior design, and stage design. Businesses employ engineers and industrial designers to develop new products. Graphic designers give products an identity that will appeal to customers. Architects design buildings to fulfill our space needs for living and working. Interior designers make offices and shopping centers comfortable environments in which to live and work.

Sometimes designers work alone, but more often they work in teams. Each member of the team has something he or she does especially well that contributes to the overall success of a design project.

Designers make decisions about the size, shape, materials, colors, and finish of a product. The product can be as small as a table lamp or as large as a building or even an entire neighborhood in an urban plan.

Are You a Designer?

You may think of designing in grand terms such as developing a video game or a line of summer clothes. But designing can also mean reorganizing a drawer, building a sand castle or a tree house, constructing shelves for books or trophies, or decorating a birthday cake. See **Figure 2-1** and **Figure 2-2**. Young designers like you can be good at generating and developing new ideas.

The list of products and services you can design is almost endless. For example, you can design a display for a school exhibition, a flower arrangement, or a flag or pennant for a club. Rearranging your room so that it reflects your tastes, painting a child's wagon, designing a birdhouse or a playhouse, or making a funny face mask are all forms of design. In each case, you design something that may make your life, or that of others, more interesting, more comfortable, or happier.

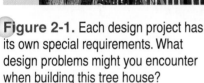

Figure 2-2. Design principles can be applied to desserts and other foods. What did the designer of these cupcakes have to think about?

Figure 2-1. Each design project has its own special requirements. What design problems might you encounter when building this tree house?

Solving a Design Problem

Designing and making a product requires the designer or engineer to use a number of skills. These skills include doing research, sketching, 3D modeling, and testing. The general process used to solve design problems is called a *design process*. If solving the problem requires a knowledge of science and mathematics, it is often called an *engineering design process*. See **Figure 2-3**.

Sometimes you will hear people talk about "the design process" as though the way in which you design a product is the same every time. This is not true. The steps a designer or engineer uses to produce a solution to a problem depend partly on the problem. For example, a dress designer works differently from an architect. The work of both is different from that of a graphic designer. However, all designers and engineers use some common steps and design process skills. See **Figure 2-4**.

Defining the Problem

The process of designing begins when one or more users have a need. In some cases, the designer may develop an entirely new product. An example of this is a global positioning system (GPS). When a product is entirely new, it is known as an *invention*. However, in most cases, the

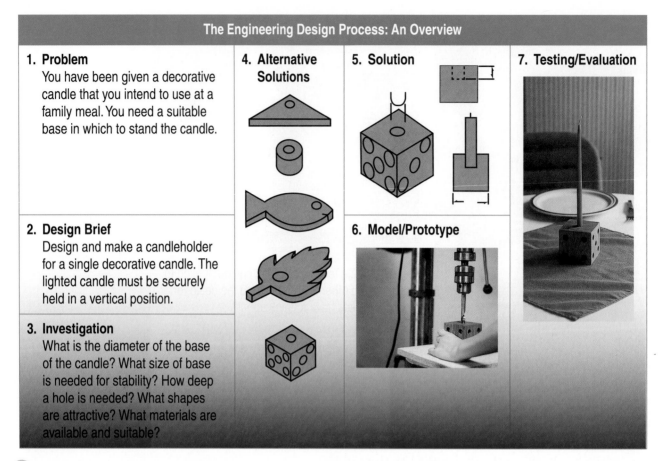

The Engineering Design Process: An Overview

1. Problem
You have been given a decorative candle that you intend to use at a family meal. You need a suitable base in which to stand the candle.

2. Design Brief
Design and make a candleholder for a single decorative candle. The lighted candle must be securely held in a vertical position.

3. Investigation
What is the diameter of the base of the candle? What size of base is needed for stability? How deep a hole is needed? What shapes are attractive? What materials are available and suitable?

4. Alternative Solutions

5. Solution

6. Model/Prototype

7. Testing/Evaluation

Figure 2-3. Each time you design, you will use some or all of these design steps.

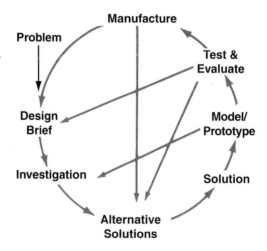

Figure 2-4. Following an orderly design process helps you create a good solution for a design problem. What other solutions can you think of for a candleholder?

designer improves an existing design. For example, cookie sheets are now coated with a nonstick material. The process of improving existing designs is called *innovation*. See **Figure 2-5**.

Before you can decide whether to use invention or innovation, you must define the *design problem*, or need, exactly. For example, imagine that you use several pens, pencils, and highlighters to do your homework. These items are currently scattered all over your desk.

Figure 2-5. Design solutions can be new inventions or innovations that improve an existing design. Are these products inventions or innovations?

Writing the Design Brief

Let's think about the need in this example. You want to design a product that will keep the writing tools together and organized. The first step is to create a statement that describes simply and clearly what is needed. Such a statement is called a *design brief*. The design brief defines the criteria and constraints that must be met by the designed product.

The design brief should be as specific as possible. A good design brief addresses some or all of the following questions:

- What sort of product (or service) is needed?
- Who it is for?
- Where will it be used?
- When will it be used?
- Where might it be sold?
- Who is likely to buy it?

An appropriate design brief for the pencil holder might be: "Design a container to hold at least two pens, three pencils, and two highlighters. Items must be easily identified and removed."

Investigating

Once you have a clear design brief, you can begin to investigate, or gather information. List all the information you think you may need. Some things to consider include:

- Function
- Appearance
- Materials to be used
- Construction
- Safety issues
- Environmental impact

Function

An object that does not *function* well fails as a design solution. The products we use should do what they are intended to do. They should also be easy, efficient, and safe to use. This is sometimes difficult to achieve. Humans vary in many ways. A simple example is that a baby needs a high chair, but an adult needs an adult-size chair. In addition, not all adults are the same: they vary in size.

The study of how a person, the products used, and the environment can be best fitted together is called *ergonomics*. Ergonomics includes these considerations:

- Body sizes—can people fit the object?
- Body movement—can everything be reached easily?
- Sight—can everything be seen easily?
- Sound—can important sounds be heard, and are annoying ones eliminated?
- Touch—are parts that a person touches comfortable?
- Smell—are there any unpleasant smells?
- Taste—are any materials toxic?
- Temperature—is the environment too hot, too cold, or comfortable?

Not all of these apply to every product. Look at **Figure 2-6**. In the design of a computer console, the seat, keyboard, and screen should adjust in various directions. Different sizes of people must be able to use the same console.

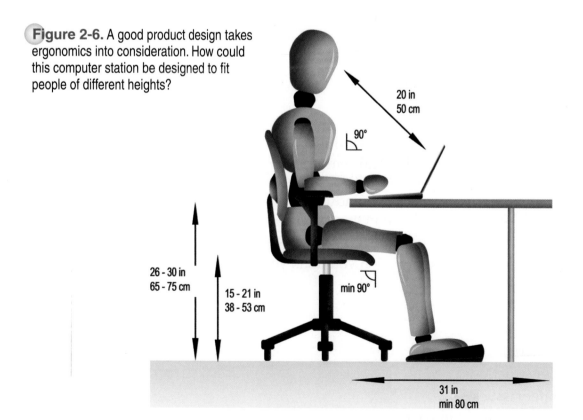

Figure 2-6. A good product design takes ergonomics into consideration. How could this computer station be designed to fit people of different heights?

20 in
50 cm

90°

26 - 30 in
65 - 75 cm

15 - 21 in
38 - 53 cm

min 90°

31 in
min 80 cm

Figure 2-7 shows two pairs of scissors. Those on the left have been designed to fit most people's hands. Their color makes them easy to see.

Appearance

As the designer, you must make decisions about how the product will look. What shapes, colors, and textures should you use? What will appeal to the identified group of users? Must the product fit with existing products?

Materials

What materials are available for the product? How much do they cost? Do they have the right physical properties, such as strength, rigidity, color, and durability? What effects will they have on the environment?

Construction

How will the parts of the product fit together? What tools and techniques will you need to cut, shape, form, join, and finish the materials?

Safety

The object you design must be safe to use. Keep in mind the age and abilities of the intended users. Your product should not cause accidents.

Environmental Impact

Are the materials used in the product or its manufacture harmful in any way to the environment? What will happen to the object when it is no longer useful? Can the product be recycled?

Sources of Information

Now that you know the questions to ask, you can begin looking for the answers. Where do you begin? Consider these sources:

- Existing solutions—look around you for similar articles, examine them, and collect pictures showing examples of other people's solutions.
- Libraries—search in your school or local library for magazines, books, and catalogs with relevant information and pictures.
- Internet—use a search engine (online software that searches the Internet for words or terms) to find possible solutions.
- Experts—seek out people in industries, schools, and colleges who have this type of problem in their daily work.

Figure 2-7. Small design changes can result in more functional products. Which of these pairs of scissors would you rather use? Why?

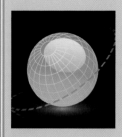

Think **Green**
Sustainable Design

As you design a new product, be sure to consider its environmental impact. A *sustainable design* is one that has little or no negative impact on the environment and society. To be sustainable, a design should be:

- Constructed of renewable natural resources.
- Free of toxic chemicals and other items that cause damage to the environment or to human, plant, or animal life.
- Easily recyclable or reusable indefinitely, or be biodegradable in common landfill conditions.

Sustainable design can be applied to small items we use every day, as well as large projects such as community planning. However, for best results, sustainability should be considered at the design stage.

Developing Alternative Solutions

After you have gathered information, you are ready to develop your own designs. It is very important that you write or draw every idea on paper as it occurs to you. Making these design sketches is an excellent way to remember, explore, develop, and communicate ideas. **Figure 2-8** shows some solutions one designer identified for the pencil holder design problem.

You may find that you like several of the solutions. Eventually, you must choose one. Usually, careful comparison with the original design brief will help you select the best. You must also consider:

- Your own skills
- The materials available
- Time needed to build each solution
- Cost of each solution

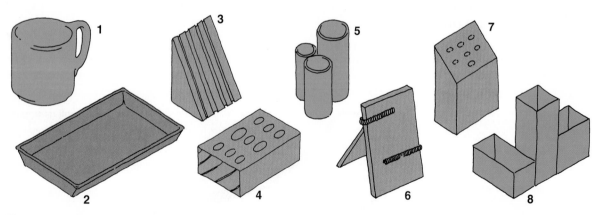

Figure 2-8. Eight possible solutions for the pencil holder design problem. Can you think of other solutions?

Choosing a Solution

Deciding among the several possible solutions is not always easy. It helps to summarize the design requirements and solutions by making a chart like the one in **Figure 2-9**.

Three solutions—numbers 5, 7, and 8—satisfy all of the design requirements listed. Which would you choose? The designer chose number 7. This design not only meets the 10 criteria listed, but also can be made economically and packaged and shipped easily.

Next, create a detailed drawing of the chosen solution. See **Figure 2-10**. This type of drawing must include all of the information needed to make the pencil holder, including:

- The overall dimensions
- Detail dimensions
- The material to be used
- How it will be made
- What finish will be required

Design Requirements	Alternative Solutions							
	1	2	3	4	5	6	7	8
Holds four pens?	✓	✓	✓	✓	✓	✓	✓	✓
Holds three pencils?	✓	✓	✓	✓	✓	✓	✓	✓
Pens and pencils separated?			✓	✓	✓		✓	✓
Are pens and pencils easily removed and replaced?	✓	✓	✓	✓	✓		✓	✓
Is container stable?	✓	✓	✓	✓	✓		✓	✓
Attractive?					✓		✓	✓
Possible to make?		✓	✓	✓	✓	✓	✓	✓
Uses appropriate materials?	✓	✓	✓	✓	✓	✓	✓	✓
Tools are avalable?		✓	✓	✓	✓	✓	✓	✓
Materials are available?	✓	✓	✓	✓	✓	✓	✓	✓

Figure 2-9. This chart allows you to evaluate the solutions at a glance.

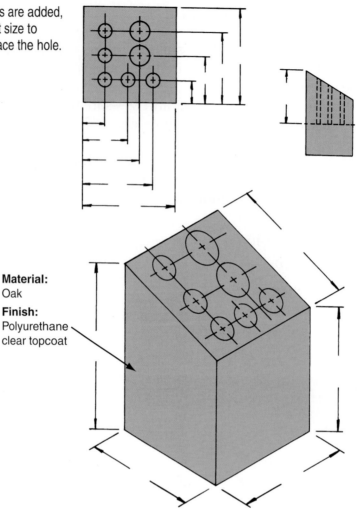

Figure 2-10. When dimensions are added, this detailed drawing will tell what size to make the holder and where to place the hole.

Material:
Oak

Finish:
Polyurethane
clear topcoat

Making 3D Models and Prototypes

Now you can choose what to do next. You can make a model and later a prototype, or you can go directly to making a prototype.

Architects, engineers, and designers use *models* to help communicate their design ideas. It is far easier to understand an idea that you can see in three-dimensional form. If the object is very large or small, a scale model is used. Building a model allows them to view the design from various angles. They can then correct any errors they see. See **Figure 2-11**.

For a simple object, such as the pencil holder, the designer probably would not make a model. He or she may go directly to a prototype. See **Figure 2-12**. A *prototype* is the first working version of the designer's solution. It is generally built at full size and is often handmade. The designer uses the prototype to help plan the steps for making the product. For example, the designer would:

- Select the materials
- Plan the steps for cutting and shaping the material
- Choose the correct tools

Figure 2-11. Using a clay model, the designers can view a product from various angles. They can correct any errors they see.

Figure 2-12. A prototype allows designers to test a design in a real-life situation.

- Cut and shape material
- Apply the finish

The steps will vary depending on the object you are making. Some products have many parts that must be assembled. The important thing is that you plan ahead.

Testing and Evaluating

Testing and evaluating a design answers three basic questions:

- Does it work?
- Does it meet the design brief?
- Will modifications improve the solution?

The question "Does it work?" is basic to good design and must be answered. An engineer designing a bridge, the designer of a baby stroller, or an architect planning a new school would ask this same question. What would happen if the holes in the pencil holder were too small to accommodate pencils? What if it were top-heavy? The holder would not work. Sometimes, however, poor design can be dangerous. If a designer makes mistakes in the design of a seat belt for a car, someone's life may be in danger!

Manufacturing

When the prototype satisfies the designer, it is time to produce a small number of samples. These samples are given to typical users who report their experiences to the manufacturer. Did it work well? How could it be improved? Is it attractive? Is it priced right? Designers use this feedback to make final changes. As they make the design changes, they must also remember that the product must be sold at a reasonable profit.

When the company is satisfied with the design, it then decides how many to make. Products may be made in low volume or mass-produced in high volume. Some items, such as specialized medical equipment or airplanes, are produced in the hundreds. Other products, such as nuts and bolts, may be produced in the millions. See **Figure 2-13.**

Other Problem-Solving Methods

The steps described in the previous section are general ones. Most designers and engineers use them to some extent. However, other problem-solving methods also exist. Examples include the I-DREAM method, experimentation, and troubleshooting.

Figure 2-13. An accurate estimate of the demand for a product is critical to planning its manufacture. How might this denim manufacturing company decide how many pairs of jeans to produce?

I-DREAM Method

I-DREAM stands for:

- Identify and accept the problem.
- Define the problem.
- Research and evaluate ideas.
- Execute the design by building a model or prototype.
- Assess whether the design meets the design criteria.
- Modify the design if it does not meet the design criteria.

You might notice that these steps are very similar to those used in an engineering design process. Sometimes steps are performed more than once or in a different order. For example, suppose you assess a solution and find that it does not meet the design criteria. To modify the design, you may need to go back to the research stage or even the define stage.

Experimentation and Troubleshooting

What should you do when you know a problem exists, but you can't put your finger on it? You may need to do some experiments. *Experimentation* means trying different ideas. You can use experimentation to define a problem or even to solve it.

You can also try to *troubleshoot* a problem. This involves systematically eliminating possible causes of the problem. Troubleshooting is a good method to use when you know what the problem is, but you do not know what is causing it. For example, suppose your computer monitor is not displaying an image. To troubleshoot this problem, you might use the process shown in **Figure 2-14**. First check to see if the monitor is plugged into an electrical outlet. If it is, you might check to see if the monitor is turned on. Then check to see if the computer is turned on, and so on. By finding the reason for the problem, you can often solve the problem.

Elements of Design

All products appeal to our senses in some way. We buy clothing that looks good. We enjoy the smooth feel of the polished wooden arm of a chair. A meal on a plate must not only look attractive but also should taste and smell good. The sound of musical chimes is preferable to the harsh sound of a door buzzer.

When you see something you like, ask yourself what is it you like about the product. Is it the color? Is it the shape or form? Is it the texture? Does its texture remind you of the beauty of things found in nature?

Line, shape and form, texture, and color are the four *elements of design*. They are what most people notice when they look at an object. When you examine any object, you will find that the elements of design have been combined to create a unique look.

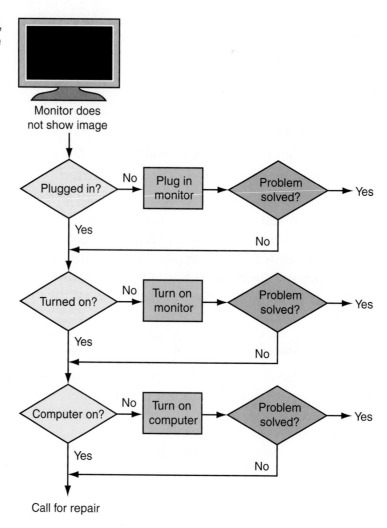

Figure 2-14. In troubleshooting, you eliminate possible causes one by one.

Line

Lines describe the edges or contours of shapes. They show how an object will look when it has been made. Lines can also be used to create some special effects in the finished product. For example, straight lines suggest strength, direction, and stability. See **Figure 2-15**. Curved or jagged lines may give a feeling of motion, grace or softness, depending on their shape. See **Figure 2-16**. Heavy lines suggest more strength than thin lines, as shown in **Figure 2-17**.

Shape and Form

All objects occupy space or possess volume. The *shape* of an object is two-dimensional, and the *form* of an object is three-dimensional. See **Figure 2-18**.

Shapes and forms may be geometric, organic, or stylized. See **Figure 2-19**. Geometric shapes can be drawn using rulers, compasses, or other instruments. Organic shapes and forms mimic nature and contain curved,

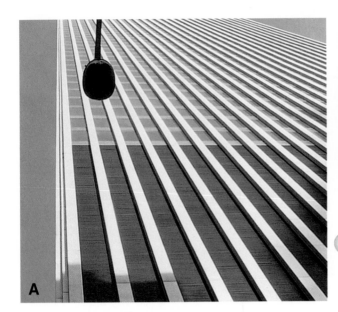

Figure 2-15. A—Vertical lines show strength. B—Horizontal lines give a feeling of stability.

Figure 2-16. A—Curved lines give a sense of grace and softness. B—Jagged lines seem harsh and unfriendly.

Figure 2-17. A—Thin lines appear fragile. B—Heavy lines indicate strength.

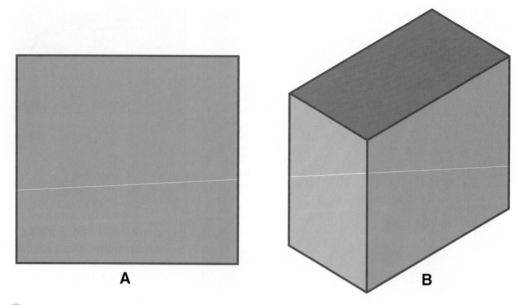

Figure 2-18. A—In design, a shape has two dimensions—width and height. B—Form has three dimensions: width, height, and depth. The shape in A was used to create the form in B.

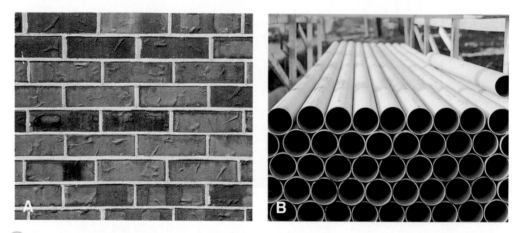

Figure 2-19. Geometric shapes: A—Bricks are rectangles (2D) or boxes (3D). B—Pipes are circles (2D) or cylinders (3D).

flowing lines. See **Figure 2-20**. Stylized shapes and forms have been simplified or streamlined, as shown in **Figure 2-21**. Organic and stylized shapes and forms are sometimes drawn freehand.

Texture

Texture refers to the way a surface feels or looks. We can describe a surface as rough, smooth, hard, slippery, fuzzy, or coarse. Sandpaper feels rough. Glass feels smooth. Rock feels hard. Ice feels cold and slippery.

A designer can choose materials according to their natural texture. She or he might also choose materials because of the way the texture can be changed. **Figure 2-22** and **Figure 2-23** show how texture can be used in wood and stone.

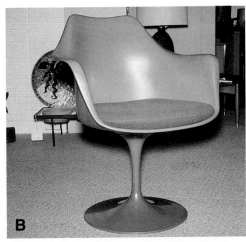

Figure 2-20. Organic shapes are found in natural and manufactured products. A—Blossoms are natural organic shapes. B—The curved, free-form shape of this chair is organic.

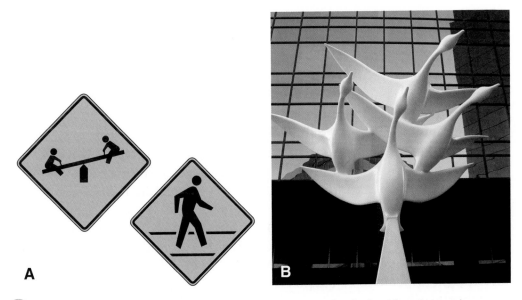

Figure 2-21. A—Stylized shapes can provide information. B—Stylized form is seen in many sculptures.

Figure 2-22. The texture of wood can be changed. A—Wood shingles on a roof. B—Wooden sculpture.

Figure 2-23. The texture of stone is attractive in a wall (A) or in the Inuit carving (B).

Color

Different colors invoke different moods. Yellow is both cheerful and exciting, while blue is associated with being calm. Green is relaxing. When people are happy, they generally prefer light colors. When they are sad or worried, they often prefer dark colors.

Isaac Newton discovered that white light is made up of rainbow colors. When a beam of white light shines through a glass prism, the path of light is bent. It splits into rainbow hues that bend at different angles. The colors can then be seen individually. These seven colors form a spectrum. See **Figure 2-24**.

We see color when light shines on objects. All objects react to light energy by absorbing certain wavelengths of light and reflecting the rest. An object that reflects all the light appears to be white. One that absorbs all the light appears to be black. Grass reflects the green wavelengths and absorbs the others.

Figure 2-24. The seven colors in white light separate when the light passes through a prism. Where do you see this happen in nature?

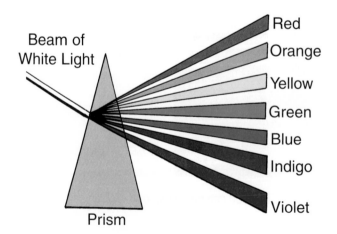

Red, yellow, and blue are called *primary colors*. If you mix equal parts of two primary colors, you obtain a *secondary color*. Red plus yellow makes orange. Yellow plus blue makes green. Blue plus red makes violet. Mixing equal parts of a primary and a secondary color creates a *tertiary color*. See **Figure 2-25**.

Designers use colors to produce certain reactions or effects. See **Figure 2-26**. Traffic signs use red to indicate danger. Yellow serves as a traffic warning. We also associate colors with objects. An apple is red. Grass is green. Charcoal is black, and milk is white.

Color can also be used to control temperature. Dark materials are best able to radiate or absorb heat. A black T-shirt, worn on a hot day, will heat up more in the sun than a white one. Houses in hot climates are painted white because white reflects sunlight.

Primary	Secondary	Tertiary
red	red + yellow → orange	red + orange → red-orange
		yellow + orange → yellow-orange
		yellow + green → yellow-green
yellow	yellow + blue → green	blue + green → blue-green
		blue + violet → blue-violet
blue	blue + red → violet	red + violet → red-violet

Figure 2-25. Mixing two primary colors produces a secondary color. Mixing a primary and secondary color creates a tertiary color.

Figure 2-26. Color can be used to create special effects. How was color used in these items?

Principles of Design

You learned earlier that line, shape and form, texture, and color are the elements of design. You can think of these as building blocks that can be put together in many different ways. The guidelines for combining these elements are called the *principles of design*. They are balance, proportion, harmony and contrast, pattern, movement and rhythm, and unity and style.

Balance

When you think of balance, think about a tightrope walker moving along a cable. She or he keeps balance using arms and a balancing pole. It is important to match or balance the mass of the body on both sides. *Balance* is also very important in design. It means that the mass is evenly spread over the space used. The three types of balance in design are symmetrical, asymmetrical, and radial. See **Figure 2-27**.

Proportion

Look at **Figure 2-28A**. Something seems to be wrong—the person is too big for the chair. Now look at **Figure 2-28B**. The person appears to

Figure 2-27. A—The designer of the Manchester Millennium Bridge combined symmetry with a creative design to provide both beauty and function. B—In an asymmetrical design, the two sides are in balance visually, but they are not mirror images. C—Like all Ferris wheels, the London Eye has radial balance. The mass moves outward in all directions from the center point.

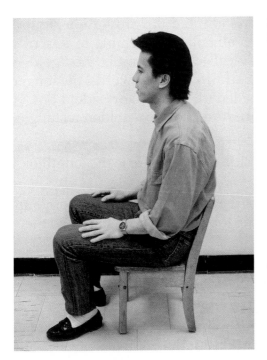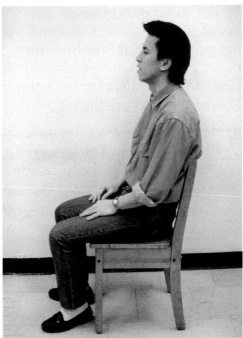

Figure 2-28. A—The person and chair are not in proportion to one another. B—The person and chair are in proportion.

be very comfortable. The relationship between the person and the chair seems to be right. The relationship between the sizes of two things is called *proportion*.

Proportion can also apply to the parts of an object. Look at the contents of the mobile home in **Figure 2-29**. Although the items are smaller than usual, their size is related to the overall size of the mobile home. They are in proportion.

Figure 2-29. Everything inside this mobile home is scaled down and in proportion to the space available.

For thousands of years, people have admired the proportions found in nature. The Greeks worked out a mathematical formula to describe the proportions found in nature. They called this formula the "golden mean." The golden mean has a ratio of 1:1.618. In other words, the longer side of the rectangle is 1.618 times the length of the short side. Look at **Figure 2-30** and use the following procedure to draw a golden rectangle:

1. Draw a base line.
2. Draw a square. The length of one side of the square is the length of the short side of the rectangle.
3. Measure halfway along the base of the square. Put the point of your compass here. Draw an arc from the top corner of the square to the base line.
4. The point where the arc touches the base line is the right-hand corner of the rectangle. Draw a vertical line upward from it.
5. Extend the top line of the square to complete the rectangle.

The golden mean also appears in the human body and in many living things. In **Figure 2-31**, the lion's proportions fit the golden mean.

Mathematics is important to designers. Still, they do not rely on mathematics alone to decide the proportions of an object. They must adjust the proportions until they look right. Look at the chest of drawers in **Figure 2-32**. Notice that the drawers at the bottom are larger than those at the top. If all the drawers were of the same height, the chest of drawers would seem to be top-heavy.

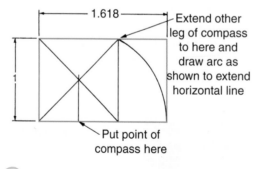

Figure 2-30. To draw a golden rectangle, begin by drawing a square.

Figure 2-32. Good proportion sometimes requires using mathematically unequal parts.

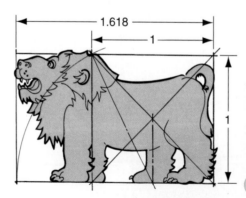

Figure 2-31. The lion's body fits the golden mean so often found in nature.

Harmony and Contrast

Observe the best figure skaters and you will notice that their movements seem to flow with the music. We say they are in harmony with the music. *Harmony* is the condition in which two things, such as color or musical notes, naturally go together.

Designers use the idea of harmony in the objects they create. Buildings and their environment should be in harmony. The dishes in **Figure 2-33** are in harmony.

Sometimes designers want to surprise you. They may want to make you feel excited about what you see. They may simply want to catch your attention. To do this, designers create an obvious difference between things. This difference is called *contrast*. While harmony makes you feel comfortable, contrast adds excitement.

The red cross on an ambulance contrasts with its white background. A jagged mountain contrasts with the calm waters of a lake. The lines and shape of an old building may contrast with those in a new office tower. See **Figure 2-34**.

Figure 2-33. These objects are in harmony. Both color and shape go well together.

Figure 2-34. Contrast can be achieved by using color (A) or lines and shape (B).

You should also consider harmony when you are selecting colors for a design. You may want the colors to be similar or to contrast. You can find similar colors next to one another on a color wheel. See **Figure 2-35**. For example, if you first choose orange, similar colors include red-orange and yellow-orange.

Suppose you want colors to contrast with each other, yet still work well together. In this case, select colors that are at the opposite side of the color wheel. For example, blue contrasts with orange. Contrasting colors are also called *complementary colors*.

Pattern

A *pattern* is a shape or form that is repeated many times in a design. See **Figure 2-36**. Patterns are found both in nature and in objects that people have designed. Sometimes they are used to make an uninteresting surface appear more attractive, as seen on the surface of the egg. At other times, the pattern may serve a particular function, as when two different colors make a pattern on a chessboard.

Rhythm and Movement

Patterns are closely related to *rhythm*, or the suggestion of movement. Some patterns, such as ocean waves, naturally create a sense of movement. See **Figure 2-37**. The rhythm of a design can be smooth and flowing or fast and dynamic. In **Figure 2-38A**, the tulip bowl creates a sense of smooth, easy movement through its use of shape and line. The spiral of the printed pattern in **Figure 2-38B** suggests a more lively feeling of movement.

Figure 2-35. Use a color wheel to choose similar or contrasting colors for a design.

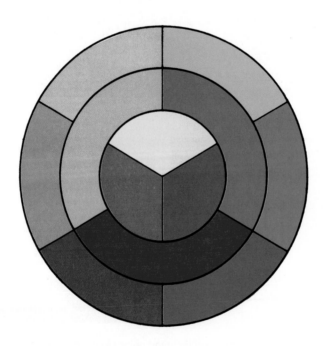

Technology Application

Art, Nature, and Technology

The shapes and patterns of technological designs are often inspired by nature. Many patterns are based on shapes that we know quite well: leaves, flowers, birds, and animals. If you look around, you may find examples of sunflower tiles on a floor, leaf patterns on a carpet, sofas and fabrics decorated with wild grasses, or a necklace of ivory pieces, each carved in the shape of a bird. Usually, the patterns are not exact copies, but abstract versions that were inspired by the objects. For example, the design for the faucet shown below may have been inspired by a tree similar to the one shown.

The shape of any object we use is influenced by our sense of touch. Have you noticed that very young children want to touch everything? They don't "know" an object until they have touched it. When we are adults, our sense of touch still influences the value we place on objects we use. For example, we prefer a door handle that is easy to grasp and operate, or a chair that feels comfortable.

Technology Activity

Design a mug for drinking your favorite beverage. Think of shape and pattern. First draw the outside shape. The shape must be easy for the user to hold. Then develop a pattern for the surface of the mug using an idea that comes from nature. Make sure the exterior shape and the surface design are in harmony with each other. Create detailed sketches to document your design.

Figure 2-36. Patterns occur in nature and in human-designed objects. A—The interior of different foods form patterns. B—The squares on a chessboard have a function. They are needed to play a game of chess. C—This egg shell has been painted in a complex pattern.

Figure 2-37. Some patterns suggest movement. Waves are an example of a rhythmic pattern in nature.

Figure 2-38. Shapes and lines create a sense of movement. A—A tulip bowl. B—A building design that suggests dynamic movement.

Unity and Style

A well-designed product must have a sense of *unity*. The parts of a design must have a sense of belonging or similarity so that visually they fit well together. See **Figure 2-39**. Not only should there be a sense of unity within an individual object but also between each object and its environment.

Style is a feature or quality that is typical of designs created by a specific person or during a specific time period. Style depends on many things:

- The availability and cost of materials: achieving the most with the least.
- The tools and techniques available to shape the materials.
- Cultural preferences.
- A knowledge of the elements and principles of design.

Figure 2-39. Each of these items has a sense of unity and the designer's unique style.

Making Design Decisions

Design principles do not provide hard and fast rules. They act only as a guide. You must make the actual design decisions. The five basic types of design decisions are conceptual, marketing, technical, constructional, and aesthetic decisions.

Conceptual Decisions

Conceptual design decisions have to do with the overall purpose of the design. What sort of product is needed? For example, what sort of product could students in your school make and sell to raise funds for a school trip? Should you design printed T-shirts, birthday cards, or jewelry? You ask a number of parents and friends which they might purchase and discover that jewelry is most popular. Choosing jewelry is a conceptual design decision.

Marketing Decisions

Who will use the product you are designing? Where and how will they use it? These questions will help you make marketing design decisions. Marketing decisions might also include where the product will be sold and the selling price. For example, you could decide to design jewelry for teens, a special friend, or a favorite aunt. You could design jewelry to give as presents, to sell in a trendy boutique, or to sell at a festival celebrating a special occasion. See **Figure 2-40**.

Technical Decisions

Technical design decisions explain how the design will work. What individual pieces are required? What materials are best for each component? What size will you make the overall product and each

Figure 2-40. Marketing decisions include how and where you will sell the item. What types of jewelry might you create to sell to people in your age group?

component? What are the safety requirements? For example, you could decide to design jewelry using natural materials such as wood and feathers. You could decide to use an easily worked metal such as copper, or recycled materials such as bottle tops. If the jewelry uses electronics, will it contain lights or sound? How will the lights be controlled? What components are needed to make sound?

Construction Decisions

Decisions about how the product will be made and assembled are construction design decisions. What tools and techniques will you need to form small pieces of copper into the shape you have chosen? How will the individual pieces fit together? What joining techniques will you use? For example, how can you join the copper pieces to create a necklace? How could they be attached to a leather cord or metal chain?

Aesthetic Decisions

Aesthetic (artistic) design decisions determine what the design will look like. What overall effect do you want? Will you make large, clunky jewelry or small, delicate jewelry? What shapes might a pendant have? Will you use natural or geometric shapes? What colors might you use? What textures could you use?

Although these five types of design decisions are described separately, they are highly interconnected. If you change one decision, you may have to change others. For example, suppose you decide to use a leather cord at first. Later you change your mind and decide to use a metal chain (technical design decision). This will affect how the decorative elements are attached (constructional design decision).

The process of designing has evolved over the centuries. Our earliest designs responded to basic needs. For example, in early times, methods of lighting a fire were vital. Fire-lighting designs were probably invented by accident in many cases. Today's designers create products or services that fulfill not only the needs, but also the wants, of a group of users.

Many of our newer designs have even been developed to create a new need or demand. New products are designed to appeal to the interests, attitudes, opinions and values of users. For example, you may be pleased with the portable music device you currently own. However, if a company offers one that can store much more music in a product half the size, you may be tempted to buy it. Articles are also being increasingly personalized. One shoe manufacturer lets you customize its soccer shoe. One automobile manufacturer lets you have your own emblem on the car instead of the manufacturer's logo. As new technologies are developed, designers will have even more freedom to create new and exciting products.

Summary

- Designers play an important role in industries such as architecture, engineering, fashion design, industrial design, and others.
- To some extent, everyone is a designer. Designs do not have to be grand schemes They can be as simple as making a flower arrangement or rearranging your room.
- Engineers and designers use specific design process skills to solve design problems.
- Problems can be solved using a design process, the I-DREAM model, experimentation, troubleshooting, or other methods. The method used depends on the problem to be solved.
- The elements of design are the qualities people notice when they look at an object.
- The principles of design are guidelines for combining design elements.
- During the design process, several different types of decisions need to be made. Decisions made in each area may affect the other types of decisions.

Reading Target

Finding the Main Idea

Create a bubble graph for each main idea in this chapter. Place the main idea in a central circle or "bubble." Then place the supporting details in smaller bubbles surrounding the main idea. A bubble graph for the first part of the chapter is shown here as an example.

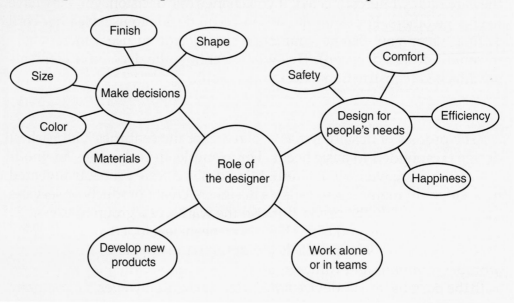

Test Your Knowledge

Write your answers to these review questions on a separate sheet of paper.

1. Why are most new products developed?
2. List the steps in a design process.
3. Compare invention to innovation. How are they similar? How are they different?
4. Explain the roles of engineering, technology, and society in invention and innovation.
5. Given a design problem, why might an engineer sketch several possible solutions?
6. What is the purpose of a prototype?
7. What word describes the study of how a person, product, and environment or surroundings can best be fitted together?
8. What specific questions would you ask if you were testing and evaluating a new wheelchair?
9. Why might a designer create a product for which there is currently no need?
10. Explain why there is no perfect design.

Critical Thinking

1. Think of one object you have seen and find attractive. Write a paragraph describing the object and explaining how the elements of design make it attractive to you.
2. Write a one-page paper comparing and contrasting the steps of the design process/ engineering design process with the steps used in the scientific method. The scientific method is described in Chapter 1. In your paper, include an introductory paragraph, one paragraph for each main idea, and a summary paragraph.
3. Imagine that you overslept this morning because your alarm clock did not wake you up. What steps would you take to troubleshoot this problem?

Apply Your Knowledge

1. Collect or draw pictures showing three natural objects. Collect three more pictures to show comparable technical objects. For example, suppose you collected a picture of a bird's nest (a natural object). A comparable technical object might be a house, because both are habitats.

2. Collect four pictures to illustrate two elements of design and two principles of design. Show your pictures to a group of four classmates. Ask your classmates to analyze the pictures.

3. Sketch as many different designs as you can showing how compact discs could be stored. Consider horizontal, vertical, and diagonal designs.

4. Often a new and exciting invention of today becomes the "old" technology of tomorrow. Think of an invention or innovation that was created more than 25 years ago. Do research and gather data to answer the following questions.

   A. What need was the product designed to meet?

   B. Is the need still important today?

   C. Is the need still being met by the same product? If so, have innovations improved the product? If not, what newer invention has replaced it?

   Write a report explaining your findings. Use data and statistics from your research to support your ideas.

5. Write an essay identifying changes in society caused by the use of inventions and innovations. Explain how new needs and wants may be caused by the use of new products.

6. Form a group with three or four classmates. Brainstorm ideas for a new product that is currently *not* needed or wanted. Use a design process to develop the idea to the prototype stage. Then create persuasive marketing tools such as color advertisements, signs, or banners. Display the marketing pieces in the classroom for a week. Then present the prototype to the rest of the class using a persuasive speech. How many of your classmates would buy your new product based on the "need" or "want" you created?

7. Carefully observe some activities in your home. Make a list of design problems that need to be resolved. Examples: storing spices in the kitchen or making sure that your baby sister doesn't fall downstairs. From your list of design problems, choose one that you will try to solve.

   A. Write a design brief that describes the criteria and constraints for the design.

   B. Sketch or describe a number of solutions to the problem and select the most appropriate solution.

   C. Use the elements and principles of design to develop the design idea.

   D. Make a list of the steps involved to build a model and/or prototype of your solution. Include information about safety precautions for using tools.

   E. Build, test, and evaluate your prototype. Be sure to follow all safety measures. Keep a good record in the form of notes and sketches. Document all changes to the prototype.

   F. Recommend further improvements to the design.

   G. Begin a design portfolio for your design ideas. Place your documentation for this design in your portfolio.

STEM Applications

1. **ENGINEERING** Make an engineering design journal in which you can store design ideas. To start the journal, record the design ideas you created for this chapter. Start each design on a new page. Include sketches drawn to scale and notes to explain your ideas. As you work through this course, add all of your design ideas to this journal.

2. **ENGINEERING** Work with a team of four other students. Brainstorm ideas for a product that will make one task easier for a person with a special need, such as a handicapped or elderly person. Use an engineering design process to build the device. Demonstrate your design in class and ask for comments and suggestions. Document your design and any changes you make. Place the documentation in your design portfolio.

3. **MATH** Look around and find an object in nature that has the proportions of the golden mean. Draw a sketch of the object. Include dimensions to show how the object satisfies the size requirements of the golden mean.

3

Communicating Design Ideas

The IDEO team uses 2D and 3D models to communicate ideas

Designers use 2D and 3D modeling to communicate and evaluate their design ideas. The team of Adam Mack, John Lai, Eleanor Morgan, Paul Silberschatz, and Brian Mason at IDEO first built a 3D model of the Aquaduct using cardboard, masking tape, and a hot glue gun. They used a *storyboard* to visualize how the Aquaduct cycle might be used. They drew detailed plans and modeled parts. Next, they built a full-size working model for testing. The next step is to involve the end users in the design process to help ensure that the product meets their needs.

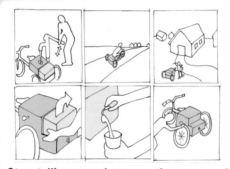

Storytelling can showcase the proposed design solution to the user.

"Prototyping allows for quick and inexpensive exploration of potential solutions to problems."
— Tom Kelley

The IDEO team created the first 3D model of the Aquaduct from readily available objects.

Reading Target

Key Terms

Objectives

Preview and Prediction

Before you read this chapter, glance through it and read only the headings of each section. Based on this information, try to guess, or predict, what the chapter is about. Use the Reading Target graphic organizer at the end of the chapter to record your predictions.

3D printing
additive fabrication
alphabet of lines
carbon footprint
circumference
communication
communication technology
computer-aided design (CAD)
concurrent engineering
construction lines
diameter
drafting
freehand sketching

isometric axes
isometric sketching
line drawings
orthographic projection
perspective sketching
rapid manufacturing
  systems
rapid prototyping (RP)
scale drawing
storyboard
symbol
views
virtual reality

After reading this chapter, you will be able to:

- ○ Explain the three basic types of communication technology.
- ○ List various forms of communication.
- ○ Communicate ideas using isometric and perspective sketches.
- ○ Draw simple objects using orthographic projection.
- ○ Demonstrate standard drawing techniques.
- ○ Explain the advantages of computer-aided design.
- ○ Summarize the principle of concurrent engineering.
- ○ Identify uses for 3D printing.

Useful Web site:
other90.cooperhewitt.org/

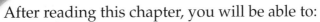

The exchange of information or ideas between two or more living beings is known as *communication*. This is a big word for a simple act. It came from the Latin, *communo*. It means to "pass along."

Communication is more than just sending a message. The message must be received and understood. If this does not happen, there is no communication. *Communication technology* is the process of transmitting and receiving of information using technical means. This chapter introduces you to the skills and equipment used in many kinds of communication.

Types of Communication

There are three types of communication. All are based on our sense of hearing and sight.

- Visual communication presents ideas in a form we can see. Thoughts are changed into words, symbols, and pictures. Stoplights, street signs, photographs, and books give out visual messages. See **Figure 3-1**.

- Audio communication consists of messages that can be heard but not seen. A buzzer tells you class is over. A doorbell at your home tells you someone is at the door. Telephones, radios, and CD players rely on audio communication. See **Figure 3-2**.

- Audiovisual communication can be both seen and heard. You are receiving audiovisual messages when you watch and listen to television, DVDs, and movies.

Figure 3-1. Some street signs communicate a clear message to all people regardless of the language they speak. What is being communicated here?

Figure 3-2. We hear messages that are transmitted using audio communication. What sources of audio communication do you use?

Forms of Communication

All forms of communication use a code or symbols. For example, *dog* and *perro* are letter symbols that communicate the idea of a certain animal. However, the same message can be given in a quite different "language." See **Figure 3-3**.

Hand Signals and Sounds

Simple movements or sounds can replace spoken and written messages. Look at **Figure 3-4A**. These are signals that most people anywhere in the world would understand. Can you think of other signals that you might use? What about the signal to be quiet?

What about sound signals? If you know any Morse code, you will recognize the dots and dashes being sent out by the ship in **Figure 3-4B** as "S.O.S.," the international distress signal. People of all languages know this signal.

Humans are not the only earth dwellers that exchange messages with sounds and body language. Sea animals, such as dolphins and whales, have systems of sound to exchange messages among themselves. Deer and beaver use their tails to signal danger.

Figure 3-3. Dogs can communicate various messages, such as when a stranger approaches, when they need to go outside, or when they are hungry.

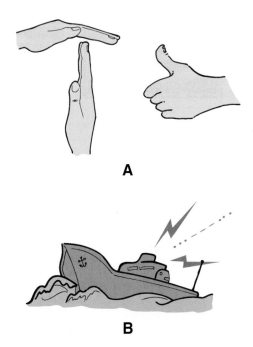

A

B

Figure 3-4. Many hand signals and sounds have well-known meanings. What message is being sent in each of these pictures?

Symbols and Signs

Simple pictures and shapes are one of the most effective methods of communication. These *symbols* can warn, instruct, and direct without using words. They "speak" in a hundred languages all at the same time. See **Figure 3-5**.

Electronic Communication

Today, people increasingly send wireless transmissions that bounce off satellites. This allows you to access information or people no matter where you happen to be.

Electronic communication takes many forms. You may decide to establish a MySpace.com™ or Facebook® profile, or send a text message to someone's phone. You could establish or contribute to a blog. You might decide to use an online chat program or check out an e-commerce site. You can communicate using an instant messaging program or check out an online community. YouTube is a site that offers amateur or professional videos that vary from funny to autobiographical.

Figure 3-5. Signs often have symbols that are easily understood at a glance. What message is communicated by each of these symbols?

Sketching

Communication is central to design and making products. Objects and ideas are often represented using lines and shapes such as the ones shown in **Figure 3-6**. These are known as *line drawings*. Designers, drafters, technicians, engineers, and architects must be able to make such line drawings. It would be impossible to explain the parts in **Figure 3-7** without drawings.

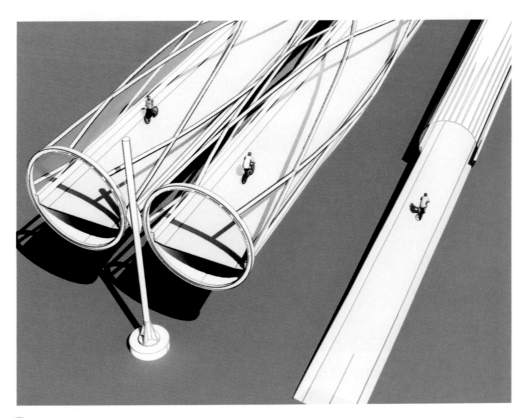

Figure 3-6. Designers use line drawings to communicate their ideas to others.

The process of creating drawings to specify the exact size, shape, and features of a design idea is called *drafting*. Drafting has always been known as the "language of industry." It prevents confusion about the size and shape of an object or structure.

The three types of drawings are isometric, perspective, and orthographic projection. These three types are summarized in **Figure 3-8**. All three may be produced freehand. This is known as sketching. All three may also be drawn using manual drafting equipment or computer-aided drafting systems.

Freehand Sketching

Freehand sketching is an essential step when designing a product. First, it allows you to record your ideas rapidly so you don't forget them. Sketching is a way of talking to yourself. Second, it allows you to share and discuss your ideas with other people. Third, it makes it easier to develop ideas. Whether you set out to design a running shoe, a sofa, or a boat, generating alternative ideas should start with small sketches. For example, the designer of the riverboard explored his early ideas by making the sketches shown in **Figure 3-9**.

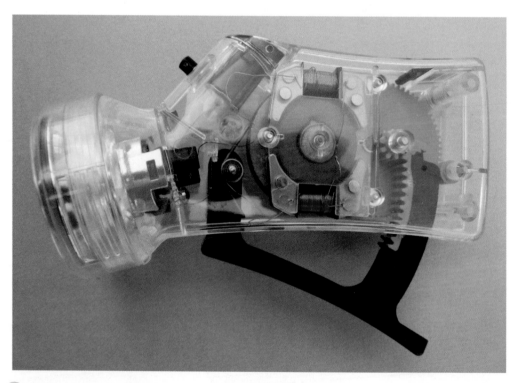

Figure 3-7. Several drawings will be needed to describe the parts of this flashlight so others can make it.

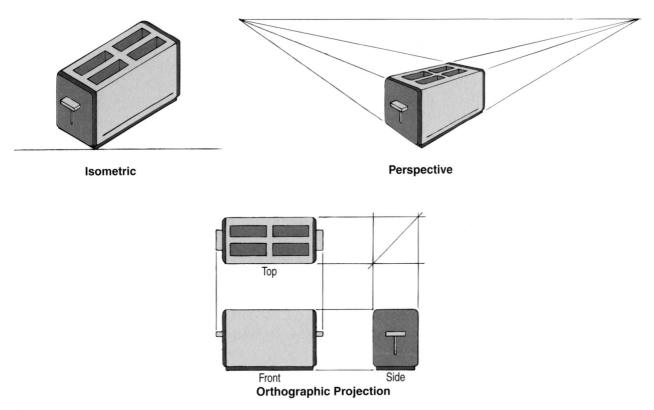

Isometric

Perspective

Top

Front   **Side**
Orthographic Projection

Figure 3-8. Different types of drawings are used for different purposes. Which type of drawing provides the most realistic image? Which provides the most information?

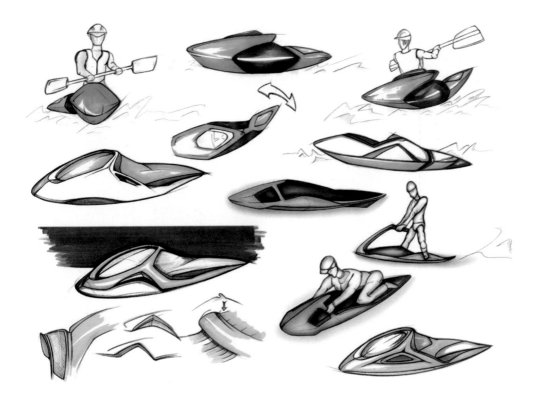

Figure 3-9. Sketches allow the designer to explore many ideas.

Isometric Sketching

In *isometric sketching*, three sides of the item are shown in a single view. Vertical lines show the height of the item. Lines representing the width and depth of the item are set at 30° from the horizontal. Refer again to **Figure 3-8**. Isometric sketches can be made on plain paper. However, when you are first learning to make isometric sketches, it is helpful to use isometric paper. This type of paper contains lines at the proper isometric angles to make sketching easier.

To sketch an isometric box that is six squares long, three squares wide, and four squares high:

1. Draw the front edge of the block. This is line 1 in **Figure 3-10**.
2. Draw lines 2 and 3 to show the bottom edges of the box. The three lines you have drawn represent the *isometric axes*.
3. Mark off the height, width, and depth of the object on the three axes, as shown in **Figure 3-11**. For this sketch, we are measuring in squares. You could also use inches, centimeters, or any other unit of measurement.
4. Draw the left and right vertical edges of the box, as shown in **Figure 3-12**.
5. Draw the top edges of the box. See **Figure 3-13**.
6. Darken the outline of the box. You may also want to add color to define the box more clearly. For a neater sketch, you may also want to remove *construction lines* that extend beyond the boundaries of the box. See **Figure 3-14**.

Whatever the shape of the object to be drawn, it is usually easiest to begin by drawing a box. In most cases, however, you will have to remove parts of the box to create the shape. See **Figure 3-15**.

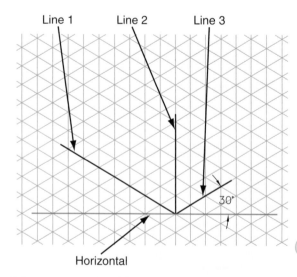

Figure 3-10. To begin an isometric sketch, make these three lines to set up the isometric axes.

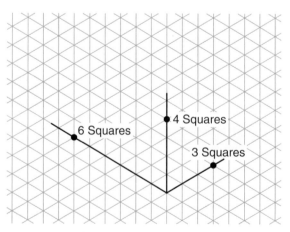

Figure 3-11. Establish the three dimensions of the object. In this case, height = 4 squares, width = 3 squares, and depth = 6 squares.

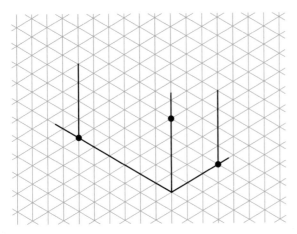

Figure 3-12. Add two more vertical lines to represent the visible vertical edges of the box.

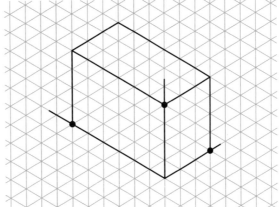

Figure 3-13. Sketch in the top edges of the box.

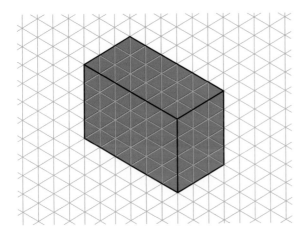

Figure 3-14. The completed isometric sketch of a box.

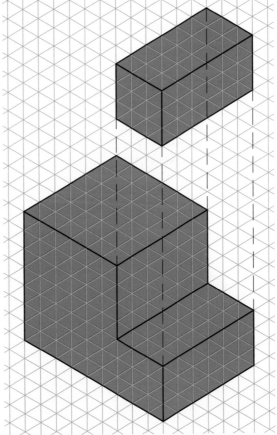

Figure 3-15. Removing part of a rectangular box creates a new shape.

Sometimes you will want to add a piece. For example, you could make a sketch of a box with a small block added to one side. This method of isometric sketching may be used to draw a simplified house.

1. Lightly construct an isometric box eight squares long, four squares wide, and six squares high, as shown in **Figure 3-16**.

2. Construct the basic shape of the house by removing a corner of the box. Add lines for the roof. See **Figure 3-17**.

3. Add details, including windows and doors, as shown in **Figure 3-18**.

4. Complete the line work by removing unnecessary construction lines. Darken the remaining lines to form the building's shape. Add color if desired. See **Figure 3-19**.

While isometric paper makes sketching easy, it has one disadvantage. It leaves grid lines on the final drawing. These could confuse someone looking at your drawing. Designers often prefer to sketch on plain paper. To make a freehand isometric sketch of a rectangular block on plain paper, use the method shown in **Figure 3-20**.

Perspective Sketching

Look at the photograph of the railroad tracks in **Figure 3-21**. Notice that the parallel lines of the tracks appear to converge. The columns on the station platform that are farther away appear shorter, and the platform seems narrower at a distance. Of course, railroad tracks don't converge, columns don't get shorter, and platforms don't become narrower.

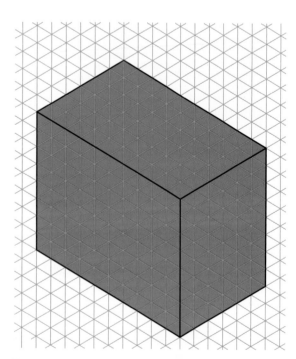

Figure 3-16. To draw a house, make a box on an isometric grid.

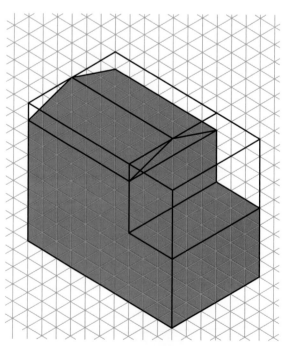

Figure 3-17. Add and remove lines to shape the house. See how points on the grid are used to draw the roof lines.

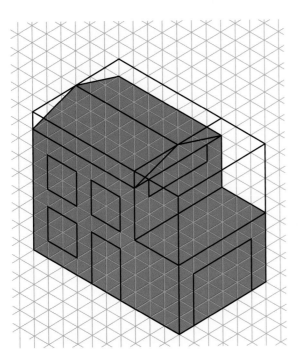

Figure 3-18. Add details such as doors and windows.

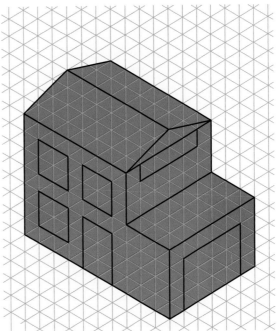

Figure 3-19. Remove construction lines to complete the house.

1.
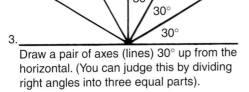
Draw a horizontal base line.

2.
Draw an axis (line 1) at right angle (90°) to base line.

3.
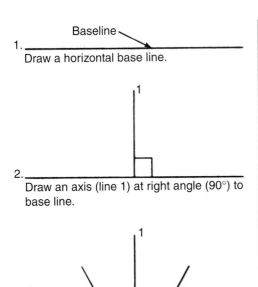
Draw a pair of axes (lines) 30° up from the horizontal. (You can judge this by dividing right angles into three equal parts).

4.
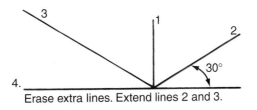
Erase extra lines. Extend lines 2 and 3.

5.
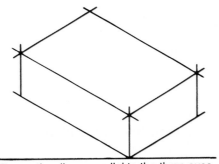
Draw other lines parallel to the three axes to complete the box.

Figure 3-20. You can make a freehand isometric sketch by following these steps.

Figure 3-21. In real life, objects at a distance seem narrower or shorter than they are close up.

Perspective sketching provides the most realistic picture of objects. The sketches are drawn to show objects as we would actually see them. Parallel lines converge and vertical lines become shorter as they disappear into the distance. Refer to **Figure 3-22** as you read the following steps for drawing a perspective sketch of a block:

1. Draw a faint horizontal line to represent the horizon. Mark two points, one at each end of the line. These are vanishing points (VP).
2. Draw the front vertical edge of the block.
3. Draw faint lines from each end of the vertical edge to the vanishing points.
4. Draw vertical lines to represent the left and right edges of the block. The length of these vertical sides will be shorter than the real object.
5. Join the top of these vertical lines to the vanishing points. Darken the outline of the object.

Orthographic Projection

You have learned that isometric and perspective sketches are quick methods of recording your ideas and communicating them to other people. They give a general idea of the shape and features of an object. Unfortunately, there are some disadvantages to isometric and perspective sketches. For example, they do not describe the shape of an object exactly because of distortion at the corners, nor do they provide complete information for the object to be made.

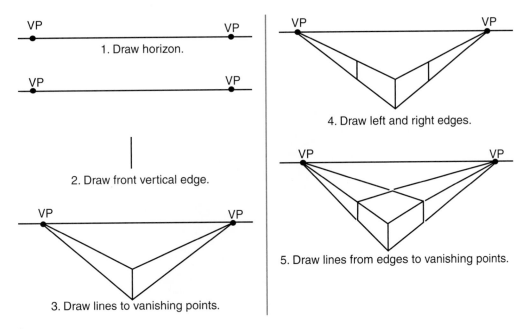

Figure 3-22. In a perspective sketch, lines become shorter as they recede into the distance.

Orthographic projection overcomes both these problems. This kind of drawing shows each surface of the object separately, as if you were looking straight at it. The viewing angle is at right angles to the surface. In this way, you see the exact shape, or view, of each surface. Complete information is usually given by drawing three *views*: front, top, and right side. To understand how a view is produced, imagine that you are the person in **Figure 3-23**. Because you are looking at the object "square on," you will only see the area that is colored red. Since this is the front of the object, this view is called the *front view*.

To produce a top view, imagine you are above the object looking down at it. The view you would see is shown in blue. The right-side view, shown in green, is drawn by looking at the right side square on. These three views are always arranged as shown in **Figure 3-24**.

Figure 3-23. An orthographic view shows only one side of an object. This person can see only the side of the block shaded red.

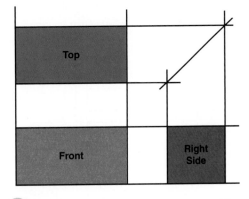

Figure 3-24. The proper arrangement for orthographic views of an object.

To draw an orthographic projection of the house in **Figure 3-25**, complete the steps described in **Figure 3-26**. It is easiest to learn to make orthographic projection drawings by working freehand on square grid paper.

Figure 3-25. The house shown in this isometric view is shown in orthographic views in Figure 3-27.

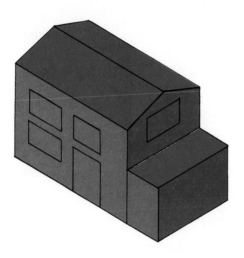

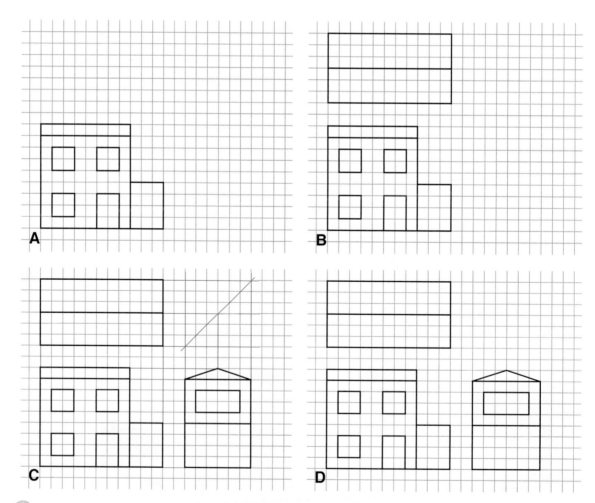

Figure 3-26. To develop an orthographic sketch, follow these steps. A—Draw the front view. B—Project the vertical lines of the front view above the drawing and draw the top view. C—Draw projection lines as shown to complete the right-side view. D—Darken object lines and erase projection lines.

Drawing Techniques

Sketches are drawn freehand, with or without grid paper. To create a more accurate orthographic drawing, you can use plain paper and drawing instruments. The instruments most often used are the T-squares, 45° and 30°/60° triangles, compass, and scale (ruler). Techniques for drawing with these instruments include:

- As a general rule, when drawing lines with a T-square, draw in the direction the pencil is leaning, as shown in **Figure 3-27**.

- When drawing vertical lines with a drafting triangle, lean the pencil away from you and draw the lines from bottom to top. See **Figure 3-28**.

- When using a 45° or 30°/60° triangle, draw lines in the directions shown by the arrows in **Figure 3-29**.

- Hold a compass between the thumb and forefinger and rotate clockwise. Lean the compass slightly in the direction of the rotation as you draw a circle, as shown in **Figure 3-30**.

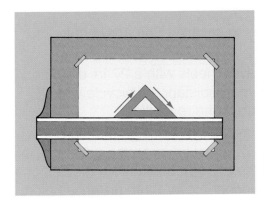

Figure 3-27. Draw in the direction the pencil is leaning.

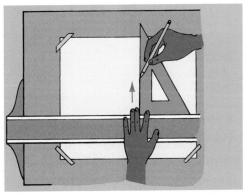

Figure 3-28. Note how the pencil is held to draw vertical lines using a triangle.

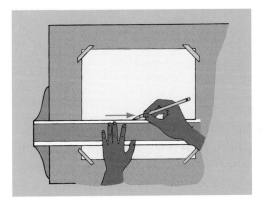

Figure 3-29. Draw sloping lines using a triangle, as shown here.

Figure 3-30. When using a compass, draw circles or arcs lightly at first. Make repeated turns to darken the line.

Math Application

Circumference of Circles

Many designs include circles or circular features, and the dimensions of these features must be included on your design drawings. This may involve mathematical calculations.

To find the length of a border that includes a circular shape, you need to find the *circumference* of, or distance around, the circle. See **Figure A**. The formula for calculating the circumference of a circle is $C = \pi \times D$ and is usually written:

$$C = \pi D$$

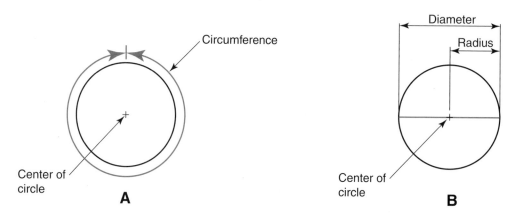

A

B

D stands for the *diameter* of the circle (the distance from one side of the circle to the other through the center). See **Figure B**. The symbol π (pronounced "pie") represents the number 3.14 (rounded to two decimal places). So, the circumference of a circle is 3.14 times its diameter.

For example, suppose your design includes a circle that is 8 inches in diameter. The circumference of the circle would be:

$$C = 3.14 \times 8 \text{ in.} = 25.12 \text{ in.}$$

Math Activity

Apply the formula explained above to calculate the answers to the following problems.

1. Most bicycles used on the road have wheels with a 27-in. diameter. Using the correct formula, calculate the distance a bicycle wheel travels on the pavement when it turns one revolution.

2. Suppose that you have created a new bicycle design that calls for 30-in. wheels. How far would a 30-in. wheel travel on the pavement in one revolution? After calculating the answer, think about the design implications. Which tire would last longer (cover more distance before failing)? Which tire would be more expensive to build? Why?

Alphabet of Lines

A number of different types of lines are used to produce orthographic drawings. Each line is used for a particular purpose and should not be used for anything else. Look at the casting in **Figure 3-31A**. The *alphabet of lines* can be used to produce the orthographic drawing of this casting shown in **Figure 3-31B**. The alphabet of lines consists of the standard line types and widths used on technical drawings.

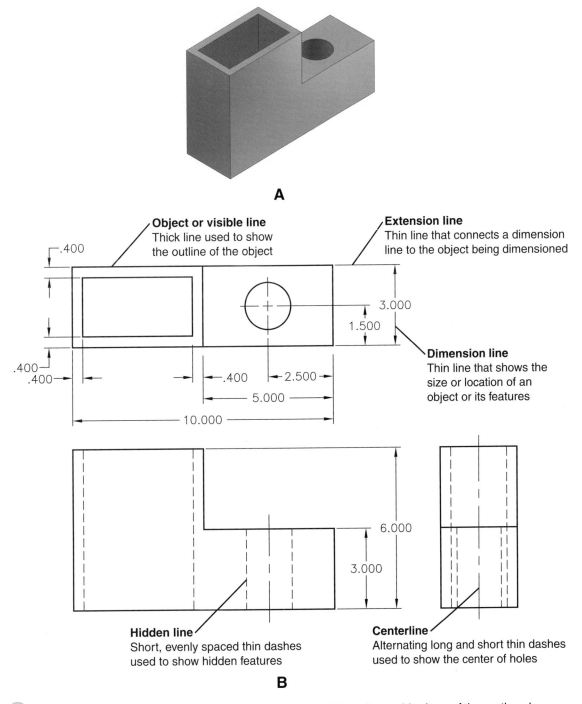

Object or visible line
Thick line used to show the outline of the object

Extension line
Thin line that connects a dimension line to the object being dimensioned

Dimension line
Thin line that shows the size or location of an object or its features

Hidden line
Short, evenly spaced thin dashes used to show hidden features

Centerline
Alternating long and short thin dashes used to show the center of holes

Figure 3-31. A—An isometric view of a metal casting. B—The orthographic views of the casting shown in A. Note the types of lines and their uses.

A few extra rules apply to hidden lines, as shown in **Figure 3-32**. For example, they almost always begin and end with a dash touching with the line where they start and end (1). However, this rule is not followed when the dash would continue a visible line (2). Dashes should join at corners (3) and (4). The dashes of parallel hidden lines that are close together should be staggered (5).

Dimensioning

Most drawings include two types of dimensions: overall dimensions and detail dimensions. To fully describe the size and shape of the view in **Figure 3-33A**, you need two overall and two detail dimensions.

If a hole is added to this view, then you must add the dimensions shown in **Figure 3-33B**. One dimension shows the size, or diameter, of the hole. The other two dimensions show the exact location of the center of the hole.

Notice the position of the dimensions in **Figure 3-33B**. Smaller dimensions are placed inside the larger, overall dimensions. This is the preferred placement for dimensions.

Scale Drawings

Ideally, objects should be shown at their full size in an orthographic projection. However, some objects are too large to fit on a sheet of paper. Others are so small that if you show them at their actual size, the details are too small to see clearly. These objects are represented on paper using a *scale drawing*. The objects in a scale drawing are larger or smaller than the object by a fixed ratio. Examples of scaled drawings are an architect's drawing of a building, a map, and an electronic engineer's schematic of a printed circuit.

If you wanted to draw a full-size front view of the skateboard in **Figure 3-34**, you would need a piece of paper larger than the skateboard. Full-size is a scale of 1:1. Each inch of the drawing paper represents 1" of the actual object. If you are working in metric, each centimeter of the drawing paper represents 1 centimeter of the actual object.

A drawing that is one-half of full size has a scale of 1:2. In this case, each inch (or each centimeter) on the drawing paper represents 2" (or 2 cm) of the actual object. Thus, the actual object would be twice the size of the views on the drawing paper.

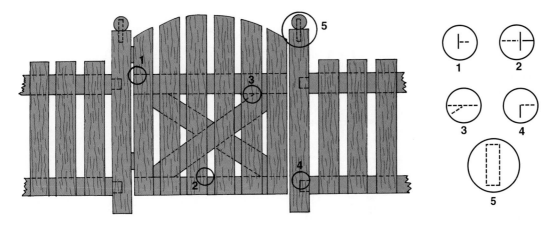

Figure 3-32. Rules for creating hidden lines in a drawing.

If an object to be drawn is very small, it may be necessary to prepare drawings to a scale larger than full size. Such a scale is referred to as an *enlarged scale*. For example, a drawing that is twice full size has a scale of 2:1. Each 2″ (or each 2 cm) on the drawing paper represents 1″ (or 1 cm) of the actual object. The parts of the compass shown in **Figure 3-35** are drawn twice their actual size.

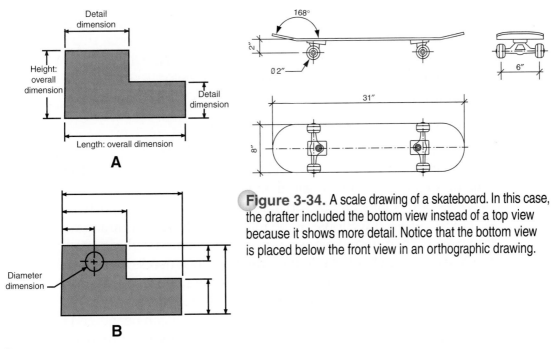

Figure 3-33. A—Overall and detail dimensions. B—Dimensioning the size and location of a circle.

Figure 3-34. A scale drawing of a skateboard. In this case, the drafter included the bottom view instead of a top view because it shows more detail. Notice that the bottom view is placed below the front view in an orthographic drawing.

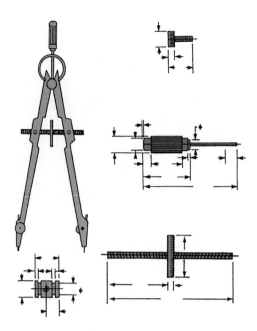

Figure 3-35. In this drawing, the compass is shown at full size (1:1). Because some of the individual parts are too small to see clearly at full size, they are shown separately, enlarged to a scale of 2:1.

Computer-Aided Design

In the past, people used the tools described previously in this chapter to create drawings. Now, however, very few drafters use drafting boards. They still need to know the types of drawings and how to construct them, but they make most drawings using computers.

This form of drawing is called *computer-aided design (CAD)*. A typical CAD system has three types of devices or parts:

- Input device—gives information or instructions to the computer.
- Processor—carries out the instructions.
- Output device—makes or displays the drawing.

Drawing commands are given by typing on a keyboard, selecting from a menu, or picking from a digitizing tablet. The designer can create the drawings, add details, and call up title blocks and other standard information. See **Figure 3-36.** Corrections can be made quickly. Also, nothing needs to be drawn more than once. Parts of a drawing that are used repeatedly can be stored in a file and loaded into the drawing when needed.

Other advantages of using a CAD system include the ability to rotate a 3D image and see it from various angles. CAD drawings can also be scaled up or down easily. In the case of a house, it can be viewed from any angle and the future owner can actually see how the finished building will look. In CAD programs designed for architectural drawing, the software can add the elevations (front and side views) automatically to a plan drawing (overhead view).

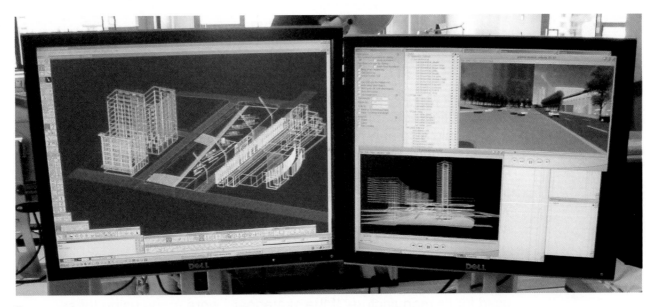

Figure 3-36. Computer-aided design can be used to develop complex drawings, such as these drawings of high-rise buildings..

The most advanced CAD systems create a virtual reality. *Virtual reality* is an artificial environment provided by a computer that creates sights and sounds in three-dimensional form. It is possible to do a virtual "walk-through" of a building to give the viewer a feeling of actually being in various rooms of the building. The viewer controls the path through the building and can look all around each room, up at the ceiling, down at the floor, or even out the windows. Virtual reality allows architects and engineers to spot errors before a building or device is built.

These advanced design systems have many advantages over drawing by hand. The computer works at high speed. The designer does not have to spend hours producing perfect line work and lettering. The CAD system makes them perfect the first time and every time, and it relieves the designer of repetitive tasks.

Some of the newer CAD programs integrate all aspects of designing, making, and supplying materials for a project. For example, an architect designing a building can find out, at any point, the cost of materials. The materials list can be sent directly to the building contractors, along with drawings and contract documents. Features like these leave more time for creative work. They also help all of the people working on the project communicate more effectively.

Other CAD programs link directly with computer-aided manufacturing software (CAD/CAM). Many types of CAD/CAM systems have been developed. In general, CAD/CAM systems allow computer-controlled machines to build parts from the information in CAD files.

Technology Application

Technical and Artistic Design

When you design a product or service, you need to keep in mind both function and appearance. Items must be solidly constructed, but they should also bring out feelings that make a potential buyer want to purchase them. For example, a potato peeler must peel potatoes easily and quickly, but it should also have an attractive shape that makes a user want to pick it up. A well-designed chair on display in a shop window must look like it can support its user and at the same time bring about a feeling that it could provide total comfort.

Technology Activity

Figure 3-34 is a technical drawing with details that would be used for its manufacture. If the skateboard were made using these measurements, it would probably work very well. However, it would not be very attractive. Your eye would not be drawn to look at it, because it lacks an interesting, appropriate design. It has a good shape and form, but it has no color or pattern. There is no sense of movement. It does not give the feeling that it would be a fast board if you were to buy it and use it.

Your task is to design a pattern that could be painted onto this skateboard. Follow these steps:

1. Draw the shape of the board to half scale on paper.

2. Ask yourself what design would give you a feeling of speed. How will this translate into the kind of pattern you will draw? Will it be geometric or a free-flowing design? Will it represent an actual object or will it be an abstract design? Will it be a traditional design or unconventional, like graffiti? It is your choice, but it must be appropriate for a skateboard!

3. Complete your pattern and present it to the class.

Concurrent Engineering

Concurrent engineering is a team effort that involves continuous communication among the entire design and production team from the very beginning of the design stage. The customer, project manager, and marketing staff also join in the process. Changes by any of these people are immediately passed on to others. Because of increased communication, concurrent engineering saves time. The product is produced quickly and meets the customer's needs.

3D Printing

Today, many industries use some form of *3D printing*. This term originally referred to small machines that could create plastic parts from a CAD file. Now, however, it can mean any system that uses *additive fabrication* to create a part from a CAD file. *Additive fabrication* is a long term that means the part is built by adding layer upon layer of plastic or metal powder. Design information is sent directly to the 3D printer. Here the data is numerically sliced into thin layers. The 3D printer then creates each two-dimensional cross section using a liquid, powder, or sheet material and bonds it to the previous layer. A complete part is built by stacking layer upon layer until the part is completed.

Specific 3D printing systems are often known by their output. For example, 3D printers that are used to create prototypes quickly during the design stage of product development are called *rapid prototyping (RP) systems*. See **Figure 3-37**. By using prototypes in real-life field tests, designers can better evaluate a product's strengths and weaknesses and avoid costly mistakes.

Figure 3-37. A student with her CAD file and 3D model of a three-dimensional puzzle she designed for the blind.

Some 3D printers can produce real parts that can be used in the actual products. These are called *rapid manufacturing systems*. This is an economical solution when only a few parts are needed. The manufacturer does not have to prepare expensive tooling. Instead, CAD information is sent directly to the 3D printer, which builds the part. A major advantage of these systems is on-demand manufacturing. The part is not manufactured until a customer orders it. This helps keep the manufacturer's costs down because no warehouse space is needed to store the part until it is purchased.

You do not have to own an expensive machine to take advantage of 3D printing technologies. 3D printers are becoming less expensive, and "personal" 3D printers are now available. See **Figure 3-38.** You can also create the CAD files and send them to a service provider for processing. The service provider has the machines capable of creating a part from your CAD files. For a fee, the service provider creates the part and sends it to you.

Figure 3-38. A personal 3D printer that builds physical models from CAD files.

End Note

Traditionally, drafters spent many hours creating exact drawings that could be used to manufacture a product. This process speeded up dramatically with the introduction of computer-aided drafting systems. Drawings could be more precise and took much less time to create.

3D printing has recently brought product design and manufacturing to a whole new level. They allow manufacturers to create a limited number of products for which demand is low. They no longer have to spend time and money creating the tooling to manufacture these products.

As 3D printer technology continues to improve, it will become more commonplace. You may even have a 3D printer in your home. In addition, scientists and technologists are considering 3D printing for applications such as organ transplants (bioprinting) and for home building, among others.

- The three basic types of communication technology are based on human hearing and sight.
- Many different forms of communication can be used to relay the same message.
- Design sketches can be isometric or perspective sketches. The purpose of a sketch determines which type you draw.
- Drawings created using orthographic projection contain all the views and information necessary to manufacture an object.
- Developing drawing techniques such as using the correct types of lines, dimensioning, and scaling drawings helps more people understand the message you want to convey.
- Freehand sketching remains an important skill for designers. However, computer-aided design (CAD) has replaced most types of manual drafting because it is faster and more accurate. Drawings and parts of drawings can also be reused, making this an efficient drafting process.
- Concurrent engineering is an efficient product development process in which all of the members of the design, development, and manufacturing teams are in communication from the very beginning of the project.
- 3D printing allows designers to experiment with prototypes. It also allows manufacturers to make low quantities of parts quickly and at a relatively low cost.

Preview and Prediction

Copy the following graphic organizer onto a separate sheet of paper. In the left column, record at least six predictions about what you will learn in this chapter. After you have read the chapter, fill in the other two columns of the chart.

What I Predict I Will Learn	What I Actually Learned	How Close Was My Prediction?

Test Your Knowledge

Write your answers to these review questions on a separate sheet of paper.

1. Name the three basic types of communication.
2. List at least three different forms of communication.
3. What is freehand sketching?
4. How many sides of a rectangular block are shown in an isometric drawing?
5. Which type of sketch provides the most realistic picture of an object?
6. What type of drawing describes the exact shape of each surface of an object?
7. What instruments are most commonly used in manual drafting?
8. Explain the process for creating a vertical line using a drafting triangle.
9. What is the alphabet of lines?
10. You have been asked to create a drawing of a printer that measures 14″ × 14″ × 8″. If you create the drawing at a scale of 1:2, how long will the overall length, width, and height lines be on your paper?
11. List the advantages of computer-aided design.
12. What is virtual reality?
13. Explain the concept of concurrent engineering.
14. What is a 3D printer?
15. What is the difference between rapid prototyping and rapid manufacturing systems?

Critical Thinking

1. Listen to a recording of whale or dolphin calls. Write an essay comparing and contrasting these calls to human speech.
2. Imagine that you are in a canoe in the middle of a large lake. You have been talking on your cell phone. While talking, you did not notice your paddle slip out of the canoe and float away. Unfortunately, you have run down the phone battery, so you can't call for help. You can see people on a dock at one end of the lake, but they do not seem to understand your shouts for help. How can you communicate a message for help?
3. Analyze the reasons for advances in drafting technology over the last 50 years. Prepare a report that answers the following questions. What factors or trends helped fuel the advances? What factors may have limited advances? Given the trends you have noted, what innovative technologies might you expect to be developed in the near future?
4. Write a persuasive paragraph explaining how the design of new communication technology does or does not depend on math and science skills.

Apply Your Knowledge

1. Create a symbol that can be used at a zoo to communicate the message: "Do not feed the animals!"

2. Design a logo (name or symbol) that you could use on your own personalized worksheets. Letters, geometric shapes, natural shapes, and simplified pictures are most appropriate for a logo.

3. Draw an isometric sketch, a perspective sketch, and an orthographic projection of a toothbrush.

4. Use a CAD system to make an isometric sketch of a tool you have used in the technology lab.

5. International Morse Code is a means of communication invented before telephone and e-mail. Check the symbols used. Then design a series of symbols that could be used to send a message by e-mail without using letters or words.

6. Research one career related to the information you have studied in this chapter. Create a report that states the following:

   - The occupation you selected
   - The education requirements to enter this occupation
   - The possibilities for promotion to a higher level
   - What someone with this career does on a daily basis
   - The earning potential for someone with this career

You might find this information on the Internet or in your library. If possible, interview a person who already works in this field to answer the five points. Finally, state why you might or might not be interested in pursuing this occupation when you finish school.

STEM Applications

1. **TECHNOLOGY** In a group of four students, create a message to be transmitted to the rest of your class. Brainstorm ideas for ways to communicate the message. Develop as many of these ideas as possible. Use both electronic and nonelectronic methods. Present your message to the class in at least five different ways.

2. **ENGINEERING** In a group of four or five students, brainstorm ideas for a creative and futuristic communication system that may help solve current or future human needs. Select your best idea and present it to the class.

3. **MATH** Select an object you use or see every day. Using a tape measure or rule, measure the dimensions of the object very carefully. Create an orthographic drawing of the object using manual drafting or CAD techniques. If the object is too large or small to be shown adequately on paper, create a scale drawing. Include dimensions.

Materials

Karim Rashid designed the Garbino2 and the Oh Chair

Karim Rashid is an industrial designer known for his work with new materials such as plastics, foams, and synthetic fabrics. Karim wants his work to inspire a sense of well-being through design. He thinks high-quality design should be accessible to everyone, not restricted to expensive, limited-run objects. He believes that an important aspect of good design is that it should be appealing to most people. His designs aim for the most simple, elegant shape that will meet the requirements of an object's purpose effectively and ergonomically.

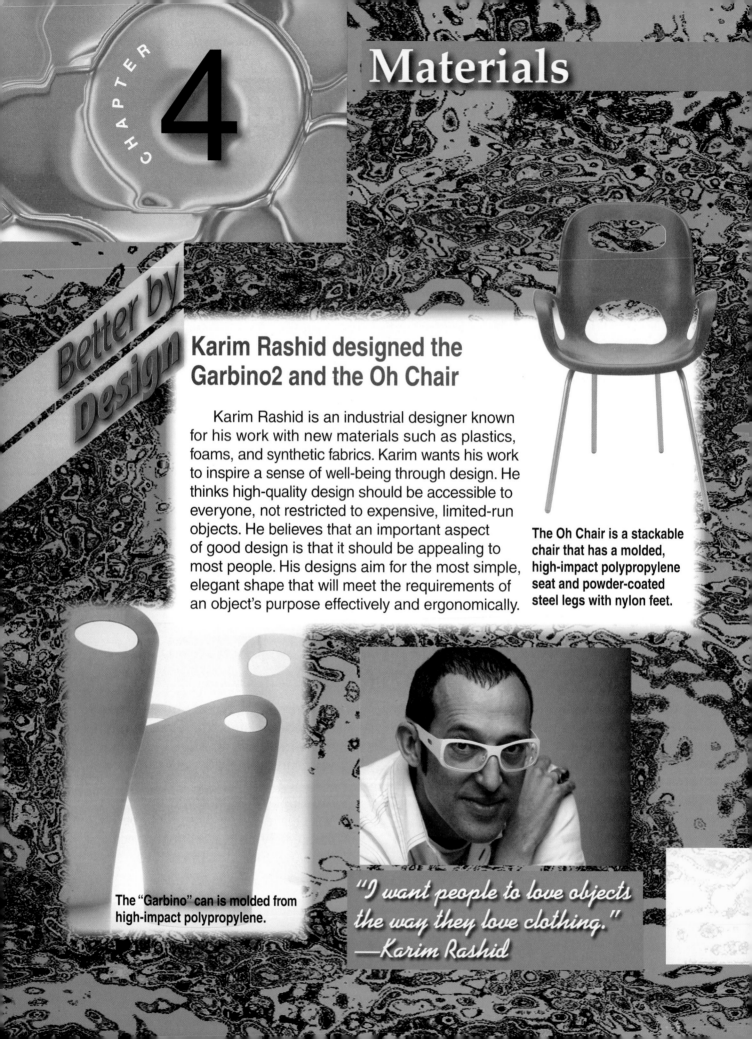

The Oh Chair is a stackable chair that has a molded, high-impact polypropylene seat and powder-coated steel legs with nylon feet.

The "Garbino" can is molded from high-impact polypropylene.

"I want people to love objects the way they love clothing."
—*Karim Rashid*

Reading Target

Key Terms

Objectives

Finding the Meaning of Unknown Words

Before you read this chapter, skim through it briefly and identify any words you do not know. Record these words using the Reading Target graphic organizer at the end of the chapter. Then read the chapter carefully. Use the context of the sentence to try to determine what each word means. Record your guesses in the graphic organizer also. After you read the chapter, follow the instructions with the graphic organizer to confirm your guesses.

acids
alloy
bamboo
bases
biobased
biodegradable

biomaterial
bioplastics
ceramics
composite
conductors
engineered wood
ferrous
hardwoods
insulators
magnetic
nanomaterials
nanotubes
nonferrous
opaque
pH scale
photosynthesis

plasticizers
polymer
primary materials
refractory material
semiconductor
sintered
smart materials
softwoods
synthetic
thermal conductivity
thermal expansion
thermoplastics
thermoset plastics
translucent
transparent

After reading this chapter, you will be able to:

- List the principle properties of materials.
- Explain the properties of acids and bases.
- Identify the types of primary (natural) materials.
- Describe how natural materials are processed into manufactured materials.
- Describe the properties of various advanced materials.

Useful Web site:
www.karimrashid.com

Designing and making products always involves the use of materials. Concrete is used for walkways, cotton for clothing, fiberglass for insulation, copper for electrical wiring, and plastic for bottles. See **Figure 4-1**.

Designers need to know about the properties of many different materials in order to choose the most appropriate material for the product being designed. The following are some examples of questions a designer might ask:

- Does the product have to withstand heat?
- Is color important?
- Should it be heavy or light?
- How strong does it have to be?
- Must it withstand bad weather?
- Does it have to conduct electrical current?
- What is the impact on the environment of using this material?

The material from which a product is made can affect how users perceive (see) the product. Most people think glass drinking utensils are more attractive than plastic. Wood has a warm appearance and feel.

The choice of materials also affects the function of a product. Most toothbrushes now incorporate one or more silicone inserts on the handle. These inserts not only feel more inviting but also help the user grip the toothbrush when the handle is wet.

Consumers are increasingly concerned with the environmental impact of materials. For example, there is an increasing demand for plastic products that are biodegradable. Some manufacturers are moving away from using plastics and are returning to wood and metal, which are less environmentally harmful.

Figure 4-1. Designers choose materials that have properties they want to use in their designs. What special property of acrylic did the designer use in the shape of this stool?

Properties of Materials

Designers and engineers judge materials by their properties. A material's properties affect how the material performs. Properties of materials can be grouped as follows:

- Physical
- Mechanical
- Thermal
- Chemical
- Optical
- Acoustical
- Electrical
- Magnetic

Physical Properties

Physical properties include a material's size, density, porosity, and surface texture. You can describe any material or product using its physical properties. Consider a pencil eraser, for example. It may be 2½" long, 1" wide, and ⅜" thick (64 × 25 × 10 mm). It is not very dense. Therefore, its mass is small. Its surface is smoother than a pine board. However, it is not as smooth as glass.

Mechanical Properties

Mechanical properties are the ability of a material to withstand mechanical forces. An elastic band will stretch and return to its original shape. A diving board will spring back. The head of a hammer will withstand sharp blows. The common mechanical properties are shown in Figure 4-2.

Thermal Properties

Thermal properties control how a material reacts to heat or cold. Materials will generally expand when heated and shrink when cooled. Some materials will also conduct heat.

Sometimes the expansion and contraction of metals causes problems. On an extremely hot day, steel railroad tracks may expand and buckle. The heat makes the molecules in the steel vibrate faster. Increased vibrations create more space between the molecules, increasing the size of the rail. The opposite happens in cold weather. In steel bridge construction, engineers add expansion joints to allow for movement.

Strength	Tension Compression Shear Torsion	*Strength* is a material's ability to withstand a mechanical force. Tension is a pulling force. A material that resists being pulled apart has tensile strength. An elastic band holding a package together is under tension. Compression is a squeezing force. A material that resists being crushed has compressive strength. The concrete pillars for a bridge are in compression due to the mass of the materials and the traffic on the bridge. Shear is a sliding and separating force. A material that resists separation has shear strength. Torsion is a twisting force. A material that resists torsion has torsional strength. A screwdriver blade must resist torsion when force is being applied to screws.
Elasticity		*Elasticity* is the ability to stretch or flex but return to an original size or shape. A material that resists elasticity has stiffness. A rubber band is elastic. A piece of glass has high stiffness.
Plasticity		*Plasticity* is the ability to flow into a new shape under pressure and to remain in that shape when the force is removed. Plasticity can be measured in two ways: 1. Ductility is a material's ability to be pulled out under tension. Chewing gum is very ductile. 2. Malleability is a material's ability to be pushed (compressed) into shape. Potter's clay is very malleable. The opposite of plasticity is brittleness. Glass is very brittle.
Hardness		*Hardness* is the ability to resist cuts, scratches, and dents. Harder materials wear less under use. Cutting tools such as knives, scissors, and drills should be hard. Diamonds are the hardest of all materials.
Toughness		*Toughness* is the ability to resist breaking. A hammer head should be tough so it will not shatter when it strikes other materials.
Fatigue		*Fatigue resistance* is the ability to resist constant flexing or bending. A springboard must have high fatigue strength.

Figure 4-2. Some common mechanical properties of materials.

Engineering Application

Material Fatigue

Most of us have experienced a moment when a light bulb suddenly burns out or a shoelace snaps. The bulb may have been turned on and off hundreds of times and the shoes could have been tied every day for months, but there comes a time when the material fails.

Engineers refer to this situation as *material fatigue*. In technical terms, it is the sudden breaking of a component after a period of cyclic loading. *Cyclic loading* is the repetitive movement that occurs in the normal use of a product. Failure often occurs due to the growth of a crack, usually at the site of a stress concentration on the surface.

The bulb burning out probably did not cause an accident unless, in the dark, you ran into a wall. However, if the cable connecting the brake handle on your bicycle to the brake mechanism snaps, or the metal rods in a reinforced concrete bridge fail, a terrible accident could happen.

Engineering Activity

Use a metal paper clip to simulate material fatigue. Follow these steps:

1. First straighten out the bends in the paper clip so it is flat.

2. Bend it back and forth, as far open and as far closed as possible, until it breaks. Count how many times you can bend the paper clip before it fails.

3. Repeat your experiment with additional paper clips, or ask your teacher to record the results of every class member's experiment. Be sure to use the same type of paper clip for all the trials. Plot the class's results on a graph.

4. Review the overall results. You will probably notice how the information is clustered. Many of you will have similar results. However, a few clips will break earlier, and a few will break later. If everyone's results are plotted on paper, or on the chalkboard, the result will be a bell-shaped curve.

5. Write a paragraph about your conclusions from this experiment. Explain how an engineer might use this knowledge to design products that do not fail. What information should be given to the consumer about possible failures?

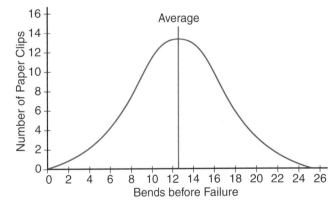

A bell-shaped curve occurs when most of the results fall within a certain range, with fewer results above and below that range. The curve for your class results may look similar to the one shown here.

This characteristic of metals, called *thermal expansion*, can be useful. A fire alarm is an example. A strip of one metal, such as brass, is joined to a strip of a second metal that has a different rate of thermal expansion, such as steel. When a fire causes this bimetal strip to heat up, one metal expands more than the other. Thus, a change in temperature causes the bimetal strip to bend. This movement closes an electrical circuit that activates the fire alarm. See **Figure 4-3**.

Thermal conductivity is a measure of how easily heat flows through a material. All metals conduct heat. Some do it better than others. Copper is a good conductor of heat. The copper bottom of a frying pan quickly conducts heat from the stove element to the pan. The metal pan then conducts this heat to the food.

Thermal insulators are materials that do not conduct heat well. Nonmetallic materials are generally thermal insulators. Plastic and wood handles on saucepans prevent heat from being conducted from the hot metal to your hand. A cooler used for preserving food on camping trips may have a casing filled with polyurethane foam to keep out heat. Foam panels or fiberglass batts are used to insulate walls and ceilings to reduce heat losses in a home. See **Figure 4-4**.

Chemical Properties

In order to understand how materials react to their environment, it is necessary to study atoms and molecules. An atom is the smallest unit of an element. Atoms are so small that several million could fit into the period at the end of this sentence. Chemical bonds between atoms form molecules and compounds. Molecules are made of specific combinations of atoms. For example, the first row (Formula) in **Figure 4-5** shows that oxygen molecules are made up of two oxygen atoms. Compounds are molecules that contain atoms from at least two different elements. Carbon dioxide is made of one carbon atom and two oxygen atoms. Water (H_2O) is the most abundant compound on earth.

Figure 4-3. Thermal expansion makes a bimetal strip bend.

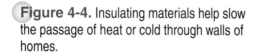

Figure 4-4. Insulating materials help slow the passage of heat or cold through walls of homes.

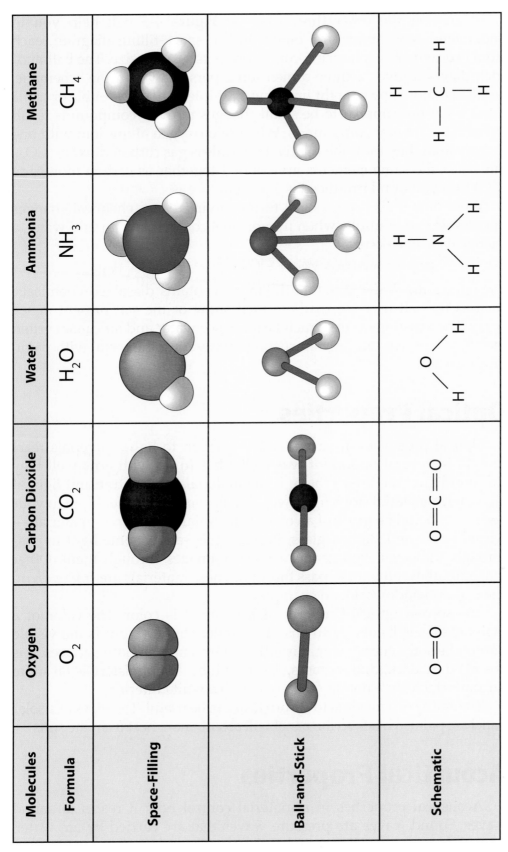

Figure 4-5. Atoms and molecules can be represented in different ways.

Examining the other three rows in **Figure 4-5** will help you to understand how atoms are joined. In the space-filling diagram, each atom is drawn as a sphere in proportion to its relative size. The ball-and-stick diagram gives a three-dimensional perspective. In the schematic, bonds are shown by straight lines and multiple bonds by double or triple lines. What may appear to be small changes to these compounds result in entirely different compounds. When two oxygen atoms join with one carbon atom they form the relatively harmless gas carbon dioxide (CO_2), but when the same elements are bonded together as carbon monoxide (CO) the compound produced is a poisonous gas.

When materials are exposed to the environment, chemical changes may occur. For example, when iron or steel contact both air and water, a chemical change occurs:

Iron + Oxygen + Water = Iron Oxide (Rust) + Water

Sometimes the water that hits the metal contains dissolved chemicals, such as the salt used on roads to melt snow or the salt present in sea spray. Then rusting occurs much faster. Since water and air cause certain metals to rust, we can prevent rusting by covering the metal with paint, oil, or grease.

Optical Properties

Optical properties are a material's reaction to light. Materials react to light in several important ways. One has to do with how well they transmit light that strikes them. Some materials cannot transmit light at all. When a material stops light, we say it is *opaque*. A roller blind in your bedroom should be made of an opaque material. *Translucent* materials—waxed paper and stained glass, for example—allow some light to pass through. However, you cannot see clear images through them. Other materials allow all light to pass through. These materials are *transparent*. Clear glass windows are an example.

The second optical property of a material is color. The color of a material affects its ability to absorb or reflect light. (Light is the visible part of the sun's energy.) Light is reflected by shiny, smooth surfaces. It is absorbed by dark, dull surfaces. A car with black upholstery is far more uncomfortable on a hot day than one with a white interior.

The ability of a material to absorb heat can be useful. The pipes of a solar panel are painted black so the panel will absorb more heat from the sun.

Acoustical Properties

Acoustical properties in a material control how it reacts to sound waves. Sound waves are pressure waves that are carried by air, water, and other materials. They are what the ears "hear."

Science Application

Chemical Symbols

There are just over 100 known substances, or elements, in the world. An *element* is the simplest form of matter and is made up of only one kind of atom or molecule. Each element is given a chemical symbol of one or two letters. Chemists, scientists, and technologists throughout the world use the same symbols, so there is never any need for translation.

For example, the letter H is the symbol for hydrogen, and the letter O is used for oxygen. These substances cannot be separated into smaller substances, but they can be combined into chemical *compounds* (substances made up of more than one kind of element). For example, the symbol H_2O represents water. It shows that water contains two molecules of hydrogen and one molecule of oxygen.

Other examples of chemical symbols are:

Ag	Silver	H	Hydrogen
Al	Aluminum	Li	Lithium
Au	Gold	N	Nitrogen
C	Carbon	Na	Sodium
Ca	Calcium	O	Oxygen
Cl	Chlorine	Pb	Lead
Co	Cobalt	P	Phosphorous
Cu	Copper	S	Sulfur
Fe	Iron	Zn	Zinc

Notice that only the first letter in a chemical symbol is capitalized.

Science Activity

The following chemical compounds are frequently used in manufacturing and processing technological designs. Name the elements contained in each compound. Also state the number of molecules of each element.

1. Calcium chloride (road salt for de-icing): $CaCl_2$
2. Ethylene glycol (car antifreeze): $C_2H_6O_2$
3. Iron oxide (rust found on steel): Fe_2O_3
4. Sodium bicarbonate (baking soda): $NaHCO_3$
5. Sulfuric acid (car battery acid): H_2SO_4

All sound energy is produced by vibrations. You can observe these vibrations if you dip a tuning fork into a glass of water. The vibrating fork will splash water out of the tumbler. See **Figure 4-6**. Sound energy travels through some materials. For example, a piece of string tightly stretched between two tin cans will carry a voice message over a short distance. Materials used in most musical instruments also transmit and amplify sound. See **Figure 4-7**.

The speed of sound through a material depends on the spacing of the molecules and how easily the molecules move. Sound travels faster in aluminum than in pine. This is true because the molecules in aluminum are closer together. They transmit sound energy more easily.

Materials vary in their ability to absorb sound. For example, acoustical tiles and heavy carpeting both absorb sound. The sound waves become trapped in the air pockets of the material. Hard materials such as the walls of a canyon reflect sound. Call out a name and it will bounce back at you as an echo.

Electrical Properties

Some materials conduct electricity, while others do not. Materials that can carry an electric current are called *conductors*. Those that do not conduct current are called *insulators*.

Figure 4-6. Sound energy is produced by vibrations.

Figure 4-7. Vibrating guitar strings cause the air inside the guitar sound box to resonate.

Metals are good electrical conductors. Some are better than others. One of the best is copper, which is often used in cables that supply electricity to lights, appliances, and machines.

Wires carrying electrical current must be insulated. They are covered with ceramic or plastic insulators.

Between these two extremes of good conductors and good insulators is a third type of material called a *semiconductor*. It allows electricity to flow only under certain conditions. Silicon and germanium are two important semiconductors. They are used in the production of transistors. Semiconductor materials are also used in devices that detect heat or light, such as the sensors in streetlights that turn them on when it gets dark.

Magnetic Properties

A *magnetic* material is one that is attracted to a magnet. The most common magnetic materials are iron, nickel, cobalt, and their alloys. Most other materials, such as wood, plastic, and glass, are nonmagnetic. A material's magnetic properties describe how the material reacts to magnetism.

Acids and Bases

Almost everything can be classified as either an acid or a base. For example, soap and the tablets we take when we have an upset stomach are bases. Household vinegar is an example of an acid.

Acids are sour-tasting chemicals. Many of the foods we eat, such as grapefruit, lemons, tea, vinegar, and yogurt, contain acid. Vitamin C is ascorbic acid, and aspirin is acetylsalicylic acid.

Some of the most powerful acids, such as nitric acid (HNO_3) and sulphuric acid (H_2SO_4), can burn holes in your skin or clothes. They can be used to etch precise patterns on glass or on silicon chips for printed circuits. Acids are used to produce car batteries, dyes, fertilizers, paper pulp, and plastics. However, even a powerful acid, when watered down, can be relatively harmless. When hydrochloric acid (HCl) is used full-strength it can dissolve metal, but in a weak form, it is in your stomach right now to help you digest food.

Bases are bitter-tasting chemicals that feel slippery or soapy to the touch. They include egg whites, ammonia, and oven and drain cleaners. Among the bases, sodium hydroxide (NaOH) is one of the most important. It is used in the manufacture of detergents, paper, and soap. It is also used in the production of acetate rayon and acetate fibers. Acids and bases are, in a way, opposites. They neutralize each other. If your stomach becomes too acidic, you may decide to take antacid tablets, made from basic chemicals, to quieten your upset stomach.

The strength of acids and bases is measured on the *pH scale*. See **Figure 4-8**. On this scale, items that are very acidic have a pH of 0 or 1, and items that are very basic have a pH of 14. Weaker acids and bases have pH values somewhere in between. A pH of 7 is neutral. It is neither acidic nor basic.

Types of Materials

Many materials exist in nature. These natural materials are called *primary materials*. Examples of primary materials include trees, clay, crude oil, and iron ore. These materials are seldom used in their natural state. Usually they are changed so that they are more useful. For example, trees are cut into boards or planks. Clay is baked into pots, crude oil is refined into gasoline, and iron ore is processed to produce steel.

Figure 4-8. The pH scale shows the relative strengths of acids and bases.

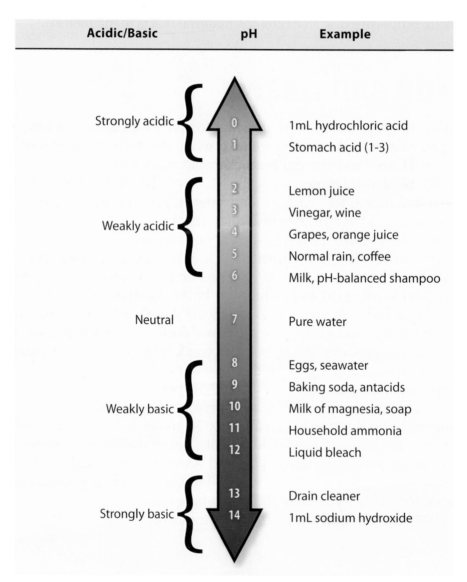

Acidic/Basic	pH	Example
Strongly acidic	0	1mL hydrochloric acid
	1	Stomach acid (1-3)
Weakly acidic	2	Lemon juice
	3	Vinegar, wine
	4	Grapes, orange juice
	5	Normal rain, coffee
	6	Milk, pH-balanced shampoo
Neutral	7	Pure water
Weakly basic	8	Eggs, seawater
	9	Baking soda, antacids
	10	Milk of magnesia, soap
	11	Household ammonia
	12	Liquid bleach
Strongly basic	13	Drain cleaner
	14	1mL sodium hydroxide

Traditionally, materials have been classified into five groups:

- Woods (including manufactured boards that use waste wood products)
- Metals
- Plastics
- Ceramics
- Composites

In addition, bamboo, which is a treelike grass used for centuries in Asia, is now being used widely. Some advanced materials have recently been developed, including biomaterials, nano-materials, and smart materials.

Woods

The two families of wood are *softwoods* and *hardwoods*. Softwood trees are coniferous, or cone-bearing. See **Figure 4-9**. Coniferous trees retain their needlelike leaves and are commonly called *evergreen trees*. Three examples of softwood trees are cedar, pine, and spruce. Hardwood trees have broad leaves, which they usually lose in the fall. See **Figure 4-10**. They are also known as *deciduous trees*. Examples are birch, cherry, and maple.

The terms *softwood* and *hardwood* refer to the botanical origins of woods, not their physical hardness. For example, balsa wood is botanically a hardwood, yet it is physically very soft.

Like all living things, trees need nutrients in order to live and grow. Their roots absorb water and mineral salts. Green plants use a process called *photosynthesis* to make food. Water containing valuable nutrients and minerals enters the tree through the roots. It passes up to the leaves where it mixes with carbon dioxide from the air. Here it is converted into sugars that are absorbed by the tree. During this process, the tree releases oxygen back into the air. See **Figure 4-11**.

Figure 4-9. Most softwood trees have needle-like leaves and bear cones.

Figure 4-10. Hardwoods are deciduous trees. Most of these trees lose their leaves in the fall.

Figure 4-11. A tree uses sunlight, water, carbon dioxide, and minerals to "feed" itself.

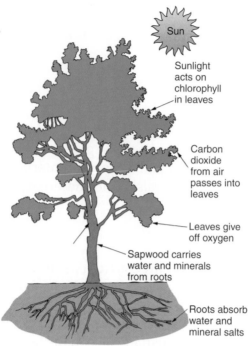

Sun

Sunlight acts on chlorophyll in leaves

Carbon dioxide from air passes into leaves

Leaves give off oxygen

Sapwood carries water and minerals from roots

Roots absorb water and mineral salts

A cross section of a tree trunk, **Figure 4-12**, shows the various parts of a tree. The outer bark protects the tree. Just inside the outer bark is the inner bark, which channels food from the leaves throughout the tree. Beneath the inner bark is a very thin, single-cell layer called the *cambium*. Here new wood cells are created, adding to the sapwood. Water moves through the sapwood to the leaves. The center part of the trunk is formed of the older, dead wood. This is the heartwood.

The internal cell structure of wood is shown in **Figure 4-13**. The cells conduct sap, store food, and provide the support system for the tree. During the spring, when plenty of water is available, larger cells are produced, which form the spring growth. During the drier seasons, the cells are smaller, heavier, and thicker-walled.

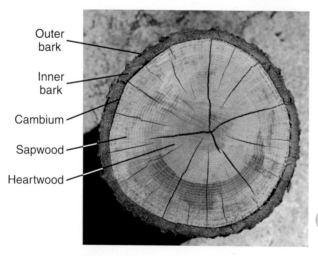

Outer bark

Inner bark

Cambium

Sapwood

Heartwood

Figure 4-12. The parts of a tree trunk.

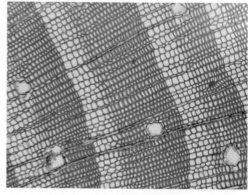

Figure 4-13. The cellular structure of trees transports food needed for growth. Growth is greatest during the spring season when more moisture is available.

The characteristics and properties of typical softwoods and hardwoods are summarized in **Figure 4-14**. To put wood to its best use, you should become familiar with these qualities.

To change the living tree into usable pieces of wood, the tree must be cut down, transported to the sawmill, sawn into boards, and dried in kilns. These steps are shown in **Figure 4-15**.

	Softwood			Hardwood		
Wood	 Spruce	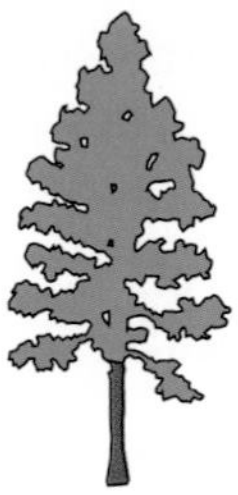 Pine	 Cedar	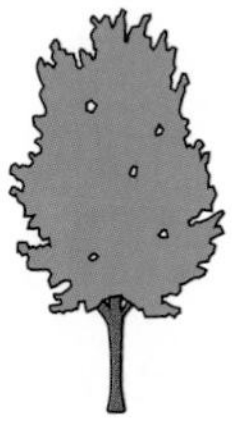 Birch	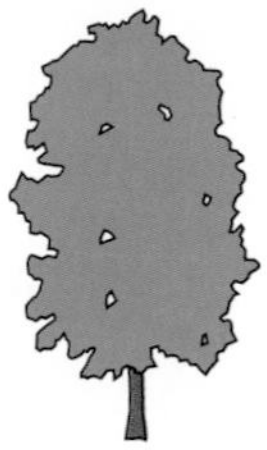 Maple	 Cherry
Resistance to decay	low	moderate	high	low	low	high
Strength	medium	low	extremely low	extremely high	extremely high	high
Color	white	white to creamy yellow	reddish-brown	light brown	golden brown	light brown
Uses	• building construction • floorboards • packing cases • newspaper	• indoor joinery • matchsticks • telegraph poles	• building construction • paneling	• plywood • furniture	• pool cues • furniture	• furniture

Figure 4-14. Characteristics and uses of common types of wood.

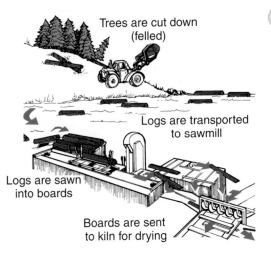

Trees are cut down (felled)

Logs are transported to sawmill

Logs are sawn into boards

Boards are sent to kiln for drying

Figure 4-15. The steps needed to convert a standing tree into finished lumber.

Wood that is used to make furniture is seasoned, or dried, to reduce its water content to about the same as the air surrounding it. Traditionally, this has been done by stacking the wood outside, which allows the moisture to evaporate. The process takes a year or more. Today, the wood is dried in a special oven called a *kiln*. Even kiln-dried wood will expand or contract as the moisture content in the air changes with the seasons.

Drying causes wood to shrink unevenly. This in turn causes the wood to warp, as shown in **Figure 4-16**. Also, wood expands and contracts as the humidity of the air changes. This is most troublesome when large pieces of solid wood are needed to make, for example, a tabletop. One way to overcome these problems is to join narrow boards edge-to-edge, as shown in **Figure 4-17**. However, this solution is time-consuming and costly. Today, the usual solution is to use manufactured boards.

Manufactured Boards

Manufactured boards include wood products such as plywood, particleboard, and other *engineered wood* products. You may also see engineered wood described as "composite wood" or "manufactured wood." In general, the purpose of engineered wood is to improve the performance or solve a problem associated with natural wood.

Plywood

Plywood is the strongest manufactured board. It consists of an odd number of plies (thin sheets of wood) glued together so the grain in each layer is at right angles to the grain in the layers above and below it. See **Figure 4-18**. Gluing the veneers in this way prevents the wood from twisting and warping.

The plies are cut from a log. The log is mounted on a huge lathe-like machine. A large blade cuts the wood, unrolling it like a reel of paper. See **Figure 4-19**.

Blockboard

Blockboard consists of a core of softwood strips faced with a veneer on each side. See **Figure 4-20**. The grain of the face veneer is at right angles to the direction of the core strips. Blockboard is strong and durable, but expensive.

Figure 4-16. Uneven expansion and contraction causes boards to warp.

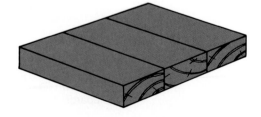

Figure 4-17. Joining narrow boards edge-to-edge reduces warping.

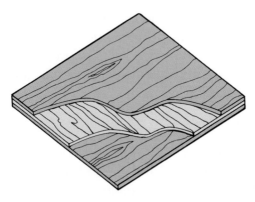

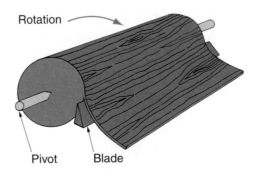

Figure 4-18. Plywood is made with an odd number of plies (layers). Placing each layer at right angles to the layer above and below it reduces warping.

Figure 4-19. A "ribbon" of wood is peeled from the log to create the plies for plywood.

Particleboard and Chipboard

Particleboard, also called *chipboard*, consists of wood chips and sawmill shavings bonded together with a synthetic resin and pressed into a flat sheet. It can be left in this form or faced with a more expensive veneer, such as mahogany or teak or a plastic laminate. It is rather brittle and difficult to join. However, it is cheap, so it is often used in the manufacture of less expensive furniture. See **Figure 4-21**.

Hardboard

Hardboard is made from wood fibers. These fibers are treated with chemicals. Then they are reformed into sheets using heat and pressure. Hardboard is usually smooth on one side and textured on the other.

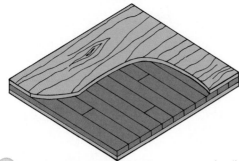

Figure 4-20. Blockboard has a core of solid pieces of softwood.

Figure 4-21. Particleboard is made from wood chips. They are glued and then pressed into a sheet. Sometimes the sheet is faced with a veneer.

Tempered hardboard can be made by impregnating standard hardboard with resin and then curing with heat. Tempering improves water resistance, hardness, and strength. Hardboard is made in thin sheets and is often used for the hidden parts of inexpensive furniture, such as the backs of cabinets or drawer bottoms. **Figure 4-22** shows several types.

MDF

MDF stands for "medium-density fiberboard." It is made from sawdust and has a smooth, even surface that is easily machined and painted or stained. It is frequently used for decorative moldings in house construction.

Bamboo

Bamboo is the fastest growing woody plant in the world. See **Figure 4-23**. It can grow over 3 feet (1 m) in a day and takes only five years to reach maturity, compared with decades for hardwoods. When bamboo is harvested (cut), the remaining plant is strengthened and continues to grow, whereas when a tree is felled it is killed.

Bamboo is a versatile material. It is most often made into boards that are stronger than most hardwoods. It is frequently used for flooring because it doesn't expand and contract like hardwood. It can also be used to make paper, furniture, and musical instruments. It can even be used in the bathroom for vanities, sinks, and bathtubs if sealed with marine varnish.

It can also be pulped and spun into the softest of fibers that can be made into fabric. Bamboo clothing, sheets, blankets, towels and even underwear and diapers are now available. See **Figure 4-24**.

Figure 4-22. Three kinds of hardboard. Sometimes a decorative face is fixed to one side.

Figure 4-23. Although bamboo is a woody substance, it is not a tree. It belongs to the same family as grass.

Figure 4-24. Bamboo can be made into fabric that is both eco-friendly and naturally antimicrobial.

Metals

Metals are inorganic materials. There are two families of metals: *ferrous* and *nonferrous*. The word *ferrous* is from the Latin word *ferrum*. It means "iron." Thus, any metal or alloy that contains iron is a ferrous metal. Metals or alloys that do not contain iron are nonferrous metals. **Figure 4-25** shows properties of common ferrous and nonferrous metals.

The internal structure of metals is crystalline. The crystals are made of atoms arranged in boxlike shapes. The atoms arrange themselves into a body-centered cubic structure with an extra atom at its center. **Figure 4-26** shows the cubic structure. Chromium, molybdenum, tungsten, and iron have this structure at room temperature. The way the atoms are combined determines the material's structure and properties.

In the face-centered cubic structure shown in **Figure 4-27**, each face of the cube has an additional atom at its center. Copper, silver, gold, aluminum, nickel, and lead have this structure.

Crystals are the basic units of metals. See **Figure 4-28**. As molten metal cools from the liquid state, its atoms bond together permanently, forming crystals. The crystals pack themselves together like the pieces of a jigsaw puzzle. Because temperature changes the way atoms are arranged, a particular metal may have different forms. Above 55°F (13°C), tin is a shiny metal. As the temperature drops, the atoms rearrange and the metallic tin begins to change into a nonmetallic gray powder. In cold cathedrals, tin organ pipes have sometimes disintegrated!

All crude (unprocessed) metals are found in the ground in the form of ore. They are mixed with rock and other impurities. The ore is extracted from the ground by mining or quarrying. It is then crushed, and the waste earth and rock are removed. The remaining ore is *sintered* (formed by heat) into pellets.

	Metal	Important Content	Melting Temp.	Resistance to Corrosion	Characteristics	Color	Uses
Ferrous	Cast Iron	93% iron 3% carbon	2200°F (1204°C)	poor	-hard -brittle -heavy	dark gray	-bodies of machine tools -engine blocks -bathtubs -vises -pans
	Mild Steel	99% iron 0.25% carbon	2500°F (1371°C)	very poor	-stronger and less brittle than iron -an be easily joined by welding	gray	-girders in bridges -tubes in bicycle frames -nuts and bolts -car bodies
	High-Carbon (Tool Steel)	0.60-1.30% carbon	2500°F (1371°C)	very poor	-not easy to machine, weld, or forge	gray	-cutting tools -drill bits -self-tapping screws -wrenches -railroad rails
Nonferrous	Aluminum	base metal	1220°F (660°C)	high	-light, soft, and malleable -conducts heat and electricity -very difficult to solder or weld	gray	-cooking utensils -foil -siding -window frames -aircraft
	Tin	base metal	450°F (232°C)	excellent	-soft -nontoxic -shiny -most often used as an alloying agent	silver	-rolled foil -tubes and pipes -tinplate -ingredient of solder -galvanizing
	Copper	base metal	1980°F (1083°C)	high	-tough and malleable -good conductor of heat and electricity -expensive -easily joined by soldering	reddish-brown	-electrical wiring -cables -water pipes

Figure 4-25. The characteristics and uses of ferrous and nonferrous metals.

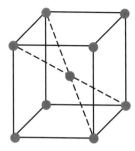

Figure 4-26. Ferrous metal has a cubic crystalline structure with one atom at the center of the cube.

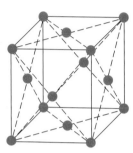

Figure 4-27. In nonferrous metals, each face of the structure has an atom at its center.

Cast Iron

Aluminum

Figure 4-28. These photographs of magnified samples of metal show some of the different crystal arrangements that can occur when metals are being formed.

The first step in making steel is to dump iron ore, coke, and limestone into the top of a blast furnace. The mixture is then heated to 2912°F (1600°C). The burning coke and very hot air melt the iron ore and limestone. The limestone mixes with the ashes and waste rock, forming a waste called *slag*. The molten iron sinks to the bottom of the furnace, and the iron and slag are tapped off separately. The iron is poured into large containers. It is then purified to become steel. **Figure 4-29** shows the entire process.

Metals are rarely used in their pure state. They are usually mixed with other metals to produce an *alloy*. In its liquid state, one metal can dissolve in another. Alloying can increase strength or hardness, inhibit rust, change color, or affect electrical or thermal conductivity. For example, mixing 10% aluminum with copper produces an alloy with approximately three times the strength of pure copper. **Figure 4-30** describes common alloys and their uses.

Gold, silver, and platinum are the rarest metals and most valuable precious metals. See **Figure 4-31.** They can be shaped easily and do not corrode.

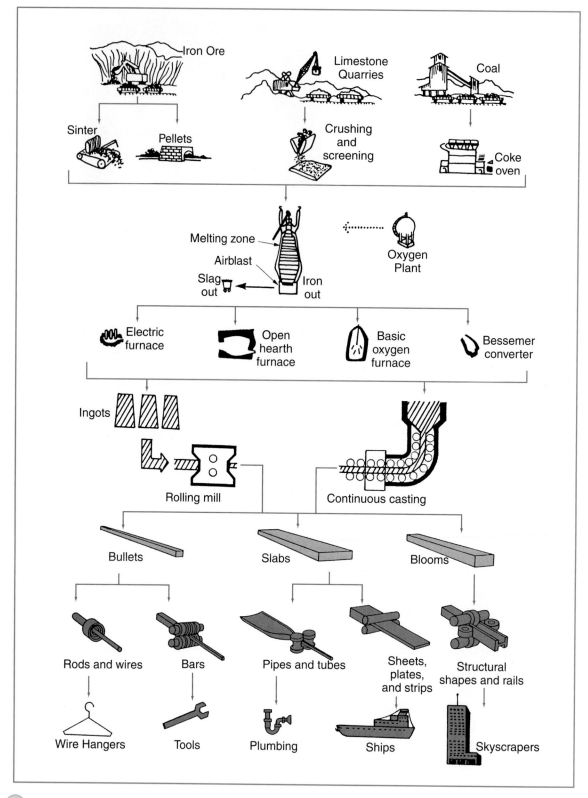

Figure 4-29. Processing iron ore into steel requires many steps. Each step must meet national standards for air and water pollution.

Alloy	Major Components	Uses
Brass	Copper and zinc	Electrical fittings, locks, hinges
Bronze	Copper and tin	Bells, castings, sculptures
Pewter	Tin and lead	Tableware, tankards
Solder	Tin and silver Tin and copper	Join metal components or wires
Stainless steel	Steel and chromium	Cutlery, sinks, surgical instruments

Figure 4-30. An alloy is a mixture of two or more metals. It combines the best characteristics of each metal.

Figure 4-31. Precious metals are combined with gemstones to make jewelry. These rings were crafted from white and yellow gold. The gems include emeralds, opals, blue sapphire, garnet, coral, and turquoise.

Gold can be hammered into sheets so thin they become transparent. It is also ductile and can be drawn into very thin wires. The purity of gold is measured in karats. Pure gold is 24-karat gold. In its pure state, gold is very soft, so it is alloyed with other metals. For example, the 14-karat gold used in jewelry and high-quality electrical connectors is 14/24 gold and 10/24 silver and copper. The gold in jewelry varies from nine carats (37.5% gold) to 18 carats (75% gold).

Silver is a little harder than gold, but is still very workable. It also has the highest electrical and heat conductivity of all metals. It is used in jewelry, electrical contacts, photography, dental alloys, coinage, and solder.

Platinum is rarer, denser, purer, and stronger than either gold or silver. One of its main uses is in car exhaust systems. The platinum converts harmful emissions into carbon dioxide and water. It is also used in jewelry and for neurosurgical and cancer treatment procedures.

Math Application

Graphing Melting Points of Metals

When you design products that contain metals, you have to take the properties of the metals into consideration. In fact, you may choose a certain metal based on its properties, such as its reflectivity (shiny or dull), weight, strength, and melting point. One way to compare the properties of several metals is to create a graph.

Graphs are used to present information in ways that can be quickly understood. The bar graph is one of the most common types of graphs. The bars on a bar graph represent different quantities of whatever is being measured. They are usually arranged in a logical order, such as lowest to highest or highest to lowest. For example, the bars on the graph below represent the electrical conductivity of four different metals. The items are arranged in order from lowest conductivity to highest.

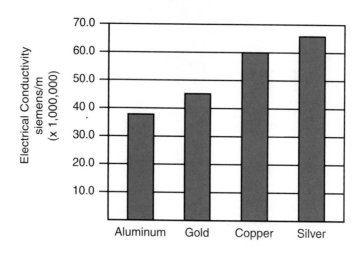

Math Activity

Follow the steps below to draw a bar graph. You will need a piece of graph paper and pencil for this activity.

Metal	°C	°F
Aluminum	660	1220
Brass	930	1710
Copper	1084	1983
Gold	1063	1945
Iron	1536	2797
Lead	327	621
Platinum	1770	3220
Silver	961	1760

1. Look at the list of melting points of various metals below. Select either the Fahrenheit or Celsius list.
2. Assemble the data in a logical sequence.
3. Choose a horizontal scale that makes it possible to draw all your data on one sheet.
4. Select a vertical scale that enables you to represent the largest amount with the tallest bar.
5. Write the temperatures on this vertical scale.
6. Write the names of the metals on the horizontal axis.
7. Draw the bars to the appropriate height for each metal.

Plastics (Polymers)

The scientific name for plastic is *polymer*. Polymeric materials are basically materials that contain many parts. "Poly" means many, and "mer" stands for monomer or unit. A polymer is a chain-like molecule made up of smaller molecular units.

Plastics are *synthetic* materials. Some of the earliest commercial polymers were made from cellulose. Cellulose is a major component of wood and cotton. Most plastics today are made from crude oil. See **Figure 4-32**.

Today, there is no single material called "plastic." Chemists can alter the mixture of components to create plastics that look and behave very differently from each other. Plastics have many important properties. These include:

- Ability to be colored
- Ease of molding
- Flexibility or rigidity
- Good electrical or thermal insulation
- Low mass (weight)
- Low cost
- Resistance to rot and corrosion
- Strength

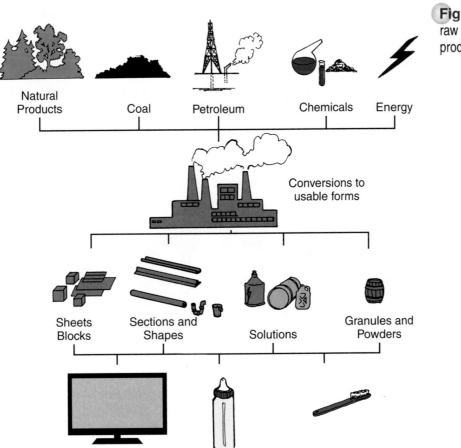

Figure 4-32. Processing raw materials into plastic products.

Natural Products   Coal   Petroleum   Chemicals   Energy

Conversions to usable forms

Sheets Blocks   Sections and Shapes   Solutions   Granules and Powders

To understand how a plastic is made, recall that atoms can combine with one another to form molecules. These molecules, joined together in chains, form the fundamental building blocks of a plastic material. Plastics include two basic categories: thermoplastics and thermosets.

Thermoplastics

Thermoplastics are materials that can be repeatedly softened by heating and hardened by cooling. The weak bonds that exist between the chains of thermoplastics can easily be broken by increasing their temperature. See **Figure 4-33**. When this happens, the chains can move past one another. The polymer becomes softer and can easily be reshaped by applying pressure.

Thermoplastics are not used to make objects that must resist high temperatures. Examples of thermoplastics include acrylic, polyethylene, polyvinyl chloride (PVC), nylon, and polystyrene. Dentists use thermoplastic to produce a detailed replica of your teeth and the shape of the tissues in your mouth. When the thermoplastic material is heated, it becomes soft and able to take up a new form. When it cools to mouth temperature, it hardens and can be removed, retaining the impression of the oral cavity.

The same plastic can take different forms. For example, PVC is used to make rigid plumbing pipes and soft vinyl shower curtains. The pliability of the material is changed by combining it with various chemicals called *plasticizers*. These act as internal lubricants, allowing the long chains of PVC molecules to slide past one another.

Thermoset Plastics

Unlike thermoplastics, *thermoset plastics* can be heated only once. See **Figure 4-34**. "Thermo" means heat, and "set" means permanent. Heat causes the material to develop cross-links between the chains. Once connected, the individual chains form a rigid structure. The plastic is no longer affected by heat or pressure. Once a thermoset has been heated and formed, its shape cannot be changed. It will burn or char before it melts.

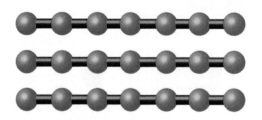

Figure 4-33. The molecular structure of a thermoplastic material has weak bonds that can easily be broken by heat.

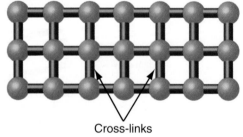

Cross-links

Figure 4-34. The molecular structure of a thermoset material is cross-linked and the bonds cannot be broken by heat.

Thermoset plastics have good heat resistance. Examples include polyester resins (often reinforced with glass fiber), phenol resins, urea resins, melamine resins, epoxy, and polyurethane. The characteristics and uses of typical thermoplastics and thermosets are summarized in **Figure 4-35**.

	Plastics	Characteristics	Uses
Thermoplastics	Polyethylene (Low Density)	-fairly flexible -soft -cuts easily and smoothly -floats -"waxy" feel -not self-extinguishing -transparent when thin -translucent when thicker -can be dyed	-detergent squeeze bottles -plastic bags -electrical wire covering
	Polyvinyl Chloride	-rigid or flexible -transparent -fairly easy to cut -smooth edges -sinks -self-extinguishing	-water pipes -records -raincoats -soft-drink bottles -hoses
	Polystyrene	-opaque -tends to crumble on cutting -very buoyant in water -burns readily -very lightweight	-packing material -insulation -ceiling tiles -disposable food containers
Thermoset Plastics	Polyester Resin	-stiff -hard, solid feel -difficult to cut -brittle -burns readily	-repair kits -car bodies -boat hulls -garden furniture
	Urea Formaldehyde	-opaque -stiff, hard, solid feel -flakes on cutting -burns with difficulty -good heat insulator	-domestic electric fittings (e.g., plug tops, adaptors, switch covers) -waterproof wood -adhesives
	Melamine	-opaque -stiff -hard, solid feel -flakes on cutting -burns with difficulty -good heat insulator -resists staining	-tableware -surfaces for counters, tables, and cabinets -expensive electrical fittings

Figure 4-35. Characteristics and uses of plastics.

Bioplastics

A new generation of materials called *bioplastics* can be processed like thermoplastics on existing equipment, with some modification. Most bioplastics are *biobased*; that is, they are made from renewable raw materials such as corn, rice, sugar cane, and soybeans. Bioplastics are ideal for packaging and for many single-use disposable items. These include food bags, gift cards, and food takeout containers, plates, utensils, and cups.

Current developments in reinforcing biopolymers with tiny plant fibers will produce materials that are longer-lasting. Major electronics companies and car manufacturers are experimenting with corn-derived compounds for use in cell phone bodies, car seat fabrics, and other components. In time, bioplastics may help overcome our dependence on plastics made from petroleum. See **Figure 4-36**.

Figure 4-36. Bioplastics used in these products are biodegradable.

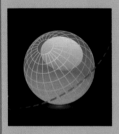

Think **Green**

Disposing of Biodegradable Products Responsibly

Some bioplastics are *biodegradable*. This means they can be degraded, or broken down, over a period of time by bacteria and other microorganisms present in soil, water, and compost. However, because landfills cover waste materials, they are not exposed to air, moisture, and sunlight, which would help the microorganisms break down the materials. Therefore, bioplastics and other organic materials rarely biodegrade in landfills. For best results, biodegradable products should be placed in home composts or in the controlled high heat and moisture environment of an industrial compost facility.

Many people are unaware of this. They think using biodegradable products is enough. How can you help raise awareness in your community about how to dispose of biodegradable products correctly?

Identifying Plastics

The most common types of plastic can be identified by the number inside the recycling symbol molded onto the base. The table in **Figure 4-37** describes the types of plastic associated with each symbol.

Symbol/Number	Name of Plastic	Description
1 PET	PET (polyethylene terephthalate)	PET is most often found in water and soft drink bottles. It is very difficult to decontaminate. It may also leach carcinogens (cancer-producing substances). PET bottles should not be reused.
2 HDPE	HDPE (high-density polyethylene)	HDPE is found in milk, laundry detergent, and cooking oil bottles, as well as in toys and grocery bags. It is one of the safest forms of plastic. Products made from HDPE can be reused.
3 V	PVC (polyvinyl chloride)	PVC is often used for food wrapping, food containers, shampoo bottles, teething rings, pet toys, and blister packaging. It can leach toxins (poisons) throughout its entire life cycle. It should not be used to reheat food in the microwave oven and should not be reused.
4 LDPE	LDPE (low-density polyethylene)	LDPE is most often found in bread bags, shrink-wraps, garment bags from dry cleaners, squeezable bottles, and plastic bags. It is less toxic than some other plastics. It can be reused by adults, but not by small children. Children are at risk of suffocation by plastic film.
5 PP	PP (polypropylene)	PP is used to make disposable diapers, carpeting, bags, and food wraps, as well as ketchup, margarine, and yogurt containers. It is one of the safest plastics, so products made from PP can be cleaned and reused.
6 PS	PS (polystyrene)	PS is used to make plastic cutlery. Meat trays, egg cartons, and foam packaging made from PS should not be reused.
7 OTHER	Other plastics	This includes acrylic and acrylonitrile butadiene styrene (ABS). However, the number is also used for layered or mixed plastics, so these products should not be reused.

Figure 4-37. The number and symbol molded into the base of the product identifies the plastic from which it is made.

Ceramics

The word *ceramics* is from the Greek *Kermos* meaning "burnt stuff." The important characteristics of ceramics are hardness, strength, resistance to chemical attack, and brittleness. They are also resistant to heat. The space shuttle is covered with ceramic tiles to protect it from the heat of friction generated by a high-speed re-entry into the earth's atmosphere. Today, the term *ceramics* covers a wide range of materials, including abrasives, cement, window glass, and the porcelain enamels on bathroom fixtures.

One of the great advantages of ceramics is the abundance of silicon and oxygen, the raw materials used to make them. Silicates include materials containing silicon and oxygen, such as sand, clay, and quartz. They are the most common minerals on earth. Silicon dioxide, SiO_2, is the basis of many ceramics and is readily obtained from common sand deposits.

Some ceramics are refractory materials. A *refractory material* is one that retains its strength at high temperatures. Refractory materials are used in linings for furnaces, kilns, incinerators, and reactors.

Most ceramics are thermoset materials. Once they have been processed and hardened, they cannot be made soft and pliable again. The one exception is glass. It can be continuously reheated and reshaped. It is, therefore, thermoplastic in nature. **Figure 4-38** lists the characteristics and uses of common ceramics.

Two of the most familiar ceramic materials are glass and cement. Their manufacture is shown in **Figure 4-39** and **Figure 4-40**. The following are some basic characteristics of ceramics:

- Strong and resistant to attack by nearly all chemicals
- Withstand high temperatures
- Stiff, brittle, and rigid
- Very stable (not likely to change shape because of heat or weather)
- High melting point
- Hardest of all engineering (solid) materials
- Raw materials are available worldwide and are consequently low in cost
- Withstand outdoor weathering from the sun, moisture, environmental pollutants, and dramatic temperature changes
- Poor electrical and thermal conductivity

Composite Materials

When two or more materials are combined, a new material, a *composite*, is formed. For example, concrete is a composite. It combines cement, sand, and gravel. Fiberglass is a composite of glass fibers and plastic resin.

Combining materials produces a new material. Each material in the composite keeps its own properties, but combining them also adds new properties.

Type	Characteristics	Uses
Stone	-hardness varies from soft sandstone to hard granite -color varies -surface appearance varies -resistant to corrosion -opaque	-building material -abrasives
Clay	-becomes workable when mixed with water -has a bonding action on drying	-house bricks -tiles -earthenware -stoneware -china -porcelain
Refractory Material	-stable at high temperatures -can withstand pressure when hot -can withstand thermal shock -highly resistant to chemical attack	-lining of furnaces to produce glass, cement, metals, bricks -insulation on spark plugs -tiles on space shuttle -firebrick
Glass	-transparent -poor tension -transmits visible light well -opaque to ultraviolet light	-fibers for insulation -fibers for filters -windows and doors -heat resistant cookware -light bulbs -drinking glasses -lenses
Cement	-correctly called Portland cement -made from limestone (80%) and clay (20%) heated in a kiln to form a solid material called *clinker* that is then ground to a fine powder -very durable -can withstand frequent freezing and thawing -can withstand wide range of temperatures	-the ingredient of concrete that binds the crushed rock, gravel, and sand together

Figure 4-38. Ceramics have characteristics that make them useful for many products. What characteristics make them particularly suitable for building materials?

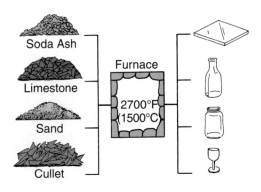

Figure 4-39. Raw materials and broken glass are used to make new glass products.

Figure 4-40. Cement is a common ceramic material.

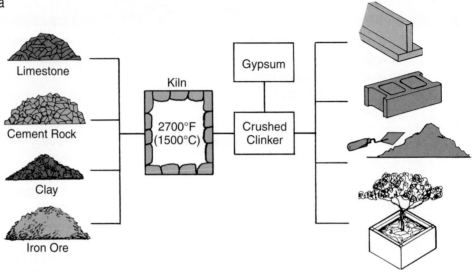

The three types of composites are layered, fiber, and particle composites. Layered composites are made up of laminations like a sandwich. Thin layers of material are tightly bonded. The tires that support you when you ride in a car or bus or land in an airplane are also layered composites. Another common use of layered composites is in the construction of hollow-core doors, **Figure 4-41.** Sheets of plywood are glued to spacers of cardboard and solid wood. The wings of some aircraft are made of layered composites. The center of the composite is an aluminum honeycomb. This honeycomb is covered with layers of fiberglass to form a sandwich.

Fiber composites consist of short or long fibers of a material such as glass or carbon embedded in a matrix of another material, such as resin, and metal. See **Figure 4-42.** One use of a fiber composite is shown in **Figure 4-43**.

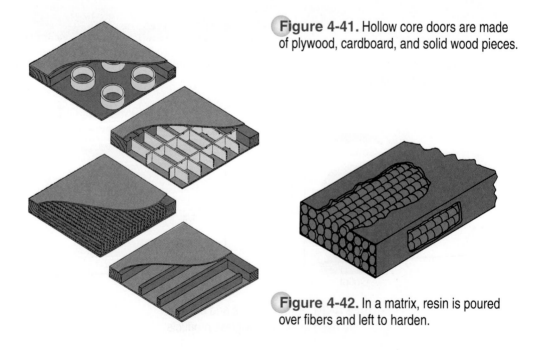

Figure 4-41. Hollow core doors are made of plywood, cardboard, and solid wood pieces.

Figure 4-42. In a matrix, resin is poured over fibers and left to harden.

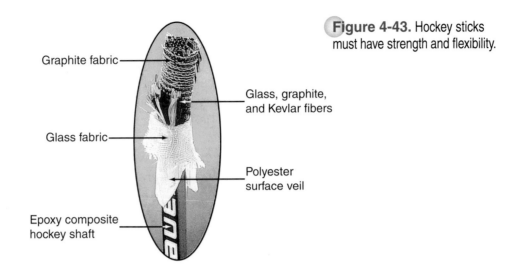

Figure 4-43. Hockey sticks must have strength and flexibility.

Graphite fabric

Glass, graphite, and Kevlar fibers

Glass fabric

Polyester surface veil

Epoxy composite hockey shaft

Particle composites, such as the particleboard shown in **Figure 4-44**, consist of particles held in a matrix. The most common particle composites are particleboard and concrete. See **Figure 4-45**.

Advanced Materials

New scientific developments are producing materials that are radically different from those that we have used in the past. These advanced materials have superior durability, elasticity, hardness, and toughness. Advanced materials include:

- Biomaterials
- Nanomaterials
- Smart materials

Figure 4-44. A particle composite consists of small particles held together by a binder. In the particleboard shown here, the particles are small pieces of wood and wood fibers, and the binder is a resin.

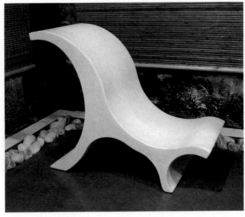

Figure 4-45. Concrete is composed of particles of stone in a matrix of Portland cement.

Biomaterials

A *biomaterial* is a material used to manufacture artificial body parts and implants that replace natural body parts. Biomaterials are biocompatible, which means they are designed to avoid rejection by our body (skin, bone, and blood). They can be natural (cellulose) or synthetic (metallic, plastic, or ceramic).

Biomaterials can be classified into two broad categories: bioactive and biopassive. Bioactive materials interact with the biological system. For example, polymers can be used as sutures that dissolve in the body. Biopassive materials do not interact with a person's biological system. For example, cobalt-chromium alloys are used as bearing materials for total joint replacements. Dental crowns and contact lenses also use biopassive materials. See **Figure 4-46**.

Nanomaterials

"Nano" means billionth, so a nanometer (nm) is one billionth of a meter. The period at the end of this sentence has a diameter of about 100,000 nm.

Materials behave differently at this small scale. The physical and chemical properties of *nanomaterials* are different from those of other materials. If you look at most materials under a powerful microscope you can see spaces, or gaps, between atoms. If these gaps line up, the material can break along the gap when pressure is applied. Certain carbon structures called *nanotubes* are free of these gaps, so they are very strong. This is because the carbon atoms are bound to each other without any spaces between them. Even though carbon is quite soft in its natural state as graphite, when its atoms are packed tightly in a nanotube arrangement, it is very hard—about fifty times stronger than steel.

Figure 4-46. This dental implant is made of biopassive materials.

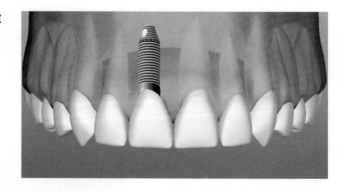

Carbon nanotubes are now being used, for example, to make paper-thin speakers. Carbon-nanotube speakers play music with sound fidelity similar to that of conventional speakers. They can even play when they are being bent and stretched. Future uses of flexible and stretchable carbon- nanotube loudspeakers include speakers on clothing, windows, flags, and video and laptop screens.

Fabrics can be coated with "nanowhiskers." These tiny surface fibers are so small that dirt cannot penetrate them. Sunscreens also use nanotechnology. They coat your skin with a nanoscopic layer that blocks out the sun's harmful ultraviolet rays.

Smart Materials

Smart materials change in response to their surroundings. For example, a T-shirt painted with a thermochromatic ink changes color in response to your body temperature. Eyeglasses made of shape memory alloys (SMA) deform but also return to their original shape when heat is applied. If you sit on them and bend them, they can be returned to their original shape undamaged.

As described earlier in this chapter, the way atoms and molecules are arranged determines a material's structure and its properties. Shape memory alloys are different because their molecules do not have a fixed arrangement. They can be rearranged. When a load is applied, they can be readily deformed. However, heating returns them to their original shape.

Smart materials can also be designed to respond to environmental stimuli such as electrical current, light, or temperature. For example, some smart materials change from a liquid to a solid when they encounter a magnetic force. Smart materials in clothes could check vital signs such as your heart rate. Sports and protective clothing can be made from a smart material that hardens on impact, helping to protect the wearer from injury.

For centuries, people have worked with and shaped materials to meet their needs. Designers need to know about the properties of materials so they can determine the material best suited for the object to be designed and made. Materials may be chosen for their physical, mechanical, thermal, chemical, optical, acoustical, electrical, or magnetic properties.

The development of advanced materials has greatly increased material design possibilities. As these materials mature, endless new products will become feasible. For example, experiments are being conducted to change the usual round shape of fibers to make them oval, rectangular, or square. When this happens, it will be possible to make a garment contract or expand so that it becomes looser or tighter. It could also become warmer or cooler depending on the season. The color of a sweater could be changed to match the shirt you decide to wear.

Other types of smart clothing will become possible when electronic components become even smaller. A smart vest could have electronic devices plugged into it. They would monitor heart rate, track body temperature and respiration, and count how many calories the wearer is burning. This information could be transmitted to a trainer or a home computer, or even stored in the vest. The information could then be played back later to a cell phone, personal computer, or wrist monitor.

As biomaterials, nanomaterials, and smart materials continue to be developed, even more fascinating applications will be found. Designers must constantly be aware of changing and emerging materials and their properties. The possibilities may only be limited by the designer's creativity.

- The principle properties of materials include physical, mechanical, thermal, chemical, optical, acoustical, electrical, and magnetic properties.
- Almost everything can be classified according to its acidity. Highly acidic substances are acids, and substances with almost no acidity are bases. Pure water is neither an acid nor a base; it is neutral.
- Traditional materials include woods, metals, plastics, ceramics, and composites.
- Advanced materials, including biomaterials, nanomaterials, and smart materials, are expanding design possibilities to include items we never imagined even 10 years ago.

Finding the Meaning of Unknown Words

Copy the following graphic organizer onto a separate sheet of paper. In the left column, record words from the chapter that you do not understand. As you read the chapter, try to guess their meanings and record your guesses in your chart. After you have read the chapter, look up each word in a dictionary. How close were your guesses? After you look up each word, go back and reread that portion of the chapter. Do you understand the chapter better?

Unknown Word	What I Think the Word Means	Dictionary Definition

Test Your Knowledge

Write your answers to these review questions on a separate sheet of paper.

1. Name and describe each of the seven properties of materials.

2. Set up a chart like the one below. Place each of the following materials in the appropriate column in the chart: polystyrene, cast iron, cedar, copper, glass, maple, aluminum, pine, cement, melamine, spruce, polyethylene, birch, porcelain, clay, steel, and concrete.

Category of Material				
Woods	Metals	Plastics	Ceramics	Composites

3. List the characteristics of hardwood and softwood trees.

4. Sketch the construction of the following manufactured boards: plywood, blockboard, and particleboard.

5. What is the difference between a ferrous metal and a nonferrous metal?

6. Is pure iron an alloy? Explain your answer.

7. Give three examples of both ferrous and nonferrous materials.

8. Copy the chart below and fill in the blank spaces.

Metal	Typical Use
Copper	
Aluminum	
High-carbon steel	
Cast iron	

9. What is the scientific name for plastic?

10. From what raw material are most plastics made today?

Test Your Knowledge (Continued)

11. Copy the chart below. List four objects you use every day that are made of plastic. In your opinion, what particular properties make plastic the most appropriate material for each object? Refer to Figure 4-35.

Object	Important Characteristics
detergent squeeze bottle	flexible, transparent, lightweight
1.	
2.	
3.	
4.	

12. What are the two most familiar ceramic materials?

13. List three advantages and three disadvantages of ceramic materials.

14. What is a composite material?

15. Briefly explain the differences between layered, fiber, and particle composites.

Critical Thinking

1. Did you know that some metals can make you sick? Some have even caused insanity. Research one of the following topics to find out more. Write a short essay answering the topic question. Explain your reasoning. Keep your essay focused and to the point, but make it interesting. Try to engage readers in the subject matter.

   A. The Romans loved sweet things, but they didn't have any sugar. As a substitute, they boiled grape juice in lead containers until it became sweet syrup. Then they added it to various foods. In small quantities, lead can result in lead poisoning. Did this lead to the fall of the Roman Empire?

   B. George III was king of England at the time of the American Revolution. It is believed that he was at least partly insane. His madness may have been the result of lead poisoning from eating sauerkraut. It was one of his favorite foods, and it was cooked in lead pots. Did his madness lead to poor decisions resulting in the loss of the American colonies?

   C. Lead carbonate was one of the main ingredients in paint until 1980. In the early to mid-1990s, babies and young children were found to have lead in their bodies. They were ingesting the paint by biting their crib rails or by touching paint dust on windowsills and then sucking their fingers. What are the ingredients in modern paints? Are they harmful to us?

2. Think about the advanced materials discussed in this chapter. Use your imagination and creativity to think of ways to apply these new materials to reduce environmental damage. Choose your best idea and develop it into a design proposal.

Apply Your Knowledge

1. Choose three objects you use every day that are made of different materials. State whether the materials used are appropriate for the item. Explain your answer.

2. Collect pictures of objects that are made of layered, fiber, and particle composites. Label the pictures to name the materials used in each composite.

3. Work with two or three classmates to collect samples of five different materials. Choose one material property. As a team, design and build an apparatus to test the materials for the property you have chosen.

4. Search the Internet to find three current applications for smart materials. Write a report describing these applications.

5. Research one career related to the information you have studied in this chapter. Create a report that states the following:

   - The occupation you selected
   - The education requirements to enter this occupation
   - The possibilities for promotion to a higher level
   - What someone with this career does on a daily basis
   - The earning potential for someone with this career

   You might find this information on the Internet or in your library. If possible, interview a person who already works in this field to answer the five points. Finally, state why you might or might not be interested in pursuing this occupation when you finish school.

STEM Applications

1. **SCIENCE** Describe the fundamental difference between thermoplastics and thermoset plastics in terms of the way their molecular chains are formed. Create models of the molecular structures to illustrate this concept.

2. **ENGINEERING** Research the materials used in tires made for specific purposes, such as winter use, summer use, or running while flat. Create a diagram that clearly shows the layers of material used in one type of tire and explain why engineers chose each material.

Bamboo is very suitable for casual clothing. What other clothing, recreational, and home products could be made using bamboo?

Better by Design

Bonnie Siefers

Bonnie Siefers is an American fashion designer and eco-designer who believes that clothing must look chic and feel luxurious. But Bonnie also believes that fashion designers have a responsibility toward the environment and must begin to use certified organic or eco-friendly fabrics (ecotextiles). So Bonnie designs apparel made from ecoKashmere®, a soft, silky fabric that incorporates the pulp of bamboo grass. The result is clothes that are soft and flowing. Also, because bamboo absorbs water rapidly, it produces fabrics that "wick" moisture at twice the rate of cotton. Bamboo-based fabrics are also highly breathable, anti-static, and naturally antibacterial, making them helpful to people with sensitive or allergy-prone skin.

Ecotextiles such as bamboo, flax, cotton, and jute are renewable, biodegradable, and sustainable.

"A pure, natural environment is vital to children of all ages. Organic [textiles] are not only gentle on the skin, but also safer for the people who make the clothes, for the farmers who grow the crops, and for the environment." —*Bonnie Siefers*

Reading Target

Summarizing Information

A *summary* is a short paragraph that describes the main idea of a selection of text. Making a summary can help you remember what you read. As you read each section of this chapter, think about the main ideas presented. Then use the Reading Target graphic organizer at the end of the chapter to summarize the chapter content.

Key Terms

adhesion
bending
casting
chemical joining
chiseling
coatings
cohesion
development
drilling
filing
finishing
forming

heat joining
jig
laminating
marking out
mechanical joining
molding
planing
sawing
shearing
volatile organic compounds
    (VOCs)

Objectives

After reading this chapter, you will be able to:
- Demonstrate responsible and safe work attitudes and habits.
- Design and shape a product using the correct tools and processes.
- Select the correct materials and methods for joining materials.
- Recall methods for applying a finish to a material.

Useful Web sites:
www.jonano.com
www.bambooclothing.co.uk/why_is_bamboo_better.html

The tools and equipment needed to make a product depend on the design and the materials to be used in the product. This chapter describes the types of tools needed to shape various types of materials. Before you actually use the tools, however, you must learn how to use them safely.

Learning to Work Safely

As you design and make products, you will use various materials, hand tools, and machine tools. General rules for safety are included in **Appendix B**. You should also follow the manufacturer's instructions carefully to help avoid injury. However, you can also take an active role in your own safety by planning carefully. To avoid hurting yourself or others or damaging tools or equipment, you need to learn about:

- Hazard identification
- Risk identification
- Risk management

This means that before using any tool or machine, you should ask yourself the following questions:

- What are the hazards?
- What are the risks?
- How will I manage these risks?

For example, when you are using a band saw, the hazards are the sharp stationary blade and moving blades. The risks are that you can cut your fingers, both on the stationary blade and even more seriously on the moving blade. Also, long hair or loose clothing can become entangled in the moving blade. You can manage these risks keeping your fingers at least four inches (10 cm) away from the blade, tying your hair back, and not wearing loose clothing or jewelry.

This approach to working safely is applicable every time you use a hand tool, a machine tool, and materials. It also applies to situations you will encounter at home or elsewhere. Working safely is your responsibility!

Shaping Materials

When you have designed your product and have chosen the material, you are ready to make the product by following several carefully thought-out steps. These steps include marking out, cutting and shaping, joining and finishing.

Technology Application

Game Design

All products are designed with safety in mind, and safety factors are foremost in the mind of designers. Power tools have guards to protect the operator from moving blades or belts. Motorized vehicles need keys and often pressure on the brake pedal to start them. A lawn mower automatically stops if the handle is released.

Safety is equally important for items used in the kitchen. Although kitchen tools and equipment are designed for safe use, there is always an element of danger and a need for the user to be cautious and alert.

Technology Activity

One of the primary safety problems in a kitchen occurs when small children are present. Children often don't know when or where danger exists. Your task is to design and make a game similar to Chutes and Ladders® to teach young children, 2 to 4 years old, the dangers of touching items found in the kitchen. Follow these steps:

1. On an 8½ × 11 sheet of graph paper, draw nine vertical lines and nine horizontal lines to create a grid of 1-inch squares. You will have a total of 64 squares.

2. Think of four things that a young child might do that could result in an accident; for example, "grabbing the blade of a sharp knife."

3. Think of four things that a young child might do to avoid an accident; for example, "staying away from a hot stove."

4. Decide on the length of four chutes and four ladders and draw them on the game, spacing them out evenly.

5. Print the accident-causing and accident-avoiding messages in small letters in the appropriate squares. The events that might result in an accident will be in the square at the top of a chute. The events that could avoid an accident will be in the square at the bottom of a ladder.

6. Design at least two player pieces to be used in the game. How will you make them? Be sure to take the age of the children into consideration when you specify the size of the pieces. They must not be a choking hazard.

7. Decide on a way to determine the number of squares a player advances.

8. Ask a friend or classmate to help you test the game by playing it.

9. Based on your experience and your classmate's comments, decide how you can improve this board game. Should you add art? How could you make the game board more durable?

Marking Out

The first step in making a product is called *marking out*. Marking out involves measuring material and marking it to the dimensions on your drawing. You should do this carefully for two reasons. First, your marks must be accurate so that the pieces fit together. Second, materials are expensive. Making mistakes wastes time and money.

Most marking out starts from a straight edge. To make a straight edge on wood, a plane is used. Plastic and metal are filed. When wood is used, this edge is the face edge, and the best side is the face side. See **Figure 5-1**. Lines should be easily seen. To be accurate, they must be thin. They should be marked as shown in **Figure 5-2**.

Sawing

Sawing removes material quickly. All saws have a row of teeth. They chip or cut away the material. The part of the tool that cuts must be harder than the material being cut. Some saws are designed for cutting wood and others for cutting metal. See **Figure 5-3**. Plastics and composites are usually cut using saws designed for cutting wood and metal.

Filing and Planing

Small amounts of material may be removed by *filing* and *planing*, as shown in **Figure 5-4**. Files are mainly used on metal. A special type of file, called a *rasp*, is used on wood. Most filing is done while the work is held in a vise. In straight or cross filing, you push the file across the work straight ahead or at a slight angle. Never run your fingers over a newly filed surface. Sharp burrs on the workpiece may cause a severe cut.

Various types of planes produce smooth, flat surfaces on wood. Planing requires that the wood piece be held in a vise. When properly adjusted, a plane should take off fine shavings from the piece of wood. Planing removes the small ridges left by a power planer or saw. This saves having to remove them with sandpaper—a slower process.

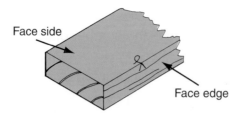

Figure 5-1. Mark lumber on both the face side and the face edge.

Face side

Face edge

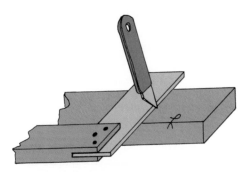

Marking wood to length
Use a try square.
Hold the handle firmly against the
  face edge of the wood.
Mark a line with a marking knife.
Always use the outside edge of the square.

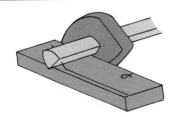

Marking wood to width
Use a marking gauge.
Press the stock of the gauge firmly
  against the wood.
Tilt the gauge in the direction you will
  push it.
Practice on scrap wood (the gauge is a
  difficult tool to use).

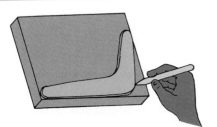

Using templates for irregular shapes
Draw and cut out the shape in paper or
  cardboard. This is called a *template*.
Hold the template to the material and
  draw carefully around it.
When drawing on wood, take careful note
  of the grain direction.
When many pieces of the same shape
  are to be made, use masonite to make
  the template.

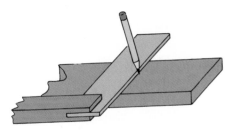

Marking metal to length
Coat the metal with marking blue.
Use an engineer's square.
Hold the handle firmly against the straight
  edge of the metal.
Mark a line using a scriber.
Always work on the outside of the square.

Marking metal to width
Use an odd-leg caliper.
Press the stepped leg of the caliper against
  the straight edge.

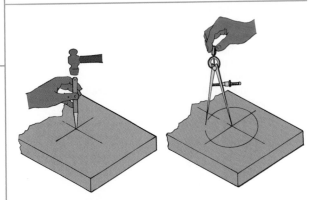

Drawing circles
Mark the center of the hole with a center
  punch on metal and an awl on wood.
Use a compass on wood.
Use dividers on metal.

Figure 5-2. Procedures for marking out.

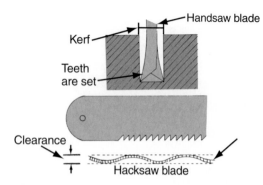

Saws need clearance
Teeth are bent in alternative directions; this is called *set*.

The kerf made by the teeth is wider than the blade thickness.

Teeth usually point away from the handle and material is cut on the push stroke.

The teeth on hacksaw blades are too small to be set. Clearance is achieved by stamping a wavy edge on the blade.

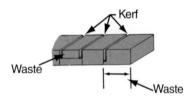

Sawing techniques
Never cut on the line; the kerf is made touching the line but on the waste side.

Stand with your hand and arm in line with the saw cut.

Use your thumb at the side of the blade when you start the cut.

Always use the full length of the blade.

At the end of the cut, support the work underneath.

Using a handsaw
Used for first cutting wood to approximate size.

There are two types: cross cut and rip.

A cross cut saw has finer teeth and is used for cutting across the grain.

A rip saw has larger teeth and is used for cutting parallel to the grain.

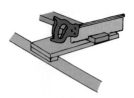

Sawing wood accurately
A tenon saw is used to make straight, accurate cuts in wood.

Always use a bench hook, held firmly in the vise.

Coping saw
Used to make curved cuts through wood and plastic.

Teeth point toward the handle; cuts on the pull stroke.

Hacksaw
Used to make straight cuts through metal and plastic.

Make sure at least three teeth are in contact with the material all the time.

Use small teeth for hard materials and large teeth for soft materials.

Junior hacksaw
Small and inexpensive.

Useful for cutting thin metals and light sections.

Abrafile (rod saw)
Used to make curved cuts through metal, plastic, and ceramics.

The blade is like a file and is held in a hacksaw frame.

The cutting edge is made of tungsten carbide particles bonded to a steel rod.

Hot wire cutter (an alternative to sawing)
Heated wire cuts straight and curved shapes in rigid foam plastic.

Must be used in a well-ventilated area as fumes are produced.

Figure 5-3. These tools are used for cutting operations.

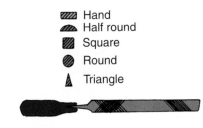

Hand
Half round
Square
Round
Triangle

Files

Used to remove small particles of metal and plastic.
Double-cut files remove metal faster but make a rougher surface.
Single cut files produce a smooth surface.
Each shape is available in many sizes and degrees of coarseness.
Always use a file with a handle.

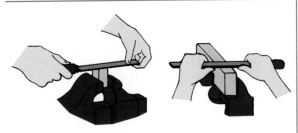

Using a file

For normal filing, hold the file at each end; it only cuts on the forward stroke (cross filing).
To produce a very smooth finish, push the file sideways (draw filing).

Cleaning a file

Small pieces of metal sometimes stick in the teeth of the file; this is known as *pinning*.
A special wire brush called a *file card* is used to clean the file.
Keeping the file clean is particularly important when filing plastic.

Rasps

Similar to coarse files.
Used for rough shaping free-form, sculptured shapes.

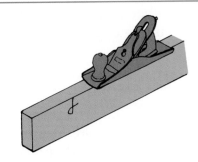

Planes

Used to remove a thin layer of wood (shaving).
A short plane, called a smoothing plane, is used on short pieces of wood.
A longer plane, called a jack plane, is used on longer pieces of wood.
Always plane in the direction of the grain.

Surforms

Are held like files and rasps but cut like planes.
Cut wood quickly but leave a rough surface.

Figure 5-4. Filing and planing tools are used to shape and smooth materials.

Shearing and Chiseling

Shearing and *chiseling* are other techniques used to shape materials. These tasks require shears and chisels. Shears, also called *snips*, cut thin metals. Chisels are designed to cut wood or metal. See **Figure 5-5**.

Snips are designed for various kinds of cuts. A straight snips cuts straight lines and large curves. For cutting curves and intricate designs, an aviation snips or a hawk-billed snips is best.

Chisels must be designed for the type of material on which they are to be used. Wood chisels have very sharp cutting edges so they will cut rather than tear the wood. Metal-cutting chisels have thicker, tougher cutting edges.

Drilling

Drilling is a process used to make holes in wood, plastic, metal, and other materials. A drill cuts while turning. Twist drills for cutting metals are made of carbon steel, high-speed alloy steel, and titanium. Twist drills are also used to bore holes in woods and plastics. **Figure 5-6** shows some types of drills and how to use them.

Cutting action of shears (snips)
Two blades are used like scissors to cut thin metal.

Shearing technique
Mainly used to cut tin plate, aluminum, and copper sheet.
To operate, use like scissors.
Alternatively one arm of the shears can be fastened in a vise while pressure is applied to the other arm.
Do not put the sheet all the way back into the shears.
Do not close the shears completely when cutting.

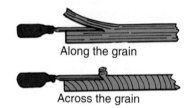
Along the grain

Across the grain

Cutting action of wood chisel
Chisel edge is straight and sharp.
Cutting action is like that of a wedge.
When chiseling along the grain, there is a greater possibilty the wood will split.

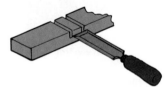

Wood chiseling techniques
Make sure wood is firmly clamped.
Keep both hands behind the cutting edge.
Never cut toward your body.
When cutting across the grain, first make two saw cuts at the limits of the groove, then chisel from both sides toward the center.

Figure 5-5. Shearing and chiseling tools are used to shape materials.

Cutting action of drill
Drills cut by rotating a cutting edge into a material.
A twist drill has a cutting edge, a spiral groove (the flute) to release the chips, and a straight shank to hold the drill in a chuck.

Drilling technique
Mark the point you want to drill.
Use a center punch on metal.
Never hand-hold work to drill; hold work in a vise or a clamp.
Place waste wood under the work to prevent damaging the bench after the drill has gone through; this also prevents the material from cracking away.
When drilling deep holes, remove the drill from the hole from time to time to avoid clogging and overheating.

Hand drill
Concentrate on keeping the drill vertical.
Turn the handle at a steady speed, trying not to wobble the drill.

Portable electric drill
Clamp small pieces of material in a vise.
Center punch the location of the hole.
Place a small piece of wood on the underside to prevent splitting.
Do not bend the drill sideways or you will snap the drill bit.

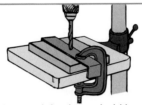

Drill press
Remember that work is always held in a machine vise or clamp.
Always wear eye protection.
The speed can be adjusted: the larger the drill bit, the slower it should turn.
Place a piece of scrap wood under the workpiece to protect the table.

Countersink drill
Used to open out the end of a hole so that a flathead screw will fit flush with the surface.

Holesaw
Used to drill large holes in wood up to 3/4 in. (18 mm) thick.
Very useful for making discs or wheels.

Figure 5-6. Hand and electric tools for drilling holes and the techniques for using them.

Bending and Forming

Bending and *forming* are two different processes. Bending sheet material is like folding paper along a straight, sharp crease. It is also quite easy to bend it along a gentle curve.

However, forming sheet material is more difficult. Forming changes the shape of sheet material into a more complex shape, such as a dome. It often involves using a mold or heating the material. Take a flat sheet of paper and try to form it into a dome. That is much harder than bending, isn't it? The sheet should bend and curve in many directions, but it tends to crease and buckle. See **Figure 5-7**.

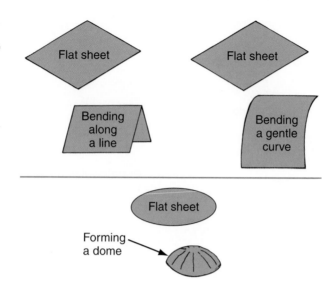

Figure 5-7. Sheet material bends or curves easily in one direction. Forming, or bending in several directions at the same time, is more difficult.

Wood

Wood does not bend easily. One way to create a curved shape from wood is to cut the curve out of a thick, solid block as shown in **Figure 5-8A**. The problem is that the curved piece would break easily because of the short grain on the curved sections. The part made this way would be weak. But there is another way: laminating. This avoids the problem of short grain.

Laminating is the process of gluing together several veneers (thin sheets of wood). See **Figure 5-8B**. These can be easily bent, glued, and held in a mold until the glue dries. The steps for laminating are:

1. Make a mold, as shown in **Figure 5-8C**. The two parts must fit together exactly.

2. Veneers vary in thickness. Calculate the number of veneers you will need to get the right thickness of laminate.

3. Clamp all of the veneers together without glue. If they don't bend easily, dampen the veneers and leave them clamped in the mold overnight. See **Figure 5-8D** and **Figure 5-8E**.

4. Completely cover the surfaces of the veneers with glue. Do not glue the outside surface of the top and bottom pieces or they will be permanently attached to the mold! Use a resin adhesive that becomes rigid when it sets. Avoid using contact cement or PVA, which remains rubbery. Once the glue has dried, the laminated wood will hold its shape. The glue keeps it from springing back to its old shape.

5. To prevent the veneers from sticking to the mold, wrap them in wax paper or a thin plastic sheet.

6. Use a thin rubber sheet between the mold and the veneers. This will take up any irregularities in the mold surface. It will also ensure even pressure over the entire surface of the mold.

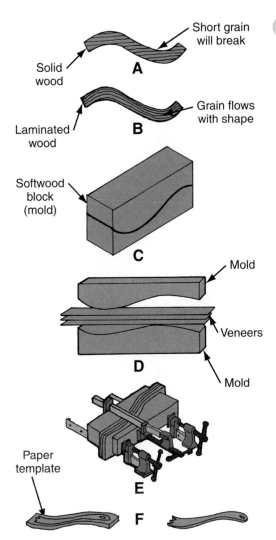

Figure 5-8. Laminating produces stronger parts than sawing the curved shapes out of solid wood.

7. Use bar clamps, **Figure 5-8E**, to squeeze the mold together until the glue is forced out along the edges of the veneer.

8. As shown in **Figure 5-8F**, make a template of the shape you want. Either glue it to the laminate or mark around it. Use the template as a pattern for cutting out the shape.

Sheet Metal

Sheet metal can be bent and folded into three-dimensional objects. A pan is a good example. Before you begin to shape the metal, you must work out a *development*. A development is a pattern. It shows where the sheet metal must be cut and folded to make the object. See **Figure 5-9A**. Mark the development on the sheet metal. Then use shears to cut the shape.

To form straight bends, follow these steps:

1. Place the marked sheet metal into folding bars. The bend line should be touching the top edge of the bar.

2. Fasten the folding bars in a vise.

3. If the metal is wider than the vise jaws, add a C-clamp.

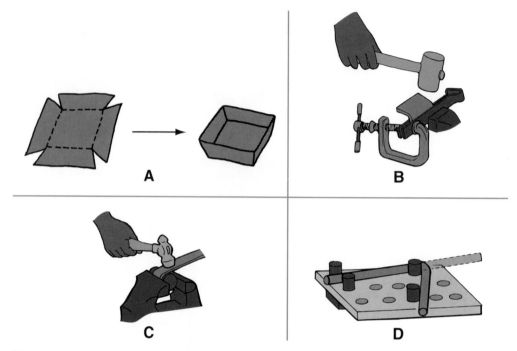

Figure 5-9. Procedures for bending metal.

4. Use a mallet with a rawhide or nylon head. Bend the metal over the bar, as shown in **Figure 5-9B**.

For some shapes, you may find a block of wood more useful than folding bars.

Lengths of mild steel not more than ¼″ (6 mm) thick can be bent fairly easily. Follow these steps to bend strip steel:

1. Clamp the metal vertically in a vise.
2. Hammer the metal from one side to bend it to the needed angle, as shown in **Figure 5-9C**.

A *jig* is a useful tool for bending metal. Small diameter rods can be bent on a peg jig, as shown in **Figure 5-9D**. Metal tubing can be bent using a similar jig.

Plastics

Thermoplastics can be bent or formed when heated to between 300° and 400° F (150° and 200° C). Heat the plastic sheet or rod along the line of the bend. The narrower the heated line, the sharper the bend will be. Refer to **Figure 5-10A**.

Using a strip heater provides heat along a straight, narrow line for a sharp bend. To use the strip heater:

1. Place the sheet of plastic on the heater. The bend line must be exactly over the heat element.

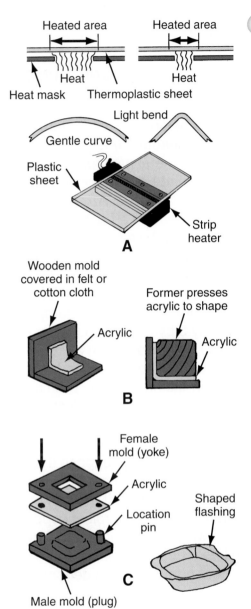

Figure 5-10. Processes for bending and forming thermoplastic sheet material.

2. For safety, wear gloves to protect your hands from the hot plastic.

3. Heat both sides of the sheet.

4. Bend the plastic to the required shape. For a 90° bend, press the plastic into a mold. See **Figure 5-10B**.

5. To produce a sharper bend, press a second mold into the corner, as shown in **Figure 5-10B**. Hold until cool.

NOTE: Wooden molds must be covered with cotton or felt material. This prevents the wood grain from marking the plastic.

Forming plastic in a mold calls for a two-part mold. To form the plastic:

1. Heat the acrylic sheet in an oven until pliable.

2. Place it over the plug. Press the yoke down on top. See **Figure 5-10C**.

3. Allow the plastic to cool before separating the mold.

Math Application

Using Mixed Fractions to Calculate Material Needed

Before manufacturers begin shaping materials, they have to decide how much material will be needed for a given production run. For example, a manufacturer of metal nameplates must determine how many nameplates can be made per linear foot of metal. This often involves working with mixed fractions. A *mixed fraction* is a number that includes both a whole number and a fraction, such as 1½.

To multiply a whole number by a mixed fraction, first change the mixed number to an improper fraction. To create an *improper fraction*, multiply the whole number part of a mixed fraction by the denominator of the fraction and add the result to the numerator. Place this number over the denominator to complete the fraction. For example, to convert 3½ to an improper fraction, multiply 3 by 2 and add the result to the 1: $3 \times 2 = 6$; $6 + 1 = 7$. The improper fraction is written as ⁷⁄₂.

After you have created the improper fractions, multiply the numerators. Then multiply the denominators. Reduce the result to lowest terms. Finally, divide the denominator into the numerator to reduce the fraction to a mixed fraction.

For example, suppose the manufacturer has an order for 50 copper nameplates. Each nameplate will be 7½″ × 1¾″. If the nameplates are blanked (stamped out) of a roll of copper that is exactly 7½″ wide, what length of copper will be needed? (Note: The improper fraction for any whole number is that number over a denominator of 1.)

Problem statement: $50 \times 1\frac{3}{4}$

Convert to improper fractions: $\frac{50}{1} \times \frac{7}{4}$

Multiply: $\frac{50}{1} \times \frac{7}{4} = \frac{50 \times 7}{1 \times 4} = \frac{350}{4}$

Divide: $350 \div 4 = 87$, with a remainder of 2

Reduce: $87\frac{2}{4} = 87\frac{1}{2}$

Math Activity

A door manufacturer needs 150 door strikes (Figure A) measuring 2¼″ × 1¾″. These will be blanked from a roll of brass, using the most economical cutting pattern. A space of ⅛″ is needed between each piece. See Figure B for examples of cutting patterns. Determine the length of brass needed if the roll of brass is:

A. 2″ wide

B. 2½″ wide

A

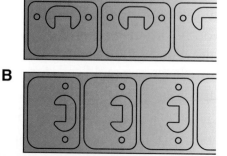

B

Casting and Molding

Pouring liquid or plastic material into a mold to shape it is called *casting* and *molding*. Every time you make ice cubes you are casting. You are making a solid shape by pouring water into a tray. The liquid takes the shape of its container.

Casting is a method of making shapes that are almost impossible to produce by sawing, drilling, or filing. Three basic materials are used for casting and molding: metals, plastics, and ceramics. When the material is poured into the mold, the process is called *casting*. When the material is forced into the mold, it is called *molding*.

Metals become liquid when they are heated above their melting point. As they cool, they solidify. Plastics are available in a liquid form. These liquids set hard through chemical action.

Ceramics include materials such as silica (sand), clay, and concrete. Silica must be melted to make glass and other products. Clays and concrete are not melted. They are mixed with liquid and poured into a mold.

Metals

Use the following steps to cast metal:

1. As shown in **Figure 5-11A**, make a pattern. A pattern is just like the finished product. Usually it is made of wood, but it could be made of some other easily worked material.

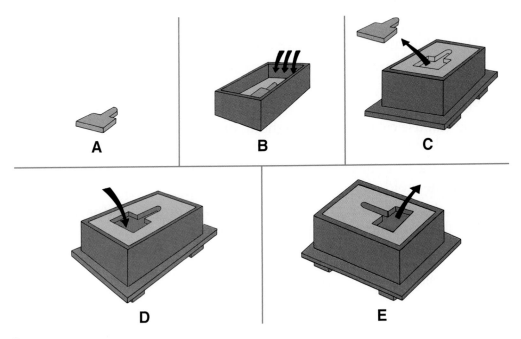

Figure 5-11. Process for casting metal.

2. Place the pattern in a molding box on a flat surface. See **Figure 5-11B**.

3. Pack molding sand carefully around the pattern. Molding sand is made from high-quality silica sand mixed with a binder such as clay to hold it together.

4. Completely fill the molding box with molding sand. Tamp it tightly around and over the pattern.

5. Cover the box with another board and turn it over. Remove the board that was on the bottom.

6. Carefully remove the exposed pattern. See **Figure 5-11C**.

7. Pour molten metal into the cavity formed by the pattern, as shown in **Figure 5-11D**.

8. Allow the casting to cool and solidify. Then remove it. See **Figure 5-11E**.

Plastics

Casting plastics has one big advantage over casting metal. The resins can be cast at room temperature. Small articles, such as paperweights, can be cast in plastic. If you wish, you can also embed decorative objects in them. Follow these steps:

1. Use a smooth mold. It will produce a smooth surface on the casting. Waxed drinking cups work well. Never use polystyrene (Styrofoam®) cups. The resin will dissolve them and may produce toxic gases.

2. Measure the amount of resin you will need. Add the recommended amount of catalyst (hardener). Mix thoroughly.

3. Pour a layer of the mixed resin into the mold and leave it to harden. This will form the top layer of the casting. See **Figure 5-12A**.

4. Place the decorative object (coin or stamp) face down on the hardened layer of resin. See **Figure 5-12B**.

5. Pour more resin around and over the object, as shown in **Figure 5-12C**. You can use clear resin throughout, or you can add pigment (color) to the last layer. This forms the base of the object.

6. When the resin has hardened, **Figure 5-12D**, remove the casting from the mold.

7. Smooth rough edges and surfaces with wet or dry sandpaper. Then polish the casting with a polishing paste.

Figure 5-12. Steps for embedding an object in plastic.

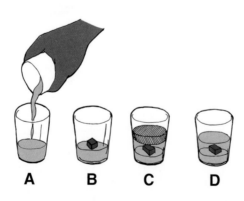

A   B   C   D

Fiberglass-Reinforced Plastic

Fiberglass canoes, racecar bodies, and crash helmets may all be made from plastic resin reinforced with glass fiber. The reinforcing glass fiber makes the shells very tough. The thermosetting resin creates a smooth, hard surface. To make a fiberglass-reinforced product:

1. Paint on a release agent over the surface of the mold, as shown in **Figure 5-13**.
2. Mix polyester resin and a catalyst in the recommended proportions. (Color may be added to the resin.)
3. Brush on a gel coat of polyester resin.
4. Add a layer of fiberglass and coat it with more resin.
5. Use a roller to make each layer take up the exact shape of the mold. Be sure to remove air bubbles.
6. Add layers of resin and fiberglass to produce the required thickness.
7. Allow the assembly to cure (set hard), and then remove it from the mold.

Joining Materials

There are many ways of joining materials. **Figure 5-14** lists some common choices for wood, metal, and plastic. You will notice that some methods, such as fastening with nuts and bolts, are good for all three. Others, such as soldering, may be used only for metal.

Figure 5-13. Molding fiberglass on a form, such as this boat, is called a *lay-up*.

	Mechanical	**Chemical**	**Heat**
Wood	Nails Screws Nuts and bolts KD (knock-down) fasteners Wedges Hinges	Glues Adhesives	
Metal	Rivets Nuts/bolts/screws KD fasteners Hinges	Adhesives	Weld Braze Solder
Plastic	Rivets Nuts/bolts/screws KD fasteners Hinges	Solvents Cements	Weld

Figure 5-14. Methods for joining wood, metal, and plastics.

Mechanical Joining

Mechanical joining is the use of physical means to assemble parts. It can be done in one of two ways. One method is to use hardware such as nails, screws, or special fasteners. Another method is to shape the parts themselves so they interlock.

Nails

Nails provide one of the easiest ways to join two pieces of wood. Nails hold the wood by friction between the wood fibers and the nail. See **Figure 5-15**. Nails can be shaped in many different ways. **Figure 5-16** shows several different types of nails. Remember these general points when using nails:

- Whenever possible, place one of the pieces to be nailed in a clamp or a vise. See **Figure 5-17A**.
- Always nail through the thinner piece into the thicker piece.
- Avoid bending the nail. Strike it squarely with the face of the hammer.
- When using finishing nails, drive the nail below the surface using a nail set, as shown in **Figure 5-17B**.
- Stagger the nails, as shown in **Figure 5-17C**. If you place them in a straight line, you may split the wood along the grain.
- To remove nails, use a claw hammer as shown in **Figure 5-17D**. Always use a block of waste wood to protect the surface of the wood.

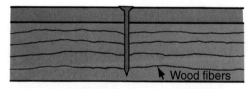

Wood fibers

Figure 5-15. When a nail is forced into wood, the compression of the wood fibers and friction between the wood fibers and the nail hold the nail in place.

Figure 5-16. Nails are made in many different shapes to serve special fastening purposes.

Type of Nail	Uses
Common 	Structural or other heavy work where head will be exposed.
Finishing	Finishing work where nail head should not be exposed.
Spiral 	Building construction. Twisted shank causes nail to thread itself into wood, increasing its holding power.
Drywall 	Fastening gypsum board to wood frames.
Concrete 	Fastening to concrete. Nail is hardened to prevent bending.
Roofing 	For wood, asphalt, and other roofing materials. Usually coated with zinc or galvanized to make them rust-resistant.

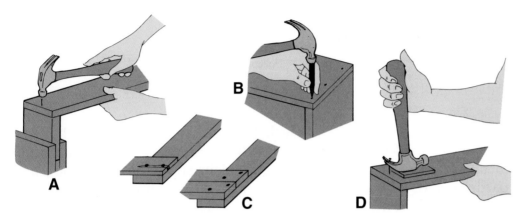

Figure 5-17. Techniques for assembly and disassembly using a hammer and nails.

Screws

Screws have greater holding power than nails. See **Figure 5-18**. They also rely on friction for their strength. When two pieces of wood are held together, the head of the screw and the grip of the screw thread pull the two pieces together. Screws can be removed more easily than nails and without damaging the material.

Wood screws are used for fastening wood to wood and metal to wood (hinges to a door). They are also used for fastening all types of hardware to furniture. To choose the correct wood screw, you must decide on the:

- Shape of head and type of slot
- Length of screw
- Thickness or gauge
- Material

The three shapes of screw heads are shown in **Figure 5-19**. Flat head screws are used when the head of the screw must be flush with or below the surface of the wood. Oval head screws can also be used when the holding power of a flat head screw is needed. The screw head will show for decoration. Round head screws are used when the object that is being fastened by the screw is too thin to be countersunk.

The common head styles or types are shown in **Figure 5-20**. Screw lengths range from ¼″ to 6″ (6 mm to 150 mm). The thickness of a wood screw is called its *gauge*. Gauge is expressed as a number. Screw gauges range from 0 to 24.

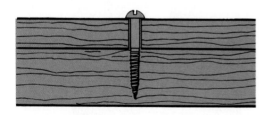

Figure 5-18. Wood fibers grip the threads of a screw. This gives the screw holding power.

Figure 5-19. Wood screws are named for the shapes of their heads. What types are shown in this photo?

Figure 5-20. Common types of screw heads. What are the advantages and disadvantages of each?

Straight slot   Phillips   Robertson

Most screws are made of steel. They are very strong, but they can rust. Brass screws do not rust, but they are weaker than steel. Sheet metal screws, **Figure 5-21**, are used to join sheet metal, plastics, and particleboard. **Figure 5-22** shows the steps to fastening hardwood parts with a wood screw:

1. Hold the two pieces together. Drill a pilot hole the length of the screw.

2. In the top piece, drill a clearance hole. This is a hole the same diameter as the screw shank (unthreaded part of the screw below the head). Note that in softwood, the clearance hole is not usually necessary. Only a pilot hole is needed.

3. If you are using a flat head screw or oval head screw, countersink the hole.

NOTE: In softwood it is usually only necessary to drill a pilot hole.

Now you are ready to install the screw. The diameters of the most frequently used pilot holes and clearance holes are shown in **Figure 5-23**.

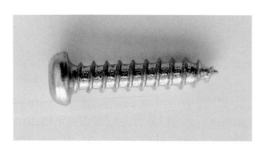

Figure 5-21. Sheet metal screws have threads that go all the way to the head.

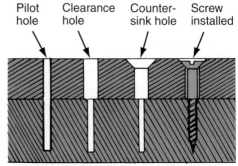

Pilot hole   Clearance hole   Counter-sink hole   Screw installed

Figure 5-22. Special holes must be drilled when fastening hardwood with screws.

Gauge No. of screw	Diameter of shank	Pilot hole	Clearance hole
4	7/64 (3 mm)	5/64 (2 mm)	7/64 (3 mm)
6	9/64 (4 mm)	3/32 (3 mm)	9/64 (4 mm)
8	11/64 (5 mm)	7/64 (3 mm)	11/64 (5 mm)
10	3/16 (5 mm)	1/8 (4 mm)	3/16 (5 mm)
12	7/32 (6 mm)	9/64 (4 mm)	7/32 (6 mm)

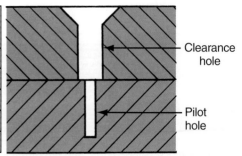

Clearance hole

Pilot hole

Figure 5-23. Pilot and clearance holes recommended for wood screws.

The most common types of wood screws and screwdrivers used in the United States are straight slot (standard) and Phillips. In Canada the most common type is the Robertson (square). Use the largest screwdriver convenient for the work. More power can be applied to a long screwdriver than a short one. Also, there is less danger of it slipping out of the slot. The tip of the screwdriver must fit the slot correctly. See **Figure 5-24**.

Nuts and Bolts

Nuts and bolts fasten metal, plastic, and sometimes wood parts together. They are quite different from wood screws. Bolt threads do not depend on gripping the fibers of the material. Bolts go completely through a drilled clearance hole. This is a hole large enough for the bolt to be pushed through. A nut threads onto the bolt end. Tightening the nut squeezes the parts together and holds them. Sometimes, threads are cut into the hole in the second piece of material, as shown in **Figure 5-25**. This takes the place of the nut.

Washers are often used under the bolt head and the nut. This protects the surfaces by distributing the load over a larger area. Lock washers prevent nuts from accidentally loosening due to vibration. Joints fastened with nuts and bolts can be taken apart and reassembled.

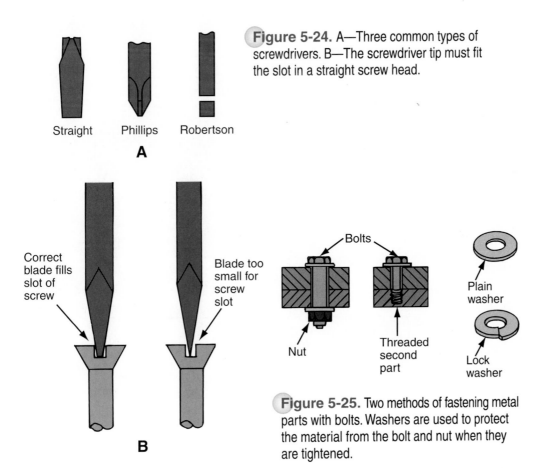

Straight   Phillips   Robertson

A

Figure 5-24. A—Three common types of screwdrivers. B—The screwdriver tip must fit the slot in a straight screw head.

Correct blade fills slot of screw

Blade too small for screw slot

B

Bolts

Nut

Threaded second part

Plain washer

Lock washer

Figure 5-25. Two methods of fastening metal parts with bolts. Washers are used to protect the material from the bolt and nut when they are tightened.

To choose the right nut and bolt, you need to decide on the following items:

- Length
- Diameter
- Shape of head
- Thread series. Some are coarse; others are fine. Some have standard threads, and others have metric threads.
- Material

Some of the choices are shown in **Figure 5-26**. Can you name any of the types shown?

Most nuts and bolts are tightened using a wrench. Machine screws are tightened with screwdrivers or Allen keys. A combination wrench is shown in **Figure 5-27**. It has one open end and one box end. The box wrench is usually preferred because it does not slip. Sometimes there is not enough room for a box end. Use an open end when clearance is a problem.

An adjustable wrench fits a range of nut sizes. As **Figure 5-28** shows, always pull on the wrench handle. Pushing can be dangerous. If the wrench should slip, you could injure your knuckles.

Figure 5-26. Nuts and bolts are made in many different shapes and sizes.

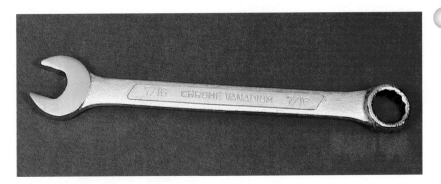

Figure 5-27. A combination wrench contains both an open-ended wrench and a socket wrench.

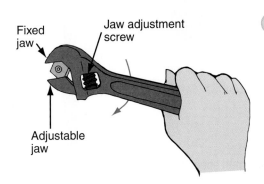

Fixed jaw

Jaw adjustment screw

Adjustable jaw

Figure 5-28. When using an adjustable wrench, always pull against the fixed jaw.

Rivets

Like nuts and bolts, rivets squeeze two or more pieces of metal or plastic together. They are either solid or pop type.

Solid rivets are usually made of mild steel. They may have round or flat heads. The four steps for installing a round head rivet are shown in **Figure 5-29**.

To use solid rivets, you must be able to reach both sides of the rivet. When this is not possible, pop rivets can be placed from one side only. They are made of a hollow aluminum head with a steel pin through it. Use the following procedure to install a pop rivet:

1. Drill a hole in the parts large enough to receive the pop rivet.
2. Push the pop rivet through the hole.
3. Slip the rivet gun over the pin.
4. Squeeze the handle to pull the pin back. This creates the rivet head on the back (concealed) side.
5. Continue squeezing until the pin breaks off. See **Figure 5-30**.

Knockdown (KD) Fasteners

Some furniture is designed for "do-it-yourself" assembly and is purchased in a flat pack. This requires special fasteners that are strong and easy to use. The fasteners require no special tools or skills. Known as "knockdown" or KD fasteners, they can be taken apart and reassembled as needed. The three most common types are shown in **Figure 5-31**.

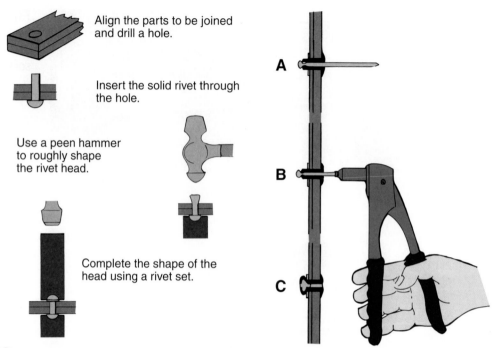

Figure 5-29. Steps to fasten sheet metal using solid rivets.

Figure 5-30. Using a pop riveter. A—Insert rivet in predrilled hole. B—Slip the tool over the pin and squeeze. C—Pin breaks off when its sleeve has expanded on the blind side of the rivet.

Movable Mechanical Joints

Some joints are made so that the joined parts can move. Think of how a door is joined to its frame. A hinge is used. The knife switch in **Figure 5-32** is another example of a movable joint. Both are pin hinges. They can only move back and forth. We say that they "move through one plane only."

A second type of movable joint is the ball and socket joint. This type of joint allows movement in more than one plane. See **Figure 5-33**. The joystick of a video game uses a ball and socket joint. A camera tripod uses a lockable ball and socket. It can move in three different directions. The drive shaft of a car uses a universal joint. It permits the joint to move up and down or left and right as the shaft spins.

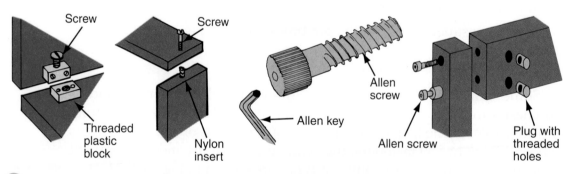

Figure 5-31. Knockdown furniture uses one of three basic fasteners.

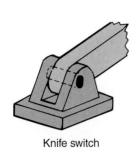

Knife switch   Door hinge

Figure 5-32. Pin hinges allow movement in one plane only.

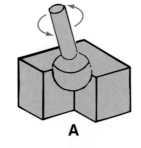

A

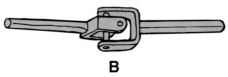

B

Figure 5-33. A—A ball and socket joint. B—A universal joint can move up and down as well as side to side.

Figure 5-34. A flexible joint can move in any direction.

A third type of movable joint is the integral or living hinge. It consists of a flexible material such as polypropylene. See **Figure 5-34**. The material itself acts as a hinge wherever it is folded.

Wood Joints

The strength of a wood joint depends on two things:

- The way the wood parts meet at the joint to provide mechanical interlocking. See **Figure 5-35**.
- The amount of surface area of the joint to be glued.

Wood joints can be grouped by type, as shown in **Figure 5-36**. One group is used on frames. The other group is used to make boxes. Frame joints are found on chairs, windows, doors, and similar products. Box joints are used to construct items such as cabinets, drawers, and storage boxes. **Figure 5-37** shows eight different joints for constructing frames or boxes.

Chemical Joining

Mechanical joints are often strengthened by *chemical joining*. Where would we be without adhesives? Furniture would disintegrate, books and shoes would fall apart, and we couldn't cap our teeth. Imagine life without stamps, tape, or Post-it® notes. Glues, adhesives, solvents, and cements are all methods of chemical joining.

Glues and adhesives are used to join woods and metals. Glues were once made from natural materials. These included animal bones, hides, and milk. They are rarely used today. Although we still use the term *glue*, it is more correct to use the term *adhesive*.

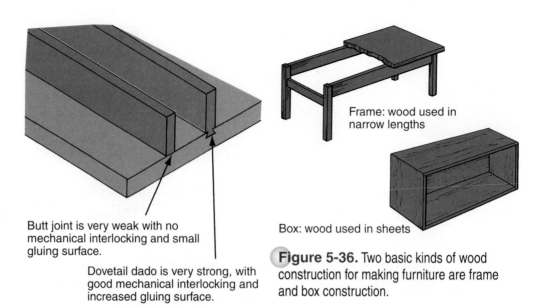

Butt joint is very weak with no mechanical interlocking and small gluing surface.

Dovetail dado is very strong, with good mechanical interlocking and increased gluing surface.

Figure 5-35. An interlocking joint is strong because one part fits into the other.

Frame: wood used in narrow lengths

Box: wood used in sheets

Figure 5-36. Two basic kinds of wood construction for making furniture are frame and box construction.

Figure 5-37. Each type of joint has advantages and disadvantages. Which of these joints has the greatest mechanical strength? Which has the largest gluing surface?

Frame Joints	Box Joints
Butt 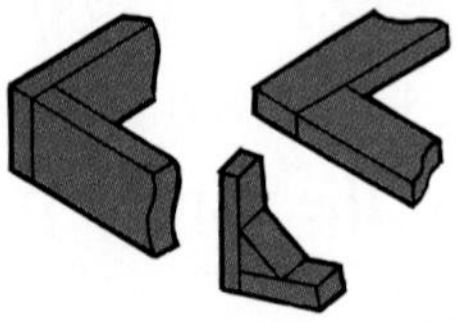	Butt 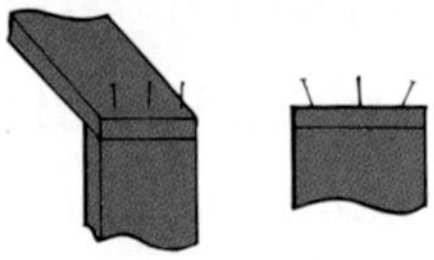
Dowel 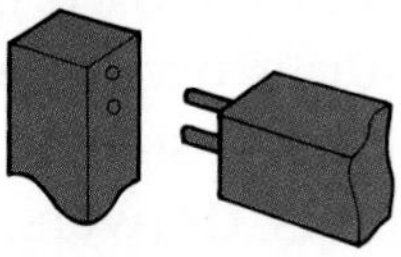	Rabbet 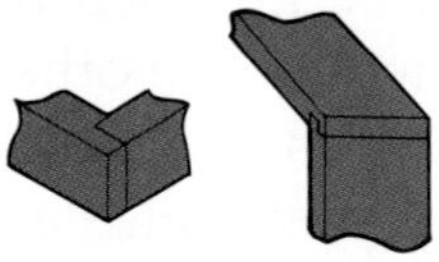
Mortise and tenon 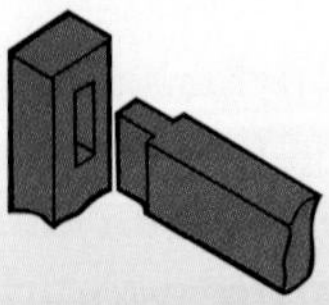	Dado 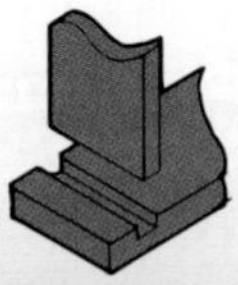
Lap 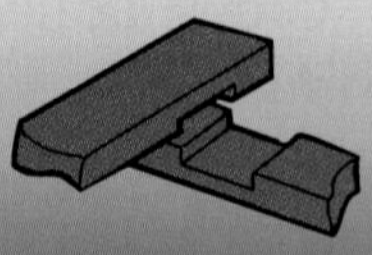	Dovetail 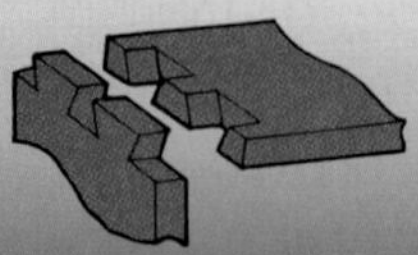

Adhesives come from petroleum products. These adhesives are of two types: thermoplastic and thermoset. One common thermoplastic adhesive is liquid white glue. Also known as *polyvinyl acetate*, it is commonly used in wood joints.

Thermoplastic adhesives harden by loss of water or solvent. They may be softened by heat and are not waterproof. Thermoset adhesives include various types of resins. Heat will not soften them, and they are waterproof.

How Glues and Adhesives Work

To act as an adhesive, the molecules that make up the glue must form strong links to one another, and the glue must stick to both surfaces being joined. This ensures that they cannot be separated when the two surfaces are pulled apart. An adhesive must flow easily to coat both surfaces and have a natural attraction, or *adhesion*, between its molecules. Adhesion is also increased when the glue hardens and tiny air bubbles get trapped. This causes a suction that has to be overcome if the surfaces are to be separated.

How Solvents and Cements Work

Solvents and cements are used to join plastics. A pure solvent softens the areas to be joined, while cements dissolve a small amount of the plastic. They penetrate deeper into the two surfaces because the solvent evaporates much more slowly. However, cement provides a stronger joint than a pure solvent.

Solvents and cements work on the principle of *cohesion*. In cohesion, the materials being joined become fluid. Then the molecules of each piece mix together. There is no foreign material in the joint. Fluid edges flow together and fuse. Cementing of thermoplastics is an example of cohesion fastening.

Figure 5-38 shows uses for different solvents, adhesives, and glues. For safety and good results, follow these general rules:

- Make sure the surfaces are clean and dry. Remove grease, paint, varnish, or other coatings.
- Carefully read the instructions and cautions.
- Secure a good fit between the two surfaces.
- Work in a well-ventilated area, especially when using solvents and cements.
- Clamp the joint until the adhesive or solvent dries.

Class	Type	Uses	Comments
Glues	Animal	Interior woodwork	Difficult to use Must be used hot Not waterproof or heat proof
	Casein	Interior woodwork	White powder mixed with water Sets in six hours Heat and water resistant
Adhesives	Polyvinyl acetate (PVA-white glue)	Wood, leather, paper	White liquid ready to use Hardens in under one hour Not waterproof
	Plastic resins (urea and phenol)	Wood	Urea: powder mixed with water Phenol: ready to use Hardens in approximately 2–6 hours Urea is water resistant: Phenol is waterproof Good strength
	Epoxy resin	Wood and metal	Two parts are mixed together Hardens in 12–24 hours Waterproof Very high strength
	Contact cement	Plastic laminates	Ready-to-use liquid Apply to both surfaces and let dry to touch Used in situations where clamps cannot be applied
	Cyano acrylate (Superglue™ or Krazy Glue™)	Nonporous materials such as glass and ceramics	Ready to use liquid Hardens almost immediately Water resistant
Solvents	Pure solvent (methylene chloride and ethylene dichloride)	Acrylics	Colorless liquid, ready to use Bonds almost immediately Waterproof
	Solvent cement	Acrylics	Colorless, viscous liquid Sets in 12–24 hours Waterproof
SAFETY NOTE: Use solvents and cements in well-ventilated areas			

Figure 5-38. Glues, adhesives, and solvents are designed for specific applications.

Solvent-Joining Acrylic Sheet

Low-viscosity solvent travels through a joint area by capillary action. This is a force that causes a liquid to rise through a solid. Properly done, solvent joining yields strong, perfectly transparent joints. It will not work at all if the parts do not fit together perfectly. To join acrylic parts using a solvent:

1. After removing the protective paper from the acrylic, hold the two pieces of acrylic in a jig, as shown in **Figure 5-39**.
2. Apply solvent along the entire joint. Work from the inside of the joint where possible. Use a hypodermic syringe or needle or an applicator bottle with a nozzle to apply the solvent.
3. Allow the joint to dry thoroughly (24 to 48 hours).
4. Remove the part from the jig.

SAFETY: Work with solvents only in a well-ventilated area!

Dipping is a second method of joining acrylic sheet material, as shown in **Figure 5-40**:

1. Set up a tray of solvent. The tray must be larger than the plastic pieces.
2. Ensure that the tray is sitting level.
3. Dip only the very edge of each plastic part into the solvent.
4. Use finishing nails in the bottom of the tray to keep the acrylic off the bottom.
5. Place the two pieces to be joined together and allow them to dry thoroughly (24 to 48 hours).

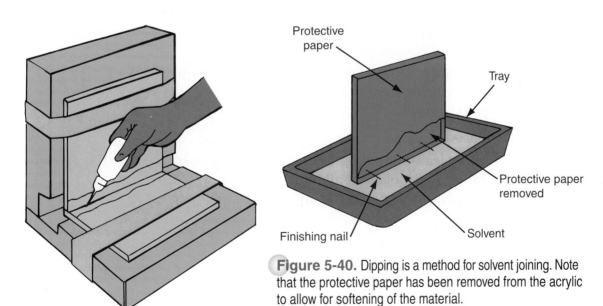

Protective paper

Tray

Protective paper removed

Finishing nail

Solvent

Figure 5-40. Dipping is a method for solvent joining. Note that the protective paper has been removed from the acrylic to allow for softening of the material.

Figure 5-39. Apply the solvent to the inside edges of the parts where possible.

Heat Joining

Heat joining is used mostly on metals. It is also used to some extent on plastics. Two types of heat joining are used on metals:

- Welding
- Brazing and soldering

Welding brings metals to their melting point. When they melt, the metals flow together. When they cool, they solidify, becoming one piece. The joint is as strong as the original metal. Welding may also be used to join plastics. This is possible with some thermoplastics such as PVC. A hot air torch heats the two parts of the joint. Heat fuses them.

Brazing and soldering work differently than welding. The heat melts the metal being used to join the parts. It does not melt the metal in the parts themselves. Brazing uses a brass alloy to join the parts. The alloy melts at 1650° F (900° C).

A mixture of tin and lead has traditionally been used in solder. However, it is being phased out because even a small amount of lead can be harmful to health. People who continue to use tin and lead solder should be sure to wash their hands after working with it. However, the best plan is to avoid lead solder entirely.

There are many different varieties of lead-free solder. One of the most popular contains 96.5% tin, 3% silver, and 0.5% copper. It is a little more costly due to its silver content. It melts at approximately 425° F, a slightly higher temperature than traditional lead solder. A high-wattage soldering iron with temperature adjustment is helpful.

To solder tinplate, copper, brass, and mild steel using lead-free solder:

1. Select a high-wattage soldering iron with temperature adjustment and several sizes of replaceable tips.

2. Choose a tip that makes the right contact with the joint to be connected. It should not be too big or too small.

3. Set the temperature at about 700° F (370° C).

4. If needed, clean the tip using a wet sponge.

5. Apply lead-free solder to the soldering iron, being sure to keep the tip wet with solder at all times. See **Figure 5-41**.

6. Allow the joint to cool slowly before moving it.

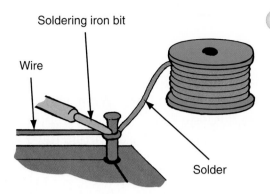

Figure 5-41. Heat the parts to be soldered before applying the solder.

Soldering iron bit

Wire

Solder

Finishing Materials

When a product is completed, its surface is usually finished. *Finishing* changes the surface by treating it or placing a coating on it. Finishing is done for several reasons:

- Protect the surfaces from damage caused by the environment
- Prevent corrosion, including rust
- Improve the appearance by covering the surface or treating it to bring out the natural beauty of the material

Converted Surface Finishes

When the surface is treated to beautify or protect, it is called a *converted surface*. The material is chemically altered to change the way it reacts to elements in the environment. The reaction of the chemical and the atoms on the product's surface provide the protective coating.

Some converted coatings are natural. Aluminum develops an oxide covering if exposed to the open air. This covering resists the natural elements.

Surface Coatings

Materials applied to a surface are called *coatings.* The most common coatings are paints, enamels, shellac, varnish, lacquer, vinyl, silicone, and epoxy. For centuries, machinery and tools have been coated with oil and grease. Paint, varnish, and enamel are widely used to protect ships, trains, cars, and bridges. Heating ducts, and sometimes nails, are galvanized (coated with zinc). Food cans are plated with tin. Many decorative objects are electroplated. A coating of nickel, chromium, copper, silver, or gold is applied to their surface.

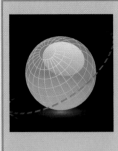

Think Green

Eco-Friendly Paints

Most people today are aware that some paints are more environmentally friendly, or "eco-friendly," than others. What you may not know is that paint companies use different definitions of the term "eco-friendly." Some paints are made of all-natural products. Others have low levels of volatile organic compounds, and still others are completely free of volatile organic compounds. *Volatile organic compounds (VOCs)* are chemicals in paint that are released as the paint dries. They often account for the strong smell of a freshly painted surface, and they can cause health problems such as dizziness and headaches.

The most eco-friendly paints are those that are free of all toxins, solvents, and odors. Tour a local home improvement center or go online to see the different "eco-friendly" paints that are available. Find out exactly what characteristics make the paint eco-friendly. Which paint would you choose?

The first step in finishing is to prepare the surfaces. They should be clean and smooth. Surfaces can be made smooth using abrasive papers or abrasive cloths, which are made in a wide range of grades and coarseness. Abrasive materials and their uses are described in **Figure 5-42**.

You should follow these three general rules when using an abrasive:

- Clean inside surfaces before assembling the project.
- Begin with a coarse abrasive. Then gradually work up to a fine grade.
- Support the abrasive whenever possible, as shown in **Figure 5-43**. A wood or cork block can be used for wood and plastic. Files can be used for holding abrasive papers while finishing metals. See **Figure 5-44**.

Figure 5-45 lists several different finishes. Some are for wood. Others are best used on plastics or metal. Finishes can be applied by wiping, brushing, rolling, dipping, and spraying.

You can apply stain and oil to wood by wiping with a cloth. Brushes work well with most finishes. They are best with liquid plastic and paint. A roller works well for painting large surfaces. Items with many curves and parts can sometimes be dipped.

Paint can be sprayed onto most shapes and materials. Aerosol spray cans are fast and easy to use on small areas. Spray guns use compressed air. They produce a high quality finish for larger surfaces. Paint dries when the solvent it contains, either water or an organic solvent, evaporates.

Material	Abrasive	Comments
Wood		*Sandpaper* was once the general name given to all abrasive papers used for smoothing wood. Today, the industry calls them coated abrasives.
	Flint paper	Crushed flint or quartz used as the abrasive Wears out quickly Cuts slowly Normally used in grades coarse to extra fine (50–320 grit)
	Garnet paper	Uses garnet as the abrasive More durable than flint paper Normally used in grades coarse to extra fine (50–320 grit)
Metal	Emery cloth	Uses emery a natural abrasive Dull black in color Normally used in grades coarse, medium, and fine (3–3/0) Oil may be added to the fine grade to give a mirror finish
Wood and Metal	Aluminum oxide	An artificial abrasive Gray-brown in color Tough, durable, and resistant to wear Normally used in grades coarse, medium, and fine (40–180 grit) Used on steel and other hard materials
Wood, metal, and plastic	Silicon carbide paper (wet-and-dry paper)	An artificial abrasive Available in three common grades: coarse (50), medium (100), very fine (400) Paper is best used wet Creates a smooth, matte finish

Figure 5-42. Coated abrasives prepare the surfaces of a material for finishing.

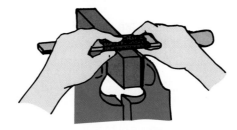

Figure 5-44. One way to use emery cloth on metals is to wrap the cloth around a file.

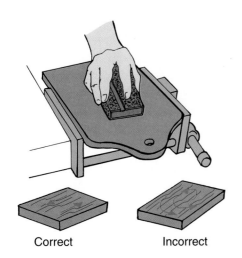

Correct   Incorrect

Figure 5-43. Always sand wood along the grain, or you will see the scratches made by the abrasive.

Material	Type	Comments
Wood	Liquid plastic (urethane)	Provides a clear coating Apply with a brush Gives a hard, water resistant, and long-lasting coating
	Stain	Changes the color of wood Cheaper wood can be stained to resemble the color of more expensive woods Applied with brush or cloth Another clear, protective finish must be applied later
	Paint	Two types: latex (water-based) and oil (linseed oil or synthetic) The surface must be primed with a primer coat Read and follow the manufacturer's directions
	Plastic laminate	Provides a decorative, durable surface The laminate is glued to a flat surface using contact cement
	Creosote and pressure treatment	Wood is immersed in a creosote or a preservative is forced into the wood under pressure Exterior use only
	Oil	Teak oil is preferred, as linseed oil requires preparation Used on handmade furniture
Metal	Paint	Surface must be completely free of oil and grease First apply a primer coat, then an undercoat, and finally a top coat
	Plastic coating	Metal is heated and dipped into fine particles of PVC that soften under the heat to form a smooth coating Useful for tool handles
	Enameling	A thin layer of glass is fused onto a metal surface For decorative work, copper is the metal used
Plastic	Polish	Surfaces are polished using very fine silicone-carbide paper followed by using a buffing attachment on a power drill Buff with light pressure to prevent melting the plastic
	Dye	Dip transparent plastic in a strong dye for a few minutes to give a tinted effect

Figure 5-45. Finishes are applied for various purposes. What criteria would you use to select a finish?

Tools are used to shape and form materials into finished products. When designing and making a product, you must know how to select and use both hand and machine tools safely. In addition to shaping and forming materials, you may need to join materials either temporarily or permanently. Materials can be joined using mechanical devices, chemicals, or heat. Finishing involves processing or coating the surface of the product.

As new materials and technologies are developed, new processes are developed for working with them. Manufacturers are looking for replacements for harmful chemicals used in the past and present, such as chromium and cadmium. A partial answer may lie in nanotechnology. For example, nanotechnology is now being applied to finishing processes. A nanocrystalline cobalt and phosphorus compound has been developed to replace chromium coatings in some cases. Other products and processes are also under development to reduce carbon emissions and the use of hazardous chemicals.

- Working safely with hand and power tools requires careful planning and attention to detail.
- To create a product, materials are first marked out. Then they are shaped using one of several processes. Examples include sawing, filing, planing, shearing, chiseling, drilling, bending, forming, casting, and molding.
- Materials can be joined using mechanical fasteners or other mechanical methods, chemicals, or heat.
- Materials are finished to protect their surfaces or to improve their appearance.

Summarizing Information

Copy the following graphic organizer onto a separate sheet of paper. For each chapter section (topic) listed in the left column, write a short, one-paragraph summary of the topic in the right column.

Chapter Section (Topic)	Summary
Learning to Work Safely	
Shaping Materials	
Joining Materials	
Finishing Materials	

Test Your Knowledge

Write your answers to these review questions on a separate sheet of paper.

1. In addition to following general safety rules, how can you help ensure your own safety when you work with tools and materials?

2. Explain how hazard identification and risk identification can help you work safely with tools and materials.

3. What tools are needed to mark a piece of wood to length?

4. Which tool is used to make straight cuts through metal and plastic?

5. To make the edge of a piece of metal smooth and flat, which tool should you use?

6. Which tool is used to make a wood surface smooth?

7. Which tool cuts aluminum and copper sheet?

8. How would you go about cutting a round hole in a piece of wood, metal, or plastic?

9. Describe the process of lamination.

10. What is the purpose of a strip heater?

11. In which process is liquid metal or plastic poured into a mold?

12. Describe the difference between mechanical, chemical, and heat joining. Give examples of each.

13. Why is a finish applied to the surface of a material?

14. Describe ways in which a finish can be applied to a material.

15. List three general rules to follow when using an abrasive to prepare surfaces for finishing.

Critical Thinking

1. A friend tells you that he has been trying to smooth the surface of the wooden table he is making, but he can't get the scratches out of the surface. What might be causing the scratches, and how can this be fixed?

2. List what you consider to be the five most dangerous activities in the technology room. What could you do to reduce the dangers?

3. When you develop a new product, you have to take into consideration all of the processes needed to convert the materials into the final product. This includes any processes needed during the design, development, manufacture, and servicing of the product. What steps should you take to ensure that someone unfamiliar with your product could reproduce these processes accurately?

Apply Your Knowledge

1. Use one of the design processes described in Chapter 2 to design and make a simple game for a young child.

2. List five objects you use at school every day. Describe how the parts of each object are joined.

3. Choose five objects, each with a different type of finish. Make a chart to show (a) the material, (b) the finish, and (c) the reason why that finish has been used.

4. List at least three tools you have in your home. State the material and process for which each is designed.

5. Research one career related to the information you have studied in this chapter. Create a report that states the following:

   - The occupation you selected
   - The education requirements to enter this occupation
   - The possibilities for promotion to a higher level
   - What someone with this career does on a daily basis
   - The earning potential for someone with this career

   You might find this information on the Internet or in your library. If possible, interview a person who already works in this field to answer the five points. Finally, state why you might or might not be interested in pursuing this occupation when you finish school.

STEM Applications

1. **ENGINEERING** Design and make an ergonomically correct chair using measurements of students in your class. Your design should do all of the following:

   - Have a seat and back.
   - Support a person weighing at least 200 pounds.
   - Be comfortable to sit in.
   - Be aesthetically pleasing.
   - Be made from corrugated cardboard.
   - Use only glue to connect the parts permanently.
   - Use as little material as possible.

2. **MATH** Choose a wood table or desk in your home or school. Measure the item carefully. Determine the type(s) of wood and other materials you would need to build it. On the Internet or in a local home improvement store, find out what standard lengths and sizes are available in this type of wood. On paper, carefully plan how to build the table with the least amount of scrap. Then make a materials list including how much of each size and type of wood is needed and the cost per piece. Also include any fasteners or adhesives necessary to build the actual table. Based on your materials list, estimate the total cost to build the table or desk.

CHAPTER 6

Structures

Better by Design

Sean Godsell designs emergency shelters

Earthquakes, floods, and hurricanes leave thousands of people needing short-term relief housing immediately. Architects around the world are tackling this problem. Some have designed inflatable concrete tents. Others have designed homes built from recycled wood pallets. Sean Godsell, an Australian architect, has designed the *Future Shack*. Made from a mass-produced, inexpensive, and durable shipping container, it has a parasol roof made from recycled plastic that packs into the container when closed and collects rainwater when open. Also packed inside the container during shipping are solar panels to generate electricity, water storage tanks, and telescoping legs to support it on uneven terrain. *Future Shack* can easily be shipped and transported by road or rail.

A shipping container provides the basis for a self-contained refugee housing unit.

The interior of the Future Shack unfolds to provide basic items such as a table and sink. What other mass-produced products could be adapted to provide emergency shelter?

"Architects have the power, if we trust them, to transform our lives."
—*Sean Godsell*

Reading Target

Creating an Outline

An outline is an orderly statement of the main ideas and major details of a text passage. Each main idea or detail is written on a separate line. Creating an outline can help you understand and remember what you read. Read each section of this chapter carefully. Then use the Reading Target graphic organizer at the end of the chapter to create an outline of the chapter.

Key Terms

abutments
arch bridge
cantilever bridge
compression
dynamic load
load
pier
reinforced concrete
shear

static load
stays
structures
strut
suspension bridge
tension
tie
truss

Objectives

After reading this chapter, you will be able to:

- Identify the loads acting on structures.
- Analyze the forces acting on a structure.
- Demonstrate how structures can be designed to withstand loads.
- Explain various bridge designs.
- Describe methods of reinforcing high-rise buildings.

Useful Web sites:
www.seangodsell.com/
www.architectureaustralia.com.au/aa/aaissue.php?article
firmitas.org/
www.architectureforhumanity.org/

Structures are all around us. We build them to live in or to cross a river. We build them to carry wires, to receive radio waves, and to transport people. Houses, bridges, and towers are not the only types of structures; airplanes, boats, and cars are structures, too.

Many structures enclose and define a space, including the homes in which we live. At times, however, a structure is built to connect two points. Examples of these structures include bridges and elevators. Other structures are meant to hold back natural forces, as in the case of dams and retaining walls. Some structures are meant to be temporary, such as the *Future Shack* described in the Better by Design feature at the beginning of this chapter. Others are permanent and need to withstand weather and various other forces over time.

Everyone has built some kind of structure. Have you constructed a ramp for a skateboard? Perhaps you built a tree house from a variety of scrap materials. Maybe you made a model crane, a dollhouse, a tunnel for a model railroad, or a sand castle on the beach. **Figure 6-1** shows two structures. What is the purpose of each?

Not all structures are made by humans. Living organisms, such as trees and our bodies, are natural structures. A giant redwood tree must be rigid enough to carry its own weight. Yet it is able to sway in high winds. The bones of a skeleton have movable joints. They permit activities such as running and lifting. **Figure 6-2** shows both natural and human-made structures.

Structures and Loads

What do all structures have in common? They all have a number of parts, which are connected. The parts provide support so the structures can serve their purpose. One important job of all structures is to support a *load*. A load is the weight or force placed on a structure. See **Figure 6-3**.

Figure 6-1. A scaffold supports workers while they build structures. Scaffolds are structures, too. They have connected parts and carry workers without collapsing.

Wasp's nest

Shells

Lighthouse

Snow Den

Figure 6-2. Structures are found all around us. Top—Some are found in nature. Bottom—Others are planned and built by humans.

Tower

Figure 6-3. Structures must be able to support the loads they are intended to carry. What loads do towers have to support?

One example of a load on a bridge is a heavy vehicle crossing it. See **Figure 6-4**. Vehicles must also carry loads, such as the weight from their own frame and the passengers they carry. See **Figure 6-5**. The load on a dam is the force of the water behind it. Both vehicles and dams must also support the materials from which they are built. This is part of the load.

Structures vary greatly in size and type. Think about the loads that each of the structures in **Figure 6-6** must withstand. What materials were used in their construction? How are the parts connected together?

All structures must be able to support a load without collapsing. A roof must not only support its own mass but also a heavy blanket of snow. A chair must carry the load of a person sitting still or fidgeting. See **Figure 6-7**. These loads are of two types: static and dynamic.

Figure 6-4. Roads, tunnels, and sidewalks help us travel from place to place by vehicle and on foot. A—Walkways can provide passages through another structure. B—Bridges span rivers, gorges, and railways lines.

Figure 6-5. Structures for transportation must carry people and support other parts of the vehicle.

Figure 6-6. The framework of each building is like a skeleton. Can you see the "skeleton" in each of these structures?

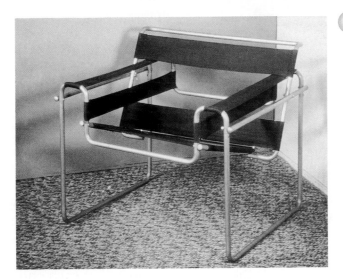

Figure 6-7. The structure of a chair must be such that it can carry the load of a person sitting on it.

Static Loads

Static loads are loads that either do not change or change slowly. They may be caused by the weight of the structure itself. Columns, beams, floors, and roofs are part of this load. Static loads also include the weight of objects placed in or on the structure. **Figure 6-8** shows an example of a static load.

Dynamic Loads

A *dynamic load* is a load that is always moving or changing on a given structure. For example, the mass of a person walking across the floor creates a dynamic load. Other dynamic loads include the force of a gust of wind pushing against a tall building and a truck crossing a bridge. See **Figure 6-9**.

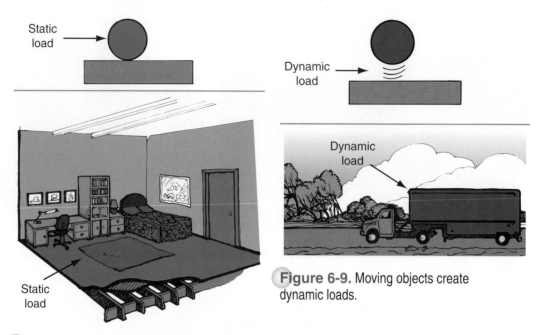

Figure 6-9. Moving objects create dynamic loads.

Figure 6-8. Objects at rest create static loads.

Forces Acting on Structures

Both static and dynamic loads create forces, which act on structures. To understand these forces and what they do, imagine a plank placed across a stream, as shown in **Figure 6-10**. When you (the load) walk across the plank (the structure), what would you expect to happen? The plank bends in the middle. The forces acting on the bridge are shown by the foam rubber in **Figure 6-11**. Notice that parallel lines have been marked on it in **Figure 6-11A**. If the foam is supported at each end and a vertical load is applied to the center of the foam, it bends. See **Figure 6-11B**.

Notice what has happened to the parallel lines. At the top edge, the lines have moved closer together. The lines at the bottom edge have moved farther apart. The top edge of the plank is in *compression* (being squeezed) and the bottom edge is in *tension* (being stretched). Along the center is a line that is neither in compression nor in tension. It has no force acting along it. This line is called the *neutral axis*.

Figure 6-10. A person standing on a plank causes the plank to bend. Bending causes compression on the top surface of the plank and tension on its bottom surface.

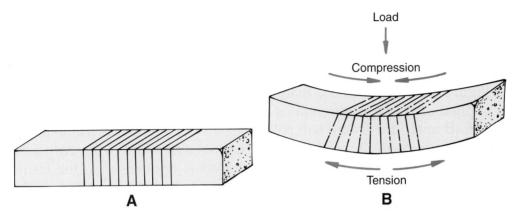

Figure 6-11. A—Foam rubber with parallel lines drawn on it will show what happens when a load is placed on a beam. B—Bending causes compression on the top surface and tension on the bottom surface.

The design and construction of structures must minimize the effects of bending. Parts must be shaped so the forces of tension and compression are balanced. These energies are then said to be in a state of equilibrium.

Designing Structures to Withstand Loads

As was shown by the foam rubber in **Figure 6-11**, the top and bottom surfaces of a beam are subject to the greatest compression and tension. These surfaces are where the greatest strength is needed. The shapes shown in **Figure 6-12** strengthen a beam along these surfaces.

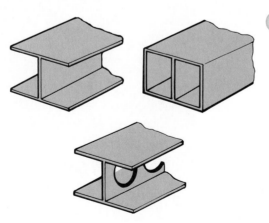

Figure 6-12. The shapes shown here can support heavy loads.

After members have been shaped to resist compression and tension, they must be connected in a way that minimizes bending. Look at the structures in **Figure 6-13**. Notice how often the triangle appears in these designs. To understand why the triangle is important in structures, look at **Figure 6-14**. The frame is made of four connected members. If a load is applied at A, the frame retains it shape. But if a load is applied at a corner (B or C), the frame will collapse. Now compare this frame to the one in **Figure 6-15**. A rigid diagonal member (running from corner to corner) has been added. Now, when a load is applied at A, the frame retains its shape. This time, however, it also retains its shape when a load is applied at corners B or C. At corner B, the load causes the diagonal to be in tension. A rigid member in tension is called a *tie*. When the load is applied at corner C, the diagonal is in compression. A rigid member in compression is called a *strut*.

What would be the effect of replacing the rigid diagonal member with a non-rigid member such as a rope, chain, or cable? See **Figure 6-16**. Would the frame retain its shape when loaded in each of the three positions? When is the rope in compression, and when is it in tension?

Pylon

Geodesic dome

Figure 6-13. Some shapes can support heavier loads better than other shapes can. What supporting shape appears most often in these two pictures?

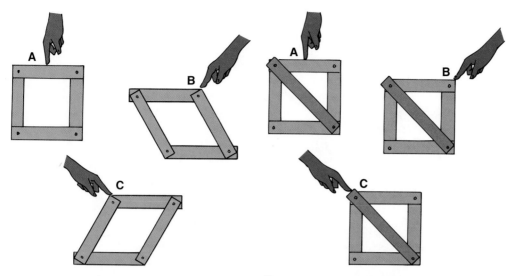

Figure 6-14. A square frame can support a load applied at point A. Why does the frame collapse when a load is applied at points B and C?

Figure 6-15. The frames retain their shape under loads at A, B, and C.

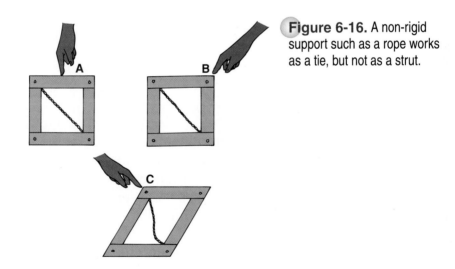

Figure 6-16. A non-rigid support such as a rope works as a tie, but not as a strut.

In addition to compression and tension, a third force acts on structures. This force is called *shear*. Shear is a multidirectional force that includes parallel and opposite sliding motions. To understand how shear takes place, imagine that you are pulling the wagon in **Figure 6-17**. Suddenly, the wheels hit a rock. The effect is a sharp jolt on the pin. This force causes the material to shear.

Figure 6-17. Shear is a force that causes one part to slide over an adjacent part.

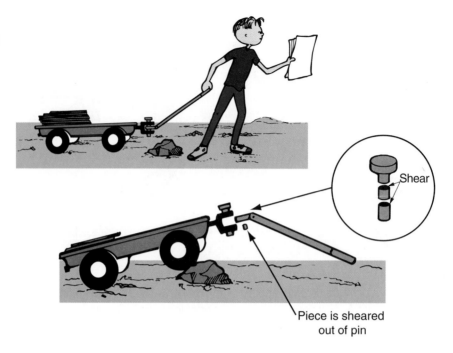

Piece is sheared out of pin

The Structure of Bridges

Let us look at how bending and the forces of compression, tension, and shear can be resisted in the design of structures. Then you will see why bridges are built the way they are.

A major problem with bridges is that they bend under a dynamic load. See **Figure 6-18.** One common way to prevent a bridge from bending is to support the center with a *pier*, as shown in **Figure 6-19.** However, it is not always possible to build piers under a bridge. Piers may not allow the passage of ships. Sometimes the river is too deep, runs too swiftly, or has a soft bed with no firm foundation. Other ways have to be found to strengthen the beam bridge.

One solution is to make the beam much thicker. This, however, would make the beam very heavy. Its own mass would make it sag in the middle.

Figure 6-18. A simple beam bridge bends easily.

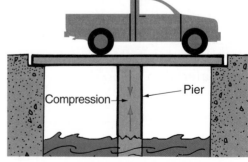

Figure 6-19. A pier supporting a beam bridge is compressed by the load on the bridge.

The beam could also be strengthened at the center where it is most likely to bend or break. See **Figure 6-20**. Once again, notice that the strongest shape is the triangle. As you saw in **Figure 6-13**, a triangle does not have to be solid. It can be a frame and still be very rigid.

Truss bridges make use of the triangle in their design, as shown in **Figure 6-21**. As the truck crosses the bridge, its mass causes the bridge roadway to bend. Member "A" moves down. This pulls down on members "B" and "C," pulling them toward the end of the bridge and carrying the forces out to the bridge supports. Most truss bridges are more complex than the simple truss. Many triangular frames are used to construct them, as shown in **Figure 6-22**.

A bridge deck can also be supported from above. Cables called *stays* provide the support. See **Figure 6-23**. Notice that the pylons are in compression and the stays are in tension.

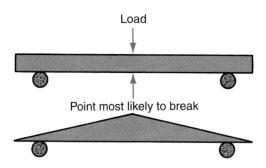

Figure 6-20. One way to strengthen a beam bridge is to make it thicker in the middle.

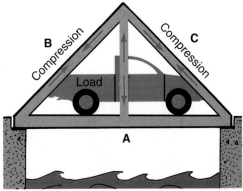

Figure 6-21. This diagram shows how a simple truss bridge works. Is the center (vertical) beam in this bridge under tension or compression?

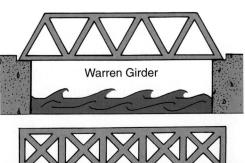

Figure 6-22. Two types of truss bridges.

Engineering Application

Testing Truss Strength

The various designs of trussed bridges are named after the engineers who first designed them. James Warren, an engineer, patented the first bridge that used equilateral triangles in 1848. An ***equilateral triangle*** is one in which all sides are the same length and all angles are the same. Today we use trusses more often in roofs of buildings than for bridges, but the reasons for using them are the same. They span distances of 10 to 100 feet (3 to 30 meters) using a minimum amount of material while providing the maximum amount of strength.

Engineering Activity

Make two models to compare the strength of a simple plank to a truss. Engineers often build scale models in order to determine the strength of their designs. Your models, like theirs, will be tested "to destruction." For this demonstration, use wood measuring 1/2 inch by 1/4 inch (12 mm by 6 mm). You may also use popsicle sticks for part of the construction.

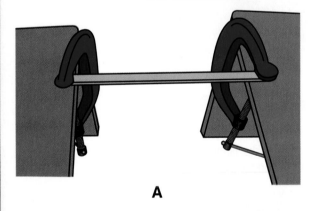

A

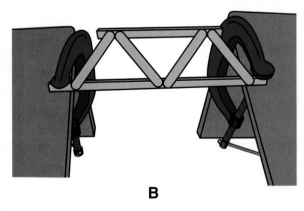

B

1. Connect a 12-inch (300 mm) length of wood between two benches and hold it in place using C-clamps, as shown in Figure A. This wood represents a simple plank.

2. Decide how you will measure the strength of the plank. How can you place greater and greater loads on the wood until it breaks? **Important:** The method of holding the weights must be suspended below the plank.

3. Gradually add weights to the wood until it breaks. Record the amount of weight it takes to break the wood.

4. Construct the truss shown in Figure B. Glue all pieces together and wait until the glue is completely dry before proceeding to the step 5.

5. Place your truss between the two benches and hold it in place using C-clamps.

6. Gradually add weights until the truss breaks. Record the amount of weight it takes to break the truss. How much stronger was the truss compared to the plank?

7. Compare your results with those of your classmates.

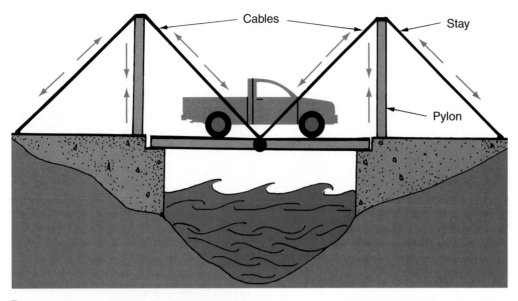

Figure 6-23. Look at the red arrows that show the forces on the pylons and stays. Describe how the mass (weight) of the truck is transferred as it travels across the bridge.

The same principle is used for *suspension bridges*. Suspension bridges are the longest bridges. See **Figure 6-24**. The bridge deck is suspended from hangers attached to a continuous cable. The cable is securely anchored into the ground at both ends. The cable transfers the mass of the deck to the top of the towers. From there, compression transfers the mass to the ground.

There are many other types of bridges. Their design follows the same general principle: reducing the amount of bending of the road deck. Two of the most common types are *arch bridges* and *cantilever bridges*.

Figure 6-24. Suspension bridges, like the Golden Gate bridge in San Francisco, rely on steel cables supported by a tower.

In an arch bridge, the compressive stress created by the load is spread over the arch as a whole. The mass is transferred outward along two curving paths. The supports where the arch meets the ground are called *abutments*. They resist the outward thrust and keep the bridge up, **Figure 6-25**.

A beam can support a load at one end provided that the opposite end is anchored or fixed. This is known as a *cantilever beam*. The principle of a cantilever is shown in **Figure 6-26**. A cantilever bridge has two cantilevers with a short beam to complete the span. See **Figure 6-27**.

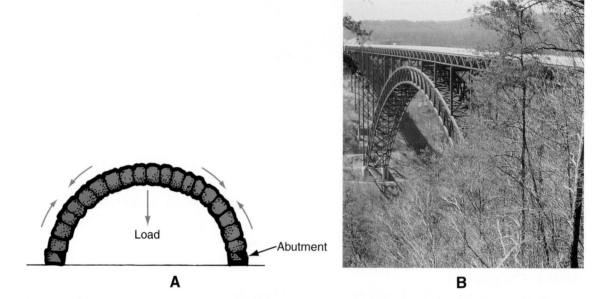

Figure 6-25. A—An arch transfers its load back to its ground support. B—The New River Gorge arch bridge in West Virginia is 3030′ (923.5 m) long.

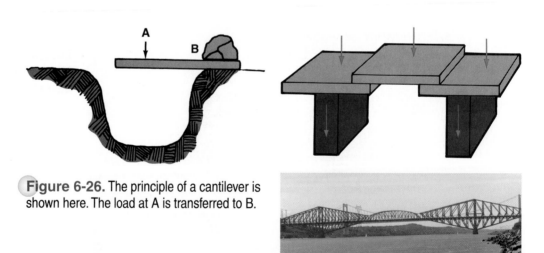

Figure 6-26. The principle of a cantilever is shown here. The load at A is transferred to B.

Figure 6-27. This bridge cantilevers its three sections from piers. In addition, it has two truss bridges supported on the tips of the cantilevers.

Bridges are made from many materials. The most common are steel and concrete. Steel is fairly inexpensive, strong under compression and tension, but needs maintenance to prevent corrosion. Steel cables made of wire rope are used to support the mass of the roadway (bridge deck) and the traffic load on it. The towers of many bridges are also made of steel. Steel trusses give rigidity to the bridge deck. They also resist bending.

Concrete is economical and resists fire and corrosion. It is strong under compression but weak under tension. To overcome this weakness, the concrete is reinforced with steel rods wherever it is in tension. The embedding of steel rods to increase the resistance to tension is the basic principle of *reinforced concrete*. See **Figure 6-28**.

The Structure of High-Rise Buildings

High-rise buildings, sometimes called skyscrapers, have only been part of the landscape for about 100 years. In 1903 the Ingalls Building in Cincinnati was built to a height of 16 stories. In 1913 the Woolworth building became the tallest building, with a height of 55 stories. Eighteen years later the Empire State Building was built to a height of 102 stories.

Figure 6-28. Concrete is made stronger by reinforcing it with steel rods.

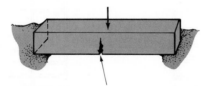

Concrete is weak in tension and cracks will occur at an unsupported center.

Reinforced concrete uses steel rods to resist tension. If these rods are stretched while the concrete is hardening, prestressed concrete is produced.

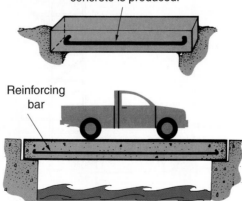

Reinforcing bar

The substructure of a skyscraper is built underground to give it a firm, solid stance and spread its weight over a larger area. Vertical steel columns are built up from this substructure. Weight from each floor is transferred by horizontal steel girders to the steel columns.

Have you noticed the shape of high-rises? While most of them are square or oblong, others are round, curved, or hexagonal. The structure of many high-rise buildings is quite different from that of houses. Houses have frames on the outside to carry the loads of roof, windows, and doors. Many skyscrapers have a stiff frame near the inner core of the building. In addition to beams and columns, the core has diagonal bars that work in both tension and compression. See **Figure 6-29A**. It supports dead loads (such as the weight of the building) and live loads (including wind). This inner frame also houses elevators and many of the pipes and ducts as they pass from one floor to the next.

If the supporting structure is placed in the center of the building, the outside walls can be made largely of glass, held in place by vertical metal struts. These walls, known as *curtain walls*, are supported by frames of steel or concrete that form part of the structure of the building.

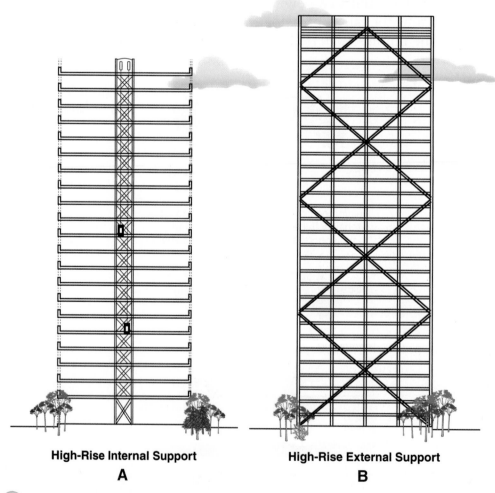

High-Rise Internal Support

A

High-Rise External Support

B

Figure 6-29. A—A central core provides the support in this high-rise. B—The exterior walls of this high-rise provide a supporting frame.

Another type of high-rise construction is shown by the John Hancock Building in Chicago. See **Figure 6-29B**. It uses the same principle of diagonal, X-shaped bracing. However, the bracing is on the outside walls of the building, making the entire building into a huge hollow tube anchored to its foundation. The Burj Khalifa, the world's tallest building with more than 160 stories, also uses this principle of construction.

Think **Green**

Green Roofs

One interesting way to make a building eco-friendly is to include a "green roof." More than just a rooftop garden, a green roof provides several advantages. It can improve both the energy performance of the building and the air quality around the building. It also provides habitat for birds and other small animals. Visually, it provides relief from the gray concrete surfaces that make up many cities.

Some green roofs include pathways for people to walk on, so the space can be used as a park or a place to grow vegetables. Others are completely covered with native grasses and plants and are designed to avoid the need for irrigation.

A good design for a green roof must take many structural factors into consideration. The weight of the soil, plants, and water or moisture must be taken into account. Drainage is also a structural issue. How will excess water be handled? The soil mix and types of plants to be used should be chosen carefully to minimize maintenance and irrigation.

More and more cities are incorporating green roofs into their municipal buildings. Check to see if any buildings in your city or county have a green roof. If not, can you identify any buildings that would be good candidates for a green roof?

Two types of loads are present in structures: static and dynamic. These loads create the forces of compression, tension, and shear. Structures must be designed to minimize the effects of these forces. The members are then connected together in a design that minimizes bending.

The iron and concrete bridges and the towering skyscrapers are the great structures of the nineteenth and twentieth centuries. They provide examples of how structures are designed to resist forces.

As new materials and processes are developed, we can build taller skyscrapers, longer bridges, and other complex structures. Engineers and designers working on the Burj Khalifa in Dubai, for example, faced many new challenges. For example, how does one wash the windows on the 160th floor of a building in an area known for high winds? Meeting these challenges and designing structures to meet the needs of people living in the 21st century are tasks for the designers and engineers of today and tomorrow.

- The two types of loads on a structure are static and dynamic loads.
- Forces that act on structures include compression, tension, and shear forces.
- Structures must be designed to withstand both static and dynamic loads.
- Bridges are designed to support the weight of the actual roadway, as well as the weight of traffic that uses the bridge and weather elements, such as snow.
- High-rise buildings can have either internal or external supports. The same triangle shape used to reinforce bridges can be used to strengthen high-rise structures.

Creating an Outline

Use the following graphic organizer to create an outline. Write your outline on a separate sheet of paper. The first section of the chapter has been outlined for you as an example. Notice that the main idea of the first section has been placed next to the Roman numeral I. The major details from that section are placed on the indented lines with letters (A, B, C, and so on). Using the first section as a pattern, outline the rest of the chapter. Try to supply at least two supporting details for each main section. Add more detail lines if necessary to describe all of a section's important details.

I. One important job of all structures is to support a load.
    A. A load is the weight or force placed on a structure.
    B. Static loads either do not change or change slowly.
    C. Dynamic loads are always moving and changing.
II.
    A.
    B.
III.
    A.
    B.
IV.
    A.
    B.
V.
    A.
    B.

Test Your Knowledge

Write your answers to these review questions on a separate sheet of paper.

1. Name three natural structures and three structures made by humans.
2. Briefly explain why all structures must be built to withstand a load.
3. Name two types of loads acting on structures. Give one example of each.
4. What forces act on the top and bottom surfaces of a beam loaded from above?
5. Where should a beam be reinforced to strengthen its ability to handle a load on its top surface?
6. Why does a triangle give greater rigidity to a structure than a rectangle?
7. What is a tie?
8. What is the difference between a tie and a strut?
9. What type of bridge uses a series of triangular frames to support the roadbed?
10. Using notes and diagrams, explain how an arch bridge resists loads.
11. Using notes and diagrams, explain the principle of a cantilever bridge.
12. From what two materials are bridges most commonly built?
13. Concrete is weak in tension. How is this problem overcome?
14. What is the purpose of the substructure of a high-rise building?
15. List two types of framing that are commonly used on high-rise buildings.

Critical Thinking

1. Search the Internet to find what you consider to be the 20 greatest engineering achievements of the 20th century. Place them in order from most important to least important, and explain your reasoning.
2. Why is the middle of a beam bridge the point most likely to break?
3. New building construction is good in many ways. For example, it allows us to incorporate new, more environmentally friendly processes and materials, and it provides employment for many people. However, current construction practices also have disadvantages, such as deforesting rainforests and contributing to global climate change. Write an essay on the ethical aspects of new construction. Should it be allowed to continue without limits or controls? Explain the role of trade-offs and compromise in this issue.

Apply Your Knowledge

1. Use books, magazines, and other sources to find illustrations of natural structures. Then find structures made by humans that closely resembles each natural structure. Create a display that compares each natural structure with a human-made structure that resembles it. Share the display with your class or school.

2. List five different types of structures. For each structure, list the loads to which it is subjected. State whether each load is static or dynamic.

3. Draw a diagram of a plank bridge with a load on it. Label the diagram to show the forces on the bridge.

4. Research one career related to the information you have studied in this chapter. Create a report that states the following:

   - The occupation you selected
   - The education requirements to enter this occupation
   - The possibilities for promotion to a higher level
   - What someone with this career does on a daily basis
   - The earning potential for someone with this career

   You might find this information on the Internet or in your library. If possible, interview a person who already works in this field to answer the five points. Finally, state why you might or might not be interested in pursuing this occupation when you finish school.

STEM Applications

1. **ENGINEERING** Using only one sheet of newspaper and 4 inches (10 cm) of clear tape, construct the tallest freestanding tower possible. Document your designing process and include the documentation in your portfolio.

2. **ENGINEERING** Using drinking straws and pins, construct a bridge to span a gap of 20 inches (508 mm) and support the largest mass possible at midpoint. Document your designing process and include the documentation in your portfolio.

3. **MATH** Explain why the triangle is one of the strongest shapes for supporting static and dynamic loads. Use sketches to show basic geometric principles in your answer.

Adrian Smith, the architect who designed the Burj Khalifa, was inspired by the hymenocallis flower. Search the Internet to find out what this flower looks like. What features of the flower provided inspiration?

CHAPTER **7**

Two coats of mud plaster are applied to the straw.

Better by Design

Rena Upitis designs and builds sustainable buildings

Rena Upitis is a Canadian educator, architect, artist, and environmentalist. She believes that people must become committed to sustainable building and environmentally respectful practices. To explore this idea, Rena designed and built Wintergreen Studios, a straw-bale building that uses recycled barn beams in its structure and locally grown straw for its walls. One wing of the building is sheltered with an earth-covered living roof. In straw-bale construction, baled straw from barley, wheat, rice, flax, rye, or oats is used to build the exterior walls. The bales are then covered with mesh that is stitched on with giant bale needles. On top of the mesh, two or three coats of plaster are applied. This construction technique provides a high insulation value, so houses are warm in the winter and cool in the summer. The house is *off-grid*. In other words, electricity is generated at the site, and the house is not connected to an electric utility.

Straw bales provide insulation in a post-and-beam structure.

"Everything I have read about energy consumption and global warming tells me that we do not have the privilege of another fifty years to think about these issues."—Rena Upitis

Connecting to Prior Knowledge

One good way to prepare yourself to read new material is to think about what you already know about the subject. You have probably seen new buildings or homes being constructed in your area. What do you know about home construction? Use the Reading Target at the end of this chapter to record your ideas, even if you are not sure of some of the facts.

adobe
batter boards
beam
closed-loop system
communication system
electrical system
floor plan
footing
foundation
heating, ventilation, and
*   air-conditioning (HVAC)*
*   system*
insulation
joist
landscaping

modular construction
off-grid
open-loop system
photovoltaic panels
plumbing system
post and lintel
prefabrication
roof truss
site
subfloor
subsystem
sustainable
system
wall studs

After reading this chapter, you will be able to:

- Identify issues that affect the structure of a house and how it is built.
- Discuss the steps involved in planning for a home.
- Describe the structural components and materials in a typical house.
- Use a systems model to explain construction technology.
- Explain how advances in construction and building systems are affecting construction technology.

Useful Web sites:
www.strawbale.com/
www.greenplanethomes.ca/about_strawbale.htm
www.wintergreenstudios.com

Modern communities rely on people with various skills and different levels of education to build quality homes and maintain them in good condition. Carpenters, electricians, cement masons, plumbers, sheet metal workers, bricklayers, and ironworkers build houses, stores, office buildings, and other commercial and industrial structures. See **Figure 7-1**. Many of these tradespeople are licensed, certified, and bonded. They often undergo an apprenticeship and join a union, which ensures their skill and competency and also secures their wages and working conditions.

Other professionals are also involved in building homes. These include civil engineers, architects, interior designers, landscape planners. In addition, inspectors design, plan, and ensure that the buildings adhere to government guidelines, regulations, building laws, and ordinances.

The Structure of a House

Many different materials have been used for homes. See **Figure 7-2**. In the past, people have built houses from the handiest material. They used what was available nearby. In North America, a mixture of clay and straw, called *adobe*, was used in the Southwest. Thick walls kept the home

Figure 7-1. Tradespeople build homes. A—Carpenters use wood to build the house frame, roof, and stairs and install windows, doors, and cabinets. B—Electricians install electrical and communication wiring, lighting circuits, and fixtures. C—Cement masons pour cement for foundations, sidewalks, and driveways.

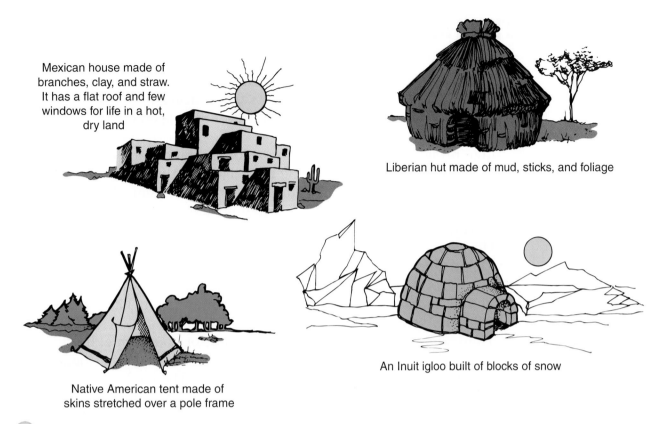

Mexican house made of branches, clay, and straw. It has a flat roof and few windows for life in a hot, dry land

Liberian hut made of mud, sticks, and foliage

Native American tent made of skins stretched over a pole frame

An Inuit igloo built of blocks of snow

Figure 7-2. Examples of homes that were built using materials found locally. What other factors determine the shape of these homes and the materials used in their construction?

warm in winter and cool in summer. Pioneers who homesteaded the Great Plains used sod cut by plow. The sod served as walls and sometimes roofs. In wooded regions of the north where lumber was plentiful, log cabins were built. Wood is still a popular building material in North America. Treatment of the wood to retard fire and decay has made the frame house more durable than ever.

Prefabrication

Building a home entirely on its site often takes a number of months. To speed up construction, some parts or even the entire house can be built in a factory. This method of building is called *prefabrication*. It is a faster method because the parts can be made on an assembly line. Workers are not affected by bad weather.

Prefabricated parts are moved to the building site for final assembly. Time is saved in the factory because of mass production methods, and time is saved on the site because much of the assembly has already been done.

A popular type of prefabrication is known as *modular construction*. Modules are basic units, such as rooms. Modules of different sizes and shapes can be combined on site.

The House Frame

The simplest framed structure is a *post and lintel* or post and beam structure. See **Figure 7-3**. The lintel is a beam simply supported on the posts. It carries the roof load. The posts are vertical struts compressed by the lintel. Post and lintel structures may be built one on top of another to frame multistory buildings.

Like bridges, houses must support loads. Static loads a house must support include the weight of the materials from which it is built and its contents. The house must also withstand dynamic loads created by weather conditions outside and the movement of people inside. To understand how these loads are supported in a house, look at **Figure 7-4**.

Horizontal forces are produced by wind. Most other loads and forces tend to push downward. Loads are usually applied to horizontal members such as *joists* or *beams*. The total load then moves downward. It transfers to columns or bearing walls and then to the foundation. Finally, the load is transferred to the soil.

Think about the loads placed on the floor of your home. If you and your friends are dancing, the mass of the people acts as a load. This load is transferred to a beam, then to a column. The load on the column is transferred through the foundation wall to the footing and into the underlying soil. See **Figure 7-5**.

Planning for a Home

Before construction of a house can begin, two major decisions must be made. One is to decide on the basic type of house. The other is to plan the house *site*, or lot. A site is the land on which a house is built.

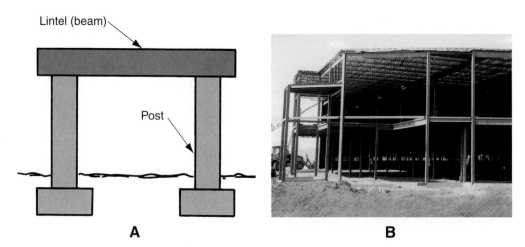

Figure 7-3. A—Heavy posts support the horizontal lintel or beam. B—Steel-framed buildings are similar to post and lintel structures.

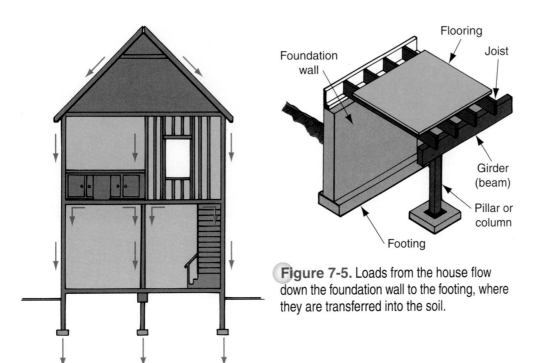

Figure 7-5. Loads from the house flow down the foundation wall to the footing, where they are transferred into the soil.

Figure 7-4. Arrows show how the load of a house is transferred to the foundation.

Four types of single-family houses are shown in **Figure 7-6** through **Figure 7-9**. The type chosen is based on a number of considerations. How much room is needed? How much land is available for a home? Which type is most appealing to the family? Will any family member have difficulty with steps?

Another consideration is how much money you wish to spend. Is the money available? Banks and other lending institutions make loans for building. However, first you have to qualify for the loan.

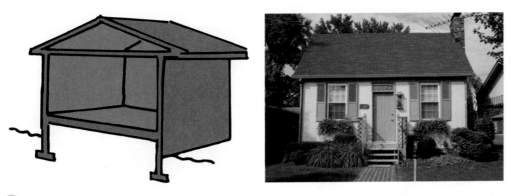

Figure 7-6. A bungalow is small and usually has only one story.

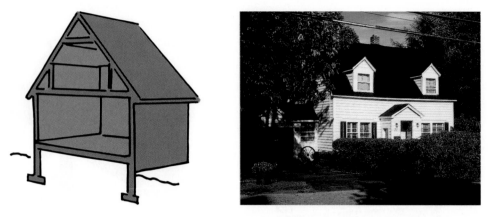

Figure 7-7. In a one-and-a-half story home, the rooms on the second level extend into the roof.

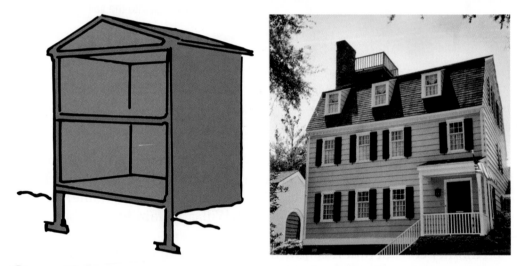

Figure 7-8. Houses may have two, three, or more stories.

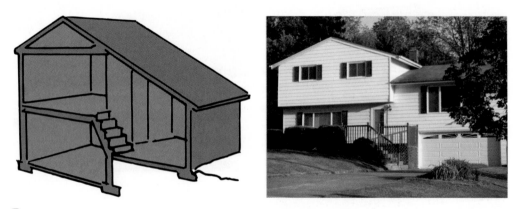

Figure 7-9. A split-level home is divided vertically. The floors of one part of the house are located midway between the floors of the other part.

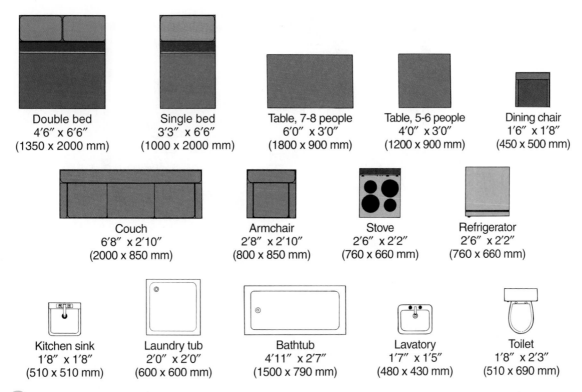

Double bed
4'6" x 6'6"
(1350 x 2000 mm)

Single bed
3'3" x 6'6"
(1000 x 2000 mm)

Table, 7-8 people
6'0" x 3'0"
(1800 x 900 mm)

Table, 5-6 people
4'0" x 3'0"
(1200 x 900 mm)

Dining chair
1'6" x 1'8"
(450 x 500 mm)

Couch
6'8" x 2'10"
(2000 x 850 mm)

Armchair
2'8" x 2'10"
(800 x 850 mm)

Stove
2'6" x 2'2"
(760 x 660 mm)

Refrigerator
2'6" x 2'2"
(760 x 660 mm)

Kitchen sink
1'8" x 1'8"
(510 x 510 mm)

Laundry tub
2'0" x 2'0"
(600 x 600 mm)

Bathtub
4'11" x 2'7"
(1500 x 790 mm)

Lavatory
1'7" x 1'5"
(480 x 430 mm)

Toilet
1'8" x 2'3"
(510 x 690 mm)

Figure 7-10. These drawings represent furniture and fixtures in a typical home. They are sometimes shown on floor plans so people can visualize how they will fit in a home.

Planning Inside Space

The second major task in designing and building a home is to plan the interior spaces. This task is often performed by an architect. What if there were no interior walls in your home and you had to divide the space into a number of rooms? How would you do it?

First of all, think about the spaces needed in a home. You must be able to fit in all of your furniture and household equipment. You should also leave sufficient space to move around. The sizes of most items of furniture and equipment are fairly standard. **Figure 7-10** shows a plan view of common items and the floor space required for them.

One way to determine the overall size and shape needed for any individual room is to try out different furniture placements. Use small pieces of card stock. Cut the card stock to represent the size and shape of each piece of furniture. Position the shapes in different ways and study each arrangement. Continue until you find a suitable arrangement, remembering to leave space for people to circulate around the room. There must also be space to open doors and pass through doorways. The result provides a possible size and shape for the room. See **Figure 7-11.** Next, fit the rooms into the overall shape of the house. Generally, rooms are grouped according to their functions: living, sleeping, and working (food preparation and dining). Rooms are grouped for a number of reasons:

- To separate noisy and quiet areas so family members can work or play without disturbing others who are resting or studying.

Figure 7-11. Rooms are often planned around furniture. What furniture items are shown by these symbols?

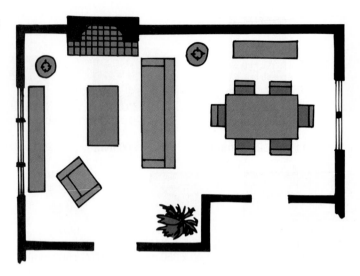

- To place bedrooms and bathrooms close to each other for convenience in washing, bathing, and dressing.
- To give direct access between kitchen and dining area for convenience in carrying hot foods from the kitchen to the table and for clearing the table.

Rooms must be connected by hallways, stairs, and doorways. In a home with good traffic patterns, you should be able to move from one area to another without passing through a third area. See **Figure 7-12**. For example, it should be possible to walk from the kitchen to the front door without going through several rooms.

When you think you have a good arrangement of rooms, you can make a drawing. An architect calls this a *floor plan*. The plan shown in **Figure 7-13** shows a floor plan for a two-bedroom apartment. **Figure 7-14** shows a floor plan for a three-bedroom bungalow.

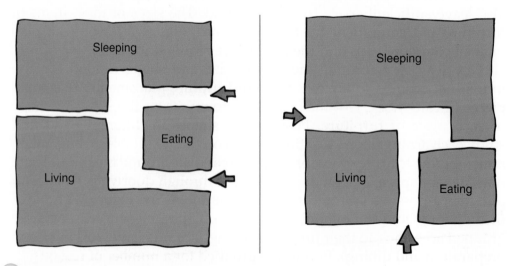

Figure 7-12. In a well-designed house, hallways allow access to all rooms without passing through other rooms.

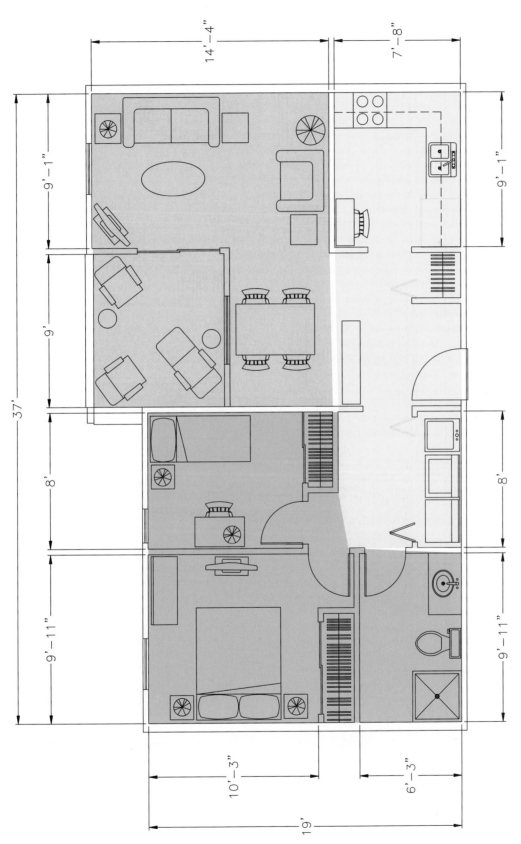

Figure 7-13. A floor plan for a two-bedroom apartment.

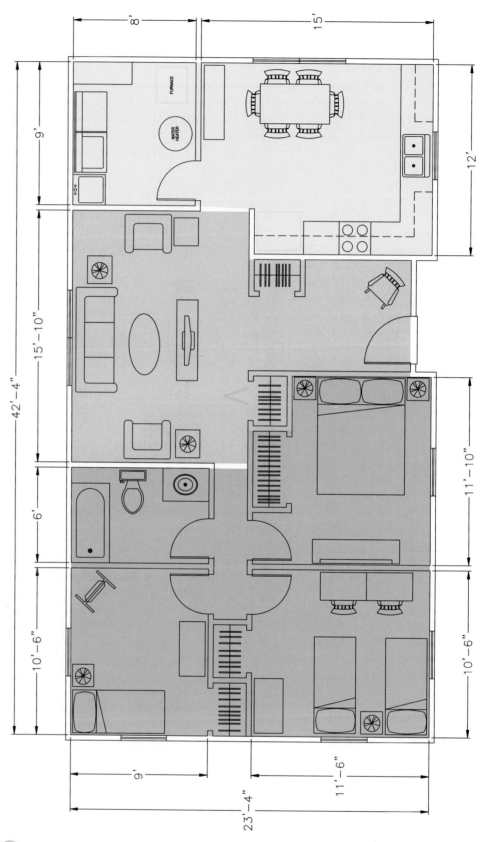

Figure 7-14. A floor plan for a three-bedroom bungalow. Why are the sleeping and living areas separated from the dining and kitchen areas?

Math Application

Floor Plan Area Calculations

The size of houses, and of rooms or areas within houses, is usually measured in square feet of surface area. An apartment may have 936 sq. ft., for example, and a mid-range house might have 1,800 sq. ft.

To find the surface area of a room, you multiply its length by its width. Therefore, if a room is 8 ft. long and 10 ft. wide, its area is 8 × 10 = 80 sq. ft. Before you multiply, make sure both the measurements are in the same unit. For example, on most floor plans, both length and width are given in feet. Because you are multiplying feet × feet, the unit of the answer is ft.^{2}, or square feet.

Math Activity

Floor plans are drawn to scale and dimensioned to show the sizes of rooms. Refer to Figure 7-14 and determine the number of square feet in each of the following areas. Assume that all closets are 2.5 ft. deep and the front entry is inset 2 ft. from the front of the house. Round your answers to the nearest square foot.

- Work area (yellow color)
- Living area (orange color)
- Sleeping area (blue color)

Finding and Preparing a Site

The land on which a house is built can be any size. City lots are usually small. Those outside the city may be as large as several acres. Planning the site is just as important as designing the home itself. It involves several important steps.

Selecting the Site

Where you locate a new home is important. You may want it to be in a certain community. Perhaps it should be close to your job, shopping centers, and schools. Or maybe you prefer a quiet wooded area.

Site Preparation

Site preparation means getting the site ready for the home. One of the first steps is to do a soil test. You need to know how well the subsoil will carry the weight of your home and whether there is hard rock underground. Rock may be expensive to remove. There may be groundwater too close to the surface. This could cause flooding in the house.

Once a soil engineer has determined that the site is suitable, a contractor will clear the site of boulders and excess soil. Grading may be needed to level a spot for the foundation. Lines and grades must be established to keep the work true and level. **Figure 7-15** shows how *batter boards* are used for this purpose. Small stakes are located at what will be the corners of the house. Nails driven into the tops of these stakes mark the four corners of the house. Straight lines between these nails indicate the outside edge of the foundation walls.

Once the four corners have been located, larger stakes are driven into the ground at least 4′ (1.3 m) beyond the lines of the foundation. The batter boards are nailed horizontally to these stakes. The boards must be level. Strong string is next held across the tops of opposite boards and adjusted exactly over the tacks in the small corner stakes. A plumb bob may be used to set the lines exactly over the nails.

Next, a saw kerf (cut) is made to mark where the string crosses the top of each batter board. Some carpenters drive a nail at this point. This is done so the strings can be removed during excavation. Later, the strings can be stretched from corner to corner, across the batter boards, to locate the corners of the building once again.

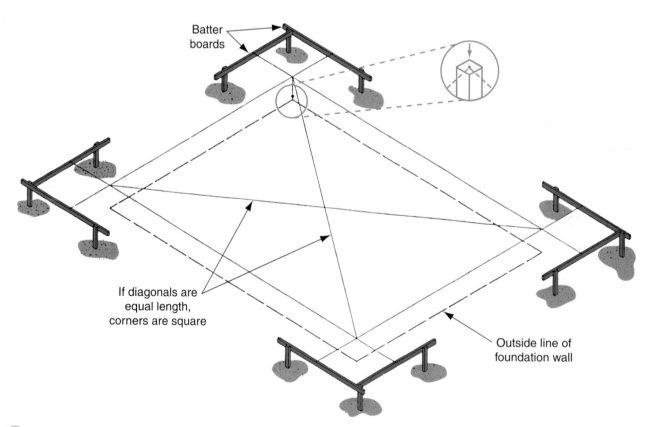

Figure 7-15. Batter boards support the lines set up to locate the building so excavating can begin for the foundation.

Structural Components and Materials

Buildings have five major components: foundation, floor, wall, ceiling, and roof. See **Figure 7-16**. The two materials most commonly used for house construction are concrete and wood.

Foundation

Most structures rest on a *foundation*. See **Figure 7-17**. Normally a foundation lies below the surface of the ground. Notice that there are two major parts to this type of foundation: the *footing* and the foundation wall.

To understand the importance of the footing, think about the reason for wearing snowshoes. See **Figure 7-18**. If you try to walk on deep, soft snow without them, you might sink down to your knees. Snowshoes spread your body mass over a larger area of the snow's surface. This prevents you from sinking.

The same principle is used to build a foundation. The load of the building is first transmitted to the foundation wall. Then it is spread over a larger area by the footing. Thus, the building is prevented from sinking into the ground.

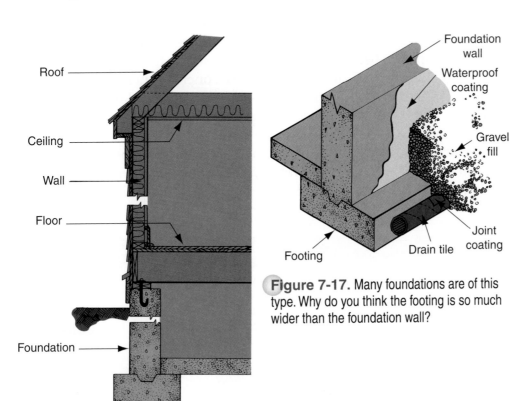

Figure 7-17. Many foundations are of this type. Why do you think the footing is so much wider than the foundation wall?

Figure 7-16. A section through a typical house shows its main structural parts.

Figure 7-18. The function of a foundation wall is similar to that of snowshoes. Why doesn't this boy sink into the snow?

In most locations, it is necessary to drain away any subsurface water to avoid damp basements and wet floors. Tile laid around the wall footings serves this purpose. These are known as drain tile, perimeter tile, or weeping tile.

The two materials most commonly used for foundation walls are poured concrete and concrete blocks. Concrete is strong in compression, so it can support heavy loads. As you read in the previous chapter, embedding steel rods or wire mesh in concrete increases its tensile strength (strength in tension). The various parts of foundation walls, their functions, and the materials used in their construction are summarized in **Figure 7-19**.

In warm climates, there is either no frost or the frost does not penetrate very far below the ground. Therefore, a combined slab and foundation is commonly used. See **Figure 7-20**.

Part	Function	Material
Foundation wall	To form an enclosure for the basement and to support walls and other building loads	Concrete Concrete blocks
Footing	To transmit the superimposed load to the soil	Concrete
Drain or perimeter tile	To provide drainage around footings	Clay Plastic
Gravel fill	To permit water to drain into the drain tile	Gravel

Figure 7-19. The parts of a foundation.

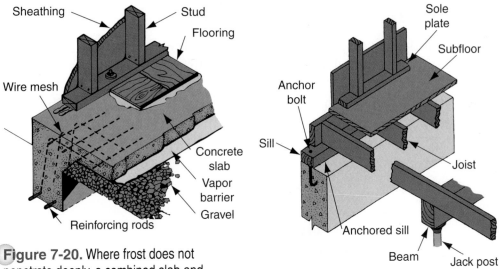

Figure 7-20. Where frost does not penetrate deeply, a combined slab and foundation can be used. In what parts of the country might you find these slab foundations?

Figure 7-21. A floor is supported by joists.

Floor

When the concrete for the foundation wall is poured, anchor bolts are set in the top. These bolts are used to fasten a sill to the foundation, as shown in **Figure 7-21**. Joists are nailed to the sill on edge, forming a framework. This framework is supported by a beam. Joists are usually made of wood nearly 2″ thick and 10″ wide or more. When joists must span a long distance, they are supported in the middle by jack posts.

The joists support a *subfloor*. A subfloor is a covering over the joists. It supports other floor coverings. The various parts of floors, their functions, and the materials used in their construction are summarized in **Figure 7-22**.

Part	Function	Material
Beam	To support joists when long distances are spanned	Wood (pine or spruce) Steel
Joist	To support a floor	Wood (pine or spruce)
Subfloor	To support finish flooring	Board or sheet material (tongue-and-groove pine, plywood)
Sill	To support joists where they meet the foundation	Wood (pine or spruce)
Jack post	To support beams	Wood (pine or spruce) Steel

Figure 7-22. Parts of a floor frame.

Walls and Finish Flooring

The subfloor is fastened to the joists. Then the walls for the first floor are laid out and built. The many parts of this type of wall are shown in **Figure 7-23**.

Wall studs provide the framework for walls and partitions. The other various parts of walls, their functions, and the materials used in their construction are listed in **Figure 7-24**.

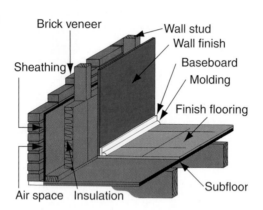

Figure 7-23. A section of a wall and floor.

Brick veneer — Wall stud — Wall finish — Sheathing — Baseboard — Molding — Finish flooring — Air space — Insulation — Subfloor

Part	Function	Material
Exterior surface	To provide protection and decoration to the outside of a building	Brick Aluminum siding Wood
Air space	To provide a barrier against the passage of moisture	
Sheathing	To reinforce studs To provide insulation	Wood Fiberboard
Wall stud	To provide a framework for walls or partitions	Wood (pine or spruce)
Insulation	To resist heat transmission	Fiberglass Polyurethane Vermiculite
Vapor barrier	To retard the passage of water vapor or moisture	Polyethylene Sheet
Interior wall surface	To cover the interior wall framing	Plasterboard Wood paneling Plaster
Finish flooring	To cover a subfloor and provide a decorative surface	Parquet Ceramic Linoleum Carpet

Figure 7-24. Parts of walls and finish flooring.

Math Application

Multiplying Decimals

Housing contractors such as bricklayers, plumbers, and roofers generally submit bids on housing projects they want to work on. The bid includes:

- Total cost of materials
- The number of hours needed to complete the job
- A labor rate, or cost per hour for the contractor to do the work

If the contractor calculates any of these items incorrectly, he may either underestimate or overestimate. Neither is good. For example, if he underestimates the cost of materials, then he won't make enough money to profit from the job. If he overestimates the cost of materials or the number of hours needed, his bid may be too high and he won't get the job.

Therefore, contractors need to be able to multiply and divide accurately, and many of their calculations involve decimals. The process of multiplying and dividing decimals is almost the same as multiplying whole numbers. The only difference is in determining the number of decimal places in the answer.

To multiply decimals, first multiply in the same way that you would with whole numbers. Next, total the number of decimal places to the right of the decimal point in both of the numbers being multiplied. Locate the decimal point by counting from the right end of the number. For example, suppose a plumber estimates that a job requires 52.75 feet of pipe. With his contractor's discount, he can buy the pipe for $1.62 per foot. How much will the pipe for this job cost?

$$\begin{array}{r} 52.75 \\ \times\ 1.62 \\ \hline 10550 \\ 31650 \\ 5275 \\ \hline 85.4550 \end{array}$$

Total of four digits to the right of the decimal point

Put decimal point four places from the right

Math Activity

Practice multiplying decimals by performing the following calculations. Then check your work using a calculator.

1. If a bricklayer can lay, on average, 150 bricks/hr., how many bricks can she lay in a 35.5-hour week?

2. A roofer used 10.75 sheets of plywood on a reroofing job. If the plywood cost $25.40 per sheet, what was the total cost of the plywood?

3. If a plumbing pipe costs $1.74 per foot, what is the cost of 25.5 feet of pipe?

Ceiling and Roof

Like floors, ceilings often require joists. A roof is made up of sloping timber called *rafters*. Roofs can also be built of a series of prefabricated trusses. Shaped like a triangle, the *roof truss* forms a framework to support the roof and any loads applied to it. Braces on the inside create further triangles to support and strengthen the rafters.

A typical ceiling and roof construction is shown in **Figure 7-25**. The various parts of ceilings and roofs, their functions, and the materials used in their construction are described in **Figure 7-26**.

Figure 7-27 shows how all the many parts of a house fit together. The next time you see a building being constructed, see how many of the parts you can identify.

Figure 7-25. Recall what you know about bridge design. Why is this type of roof called a truss roof?

Part	Function	Material
Joist	To support a ceiling	Wood (pine or spruce)
Insulation	To resist heat transmission	Fiberglass Polyurethane Vermiculite
Vapor barrier	To retard the passage of water vapor or moisture	Polyethylene Sheet
Interior surface	To form the ceiling	Plasterboard plaster
Roof truss	To form a framework for the roof and to support loads applied to it	Wood (pine or spruce)
Exterior finish	To provide protection from rain, snow, and other weather conditions	Asphalt Wood Shingles Tar and gravel

Figure 7-26. Information about the parts of a ceiling and roof.

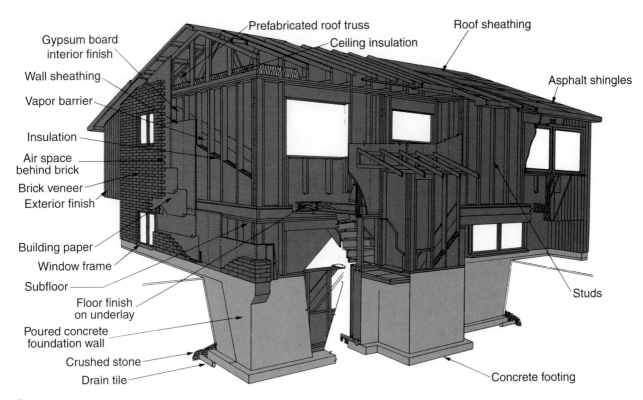

Gypsum board interior finish
Wall sheathing
Vapor barrier
Insulation
Air space behind brick
Brick veneer
Exterior finish
Building paper
Window frame
Subfloor
Floor finish on underlay
Poured concrete foundation wall
Crushed stone
Drain tile

Prefabricated roof truss
Ceiling insulation
Roof sheathing
Asphalt shingles
Studs
Concrete footing

Figure 7-27. The parts of a house and how they fit together.

Finishing the House

The final stages in building a house include trimming, painting, decorating, and landscaping. Trimming involves covering rough edges and openings with moldings. For example, a baseboard is the trim used to cover the small space between a wall and a floor. Painting protects and improves the appearance of interior and some exterior surfaces. Wallpaper is also a common interior decorating material. Paneling or tongue-and-groove boards are also used to provide a finished surface for walls.

Landscaping, **Figure 7-28,** is a design for the exterior space that surrounds a home. It involves planning the location of lawns, hedges, trees, shrubs, and plants. The plan shows the location of accesses such as driveways and walkways. It also shows special features such as patios, fences, walls, and plant boxes.

After accesses and features have been built, topsoil is added to the site. Topsoil is a layer of rich earth. It is needed so that trees, shrubs, lawns, and plants can grow. Then the plants are added to finish the landscape.

Maintaining a House

Even after a house is built, the work is still not complete. All houses, even new ones, require maintenance. Some work, such as cleaning, is required weekly. Heating systems require annual check-ups. Some maintenance is done only when the need arises, such as when a light bulb burns out.

Figure 7-28. A well-landscaped house has flowers, shrubs, trees, ground cover, and walkways that are in harmony.

Materials and equipment age, wear out and break down. That is why people should learn to maintain their homes. Typical home maintenance jobs include:

- Replace old carpet
- Repaint walls and ceilings
- Replace blown fuses
- Rewire circuits to meet updated building codes
- Repair concrete cracks
- Change air filters in furnace
- Install more insulation in walls or ceilings

You can save money by doing some of these maintenance jobs yourself instead of hiring a contractor. It can also bring you personal satisfaction for a job well done, whether you are working on your own house or helping a friend. There are other important reasons for keeping your home well maintained. If you plan on selling your house, a home inspector will come and check everything from electrical, plumbing, drywall, roof, carpentry, and furnace, hot water heater, and exhaust ducts. You will be required to repair or replace whatever is not up-to-date with the code.

Systems in Structures

What happens when you telephone a friend? After lifting the receiver or pressing a button, you dial a number. Signals travel to a central location where automatic switching equipment sends your call to your friend's house or cell phone. Your friend answers and your voices are carried over the lines or airwaves. At the end of the conversation, the caller disconnects. The telephones, cables, and automatic switching equipment are part of a *system*.

Some systems are very large. Others are quite small. The sun and the planets that revolve around it form the solar system. The skeletal system of your body is made up of more than 200 bones. Together, they support the body's mass. They also give the body shape and protect important organs.

The fuel system of a car is a system. It pumps gasoline from a fuel tank, through fuel lines, to the injectors and into the cylinders of the engine. The fuel system is connected to other car systems, such as the brake system, to make the car operate smoothly. This is an example of how technological systems can work together to accomplish a task.

To work successfully with any kind of system, it is important to think about how each part of the system relates to all the other parts. A problem in any part of a system can affect the entire system.

Types of Systems

Every system is a series of parts or objects connected together for a particular purpose. There are two types of systems: open-loop and closed-loop. See **Figure 7-29**.

Open-Loop System

A portable space heater without a thermostat is an example of an *open-loop system*. When it is plugged in and switched on, the heating element warms the air passing over it. It continues to heat the room until switched off. There is no method of controlling whether there is too much or too little heat.

Closed-Loop System

In a *closed-loop system*, the same heater would be connected to a control mechanism such as a thermostat. When the room air reaches the temperature you set on the thermostat, the heater shuts off. It will switch itself on again when the temperature falls below the set limit.

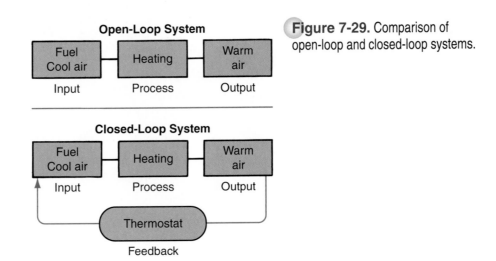

Figure 7-29. Comparison of open-loop and closed-loop systems.

In both cases, the systems contain input (cool air and fuel), process (burning the fuel), and output (warmed air). Input, process, and output are characteristics of all systems. However, in a closed-loop system there is also a feedback device, which provides control.

Control of our environment is a major reason why technological systems have been created. Controls in technological systems, such as the thermostat in a heater, allow us to change the system. For example, the thermostat uses information about the air temperature to operate the heater and maintain a comfortable temperature.

In our homes, there are four major systems:

- Heating, ventilation and air conditioning (HVAC)
- Electrical
- Plumbing
- Communication

Heating, Ventilation, and Air-Conditioning Systems (HVAC)

Figure 7-30 illustrates the major parts of a forced air *heating, ventilation, and air-conditioning (HVAC) system*. Some heating systems use water to carry heat throughout a house. However, forced air is a more popular way to move heat from a furnace to various rooms.

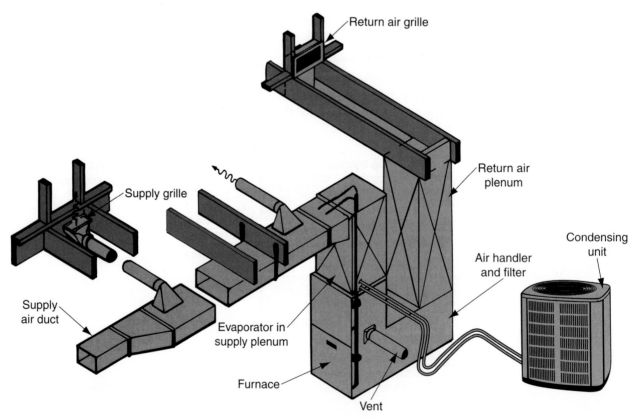

Figure 7-30. Central air conditioning and heating system. Air is conditioned and carried to all rooms of a house.

Cool air enters the bottom of the furnace. Here the filter traps dirt. A blower forces the filtered air up into a compartment, called a *heat exchanger*. The exchanger has passageways that are heated by electricity or the combustion gases from burning oil or gas. The blower forces the warmed air through a network of ducts into each room. Cooler, heavier air sinks to the floor and flows through return air ducts leading back to the furnace. Control switches turn on the blower and the supply of heat. Thus, the furnace controls the temperature of the circulating air.

An important part of any HVAC system is the *insulation*. Insulation is material installed in walls and ceilings to help control heat loss in winter and heat gain in summer. Some common forms of residential insulation are blanket, batt, rigid, and loose fill. Insulation is commonly made from fiberglass, Styrofoam®, treated paper, and a variety of other materials.

The HVAC system is, in fact, composed of several subsystems. A *subsystem* is a smaller system that operates as a part of the larger system. A home HVAC system contains some or all of the following subsystems.:

- Heater to produce heat
- Air conditioner
- Blower unit to push the heat through the ductwork
- Network of ducts to carry the conditioned air
- Thermostat to provide continuous feedback
- Humidifier
- Electronic filter
- Heat pump

Electrical System

An *electrical system* supplies electricity for light, heat, and appliances. Electricity is carried throughout the home by a number of separate circuits. A circuit is a pathway for electrical current. Each circuit normally has a cable with three wires (positive, negative, and ground) running inside the walls and ceilings.

A circuit carries current from a power source, as shown in **Figure 7-31**. Electric current travels to lights, motors, or heaters and back to the source. To supply these circuits, electricity from a utility company's wires must pass through a meter and a service panel. The service panel distributes the power among the separate home circuits. Lamps, television sets, and small appliances are connected to 120 volt, 15-ampere circuits. Appliances, such as refrigerators, toasters, and power tools, are connected to 120 volt, 20-ampere circuits. Separate 240 volt, 30-ampere circuits are provided a clothes-dryer or an electric range.

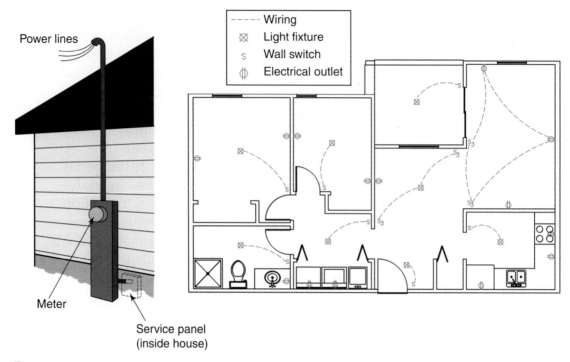

Figure 7-31. The parts of an electrical system.

Plumbing System

The *plumbing system* consists of major subsystems: the water distribution system and the drainage system. The water distribution system brings (drinkable) water into the home. See **Figure 7-32.** It is piped directly to all the faucets and outlets, such as sinks, toilets, baths, and washing machine. It is also piped to the hot water heater. From this heater, heated water goes to all hot-water faucets, the washing machine, and the dishwasher. The used water is drained from the house and disposed of by the drainage system.

Communication Systems

The *communication systems* in a home may include the telephone system, the radio and television broadcasting system, an Internet connection, a home PC network, and the cable or satellite television system. Telephone service is provided to most homes by copper wires or fiber optic cables. A nationwide switching system enables the telephone to be connected to any other telephone. The same copper or fiber cables can be used for data transmission so that any home can have access to a national or international computer network.

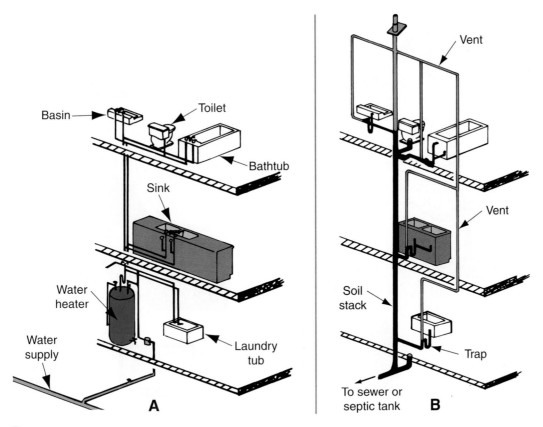

Figure 7-32. Residential plumbing system. A—Water distribution subsystem brings fresh water to different rooms in the home. B—Drainage system carries away wastewater.

Radio and television signals are received in each home. The programs may be broadcast from local stations. Satellites, antennas, and cables allow you to receive live and instantaneous radio and television coverage of events from around the world. See **Figure 7-33.**

Advances in Construction and Building Systems

Advances in other areas of technology continually affect construction and building systems. "Smart buildings" are now possible with new microelectronics. Environmental concerns have prompted other new technologies, including "green" materials, processes, and structures.

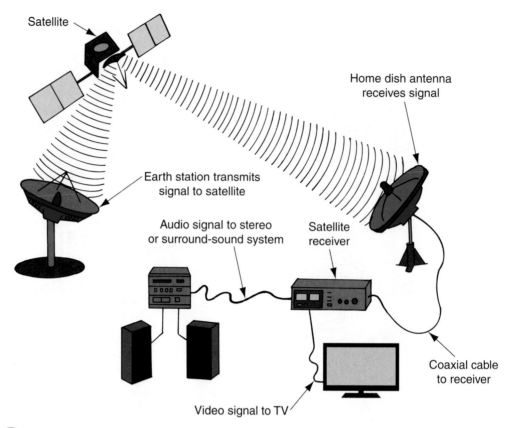

Satellite

Home dish antenna
receives signal

Earth station transmits
signal to satellite

Audio signal to stereo
or surround-sound system

Satellite
receiver

Coaxial cable
to receiver

Video signal to TV

Figure 7-33. A telecommunication system can be linked to a satellite in space. Signals can be received in an instant in spite of great distances.

Automated Systems: Smart Buildings

In the early 1970s, the microprocessor was developed. Complete electronic circuits were etched on a tiny slice of silicon called a *microchip* or *chip*. Over the years, these chips have become smaller and more powerful. The microchip is now so small that it can be embedded into almost any device. This enables automatic control, programming, and connection to just about anything.

Think about the center of the home: the kitchen. A refrigerator could suggest what meal might be cooked based on what is stored inside and display the ingredients on its video screen. Cupboards could be designed so that when food is consumed, it is automatically reordered from an online grocer. Microwave ovens could scan the bar code on the food packaging and set themselves to the correct power level and cooking time. Such devices are not fantasies. They could be made right now using existing technologies. Imagine a "kitchen command center" that would include a microwave with a flat computer screen enabling online shopping, banking, and e-mail!

"Green" Buildings

Builders are becoming more environmentally conscious and are seeking alternatives to material-intensive standard practices. Pressure from society to conserve natural resources has resulted in wider acceptance and use of "green" alternatives. One green building rating system is the Leadership in Energy and Environmental Design (LEED) system of the U.S. Green Building Council.

A green home provides the following benefits:

- Uses less energy, water, and natural resources
- Creates less waste
- Is healthier and more comfortable for its occupants

Benefits of a LEED home include lower energy and water bills, reduced greenhouse gas emissions, and less exposure to mold, mildew, and other indoor toxins.

In building a green home, an initial concern is how to place a building on its lot. The windows are oriented to make use of available daylight, minimize artificial lighting and cooling requirements, and increase the R-value of the insulation.

Sustainability, sometimes, called *sustainable development*, is an important part of LEED-certified buildings. Sustainability means using resources to meet current human needs, while preserving the environment so that these needs can continue to be met in the future. It means considering the potential for long-term maintenance of human well-being. This, in turn, depends on the well-being of the natural world and the responsible use of natural resources.

Commercial LEED-certified buildings make use of air source heat pumps that extract and recycle up to 80 percent of waste heat from the building's air stream before it leaves the building. Other sustainable features include geothermal heating, low-flow washroom fixtures, rainwater

Think **Green**

Sustainable Lumber

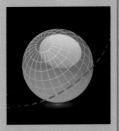

One way to incorporate green building practices into a building project is to use certified sustainable lumber. A number of forests around the world are now being managed sustainably. Although they represent only a small percentage of the forests being harvested, the number is growing. It is now possible to specify certified sustainable lumber for residential and commercial construction projects.

Not all "certified" lumber is equally sustainable, however. Look for well-established certification labels such as those offered by the Sustainable Forestry Initiative (SFI) and the Forest Stewardship Council (FSC). These two independent groups certify that lumber was planted, grown, cut, and renewed in a way that protects the forest's long-term health. They also guarantee that the "chain of custody" from the forest to the wholesale or retail store includes only companies that use responsible environmental practices.

collection, and use of fly ash (a waste material from power plants) in concrete.

Solar panels, or *photovoltaic panels*, harness the sun's energy. Solar panels that can generate electricity from the sun and also provide shade are a favored method. They can be made in many forms, including opaque or clear glass, asphalt-like shingles, and other elements that replace normal parts of a building shell. The electricity produced, even under cloudy conditions, generates current for lights and small appliances. Inside the building, sensors can measure the amount of natural light coming into the building and control the amount of electrical lighting.

Rooftops are also "going green." See **Figure 7-34**. Roof surfaces provide space for gardens, including grass, flowers, and shrubs. Roof gardens help control the temperature inside the house, and the plants conserve water and clean the air. The design of green roofs is more complicated than ordinary roofs. They must have several layers that provide the right structural support, waterproofing, and drainage. Nevertheless, they are starting to become popular. For example, Chicago has more than 2.3 million sq. ft. (214,000 sq. m) of "green" rooftops.

Figure 7-34. In addition to their other advantages, green roofs provide a space to grow vegetables and herbs.

The structure of a house must support the loads of the materials from which it is built plus its contents. It must also support the dynamic loads created by the weather conditions outside and the movement of people inside. In the past, the structure of homes was influenced by locally available materials.

Homes today contain four basic systems: HVAC, plumbing, electrical, and communication. Green building materials are being used to reduce waste and increase energy efficiency. Straw bale homes use one type of green building materials. They have insulation values that are more than double that of standard wood frame homes so they use less than one half of the heating and cooling energy.

Future homes will incorporate wireless communication. Perhaps when a visitor rings the doorbell, his or her photo will be taken and beamed to the homeowner's cell phone, whether he or she is at home or far away. The owner could choose to unlock the door or just chat on the phone. A house system, operated by a universal remote, will coordinate lights, security, and the HVAC. Using light-emitting diodes (LEDs), the color of interior walls might be changed to match the owner's mood or that of the visitor's. In the laundry room, washers and dryers will be able to read radio frequency identification (RFID) tags on clothing and make adjustments to clean the clothes appropriately. Similarly, in the refrigerator, RFID tags may keep tabs on the freshness of foods and notify the owner of any expiration dates.

- In the past, the structure and materials used in a house depended on climate and locally available materials. Prefabrication techniques allow parts of a house to be prepared off-site and then assembled at the building site.

- Planning for a home includes planning the inside space and finding and preparing a building site.

- The five major components of a house are the foundation, floor, walls, ceiling, and roof.

- Types of systems in residential building construction include HVAC, electrical, plumbing, and communication systems. These systems can be open-loop systems or closed-loop systems.

- Advances in various fields can be put to use in "smart buildings." Builders are beginning to use technology and new—and sometimes very old—techniques to make buildings "greener" and more efficient.

Connecting to Prior Knowledge

Copy the following graphic organizer onto a separate sheet of paper. Allow space in each row for one or more sentences. Before you read the chapter, write sentences in the first column to record your current knowledge about home construction. As you read through the chapter, record new concepts you learned by reading the chapter. When you are finished, review the chart to see how much you have learned. If you have questions about topics in the chapter, record your questions at the bottom of the chart. Ask your questions in class or do research on your own to find the answers.

What I Know (Or Think I Know) about Home Construction	What I Learned in This Chapter
Structure of a Home:	
Planning for a Home:	
Components and Materials for a Home:	
Systems Used in a Home:	
Recent Advances in Home Construction:	
Further Questions:	

Test Your Knowledge

Write your answers to these review questions on a separate sheet of paper.

1. What factors determined the use of a particular material when early settlers built their homes?
2. Sketch and name the parts of a post and lintel structure.
3. List two static and two dynamic loads that a typical house must withstand.
4. List four types of single-family houses.
5. How could you plan alternative arrangements of furniture in a room?
6. Why are rooms in a home grouped by function?
7. Describe the similarity between the footing of a foundation wall and a pair of snowshoes.
8. Why is a portable space heater without a built-in thermostat an example of an open-loop system?
9. What do all systems have in common?
10. What is the difference between a smart building and a green building?

Critical Thinking

1. What characteristics do reinforced concrete buildings have in common with post and beam structures?

2. The R-2000 standard is a Canadian energy efficiency standard that has been adopted by other countries, including parts of the United States. Research this standard to find out more about its requirements. Then suggest ways of bringing existing homes into compliance with the R-2000 standard. Do not limit your list to methods currently in use. Instead, use your imagination to think of as many "green" solutions as possible.

3. "Green" structures and building procedures are becoming more and more common. Think about how cultural, social, economic, political, and historical factors, as well as the environment, have contributed to the popularity of "building green." What role has technology played? Write an essay about the impact of these factors on green construction.

Apply Your Knowledge

1. Draw and label the parts of a house foundation.

2. Collect four pictures to illustrate the types of single-family homes described in this chapter.

3. Building laws and codes control many aspects of the systems in a home. The building codes that apply in your area may be national, state, or local codes, or a combination of all three. Find out what building codes apply in your area. If possible, obtain a copy of local codes and read the requirements. What requirements are specific to your local area? Why might local codes not need to be included in national requirements?

4. Use the systems model shown in Figure 7-29 to describe how a spaghetti dinner is prepared.

5. Select a technological system in which two or more systems work together. Evaluate the qualities of the system. Analyze the benefits of the system, any drawbacks it may have, and its overall efficiency. Report your findings to the class.

6. Give one example of a device in a typical home that uses an open-loop system and one example that uses a closed-loop system. Draw a system diagram of each.

7. Construct a scale model of the floor and walls of one room in your home. Think of the furniture you would like to choose if the room were empty. Cut blocks of Styrofoam® or cardboard to represent the furniture. Be sure to use the same scale you used to create the room. Position the furniture in the room.

Apply Your Knowledge *(Continued)*

8. More and more new houses and apartments are "smart homes," loaded with appliances connected to the Internet. Which devices can be connected to a local network, and for what purpose?

9. Research one career related to the information you have studied in this chapter. Create a report that states the following:

   - The occupation you selected
   - The education requirements to enter this occupation

   - The possibilities for promotion to a higher level
   - What someone with this career does on a daily basis
   - The earning potential for someone with this career

You might find this information on the Internet or in your library. If possible, interview a person who already works in this field to answer the five points. Finally, state why you might or might not be interested in pursuing this occupation when you finish school.

STEM Applications

1. **TECHNOLOGY** Make a list of common household devices that are not currently automated. Describe which of these might be automated in the future and how their function could be changed.

2. **ENGINEERING** Design a floor plan for a new home. The client wants a home with no more than 1700 square feet. She wants 3 bedrooms and two bathrooms. Design the home so that each room is large enough for its intended purpose. Label each room with its purpose (such as KITCHEN or BEDROOM) and the size of the room (such as 8′ × 12′). Check the total square footage of the house to make sure it is within the client's 1700-sq. ft. limit. Do not forget to include the square footage of hallways. Use graph paper or a CAD system to draw the floor plan after you have finalized the design.

3. **SCIENCE** One test that is commonly performed before a site is approved for building is a water (soil) percolation test. This test shows how well water drains from the soil. It is particularly important if the planned home will have a septic tank. The percolation rate depends on the type of soil. To determine which type of soil has the best percolation rate, collect samples of the following types of soil:

   - Sandy
   - Rocky/gravel
   - Clay

Design a simple test to determine which type of soil drains water the fastest. Record the time it takes water to run through the soil and the amount of water that is retained by the soil. Document your test procedure and results in a lab notebook. Include an analysis of your findings.

Not all bridges are large structures. Smaller bridges, such as this footbridge, must also be designed for strength and safety. What materials did the designers use in this bridge? How is the bridge supported? Where would a bridge like this be appropriate?

Machines

In this walk-behind lawn mower, the gas tank is placed with the engine for a more unified appearance.

Better by Design

Charles Harrison designs machines used around the home

Charles "Chuck" Harrison is one the most productive and respected American industrial designers of his time. He has been involved in the design of more than 750 consumer products that have improved the life of millions. Harrison helped design the portable hair dryer, toasters, stereos, lawn mowers, sewing machines, and the see-through measuring cup. He worked on power tools, fondue pots, and stoves. Among his most important designs was the first plastic garbage can. The tough plastic can was lighter and more durable than the traditional metal can. The round container evolved shortly into the familiar square green hulk with two wheels and raccoon-proof lid.

Harrison designed this hedge trimmer with plastic parts to minimize its weight.

"My best efforts resulted in products that did their job as expected—you look at it, right away guess what it is supposed to do, and that's exactly what it does."

Finding the Meaning of Unknown Words

Before you read this chapter, skim through it briefly and identify any words you do not know. Record these words using the Reading Target graphic organizer at the end of the chapter. Then read the chapter carefully. Use the context of the sentence to try to determine what each word means. Record your guesses in the graphic organizer also. After you read the chapter, follow the instructions with the graphic organizer to confirm your guesses.

friction   *pneumatics*
gear   *power*
hydraulics   *pressure*
inclined plane   *pulley*
lever   *screw*
linkage   *torque*
lubrication   *velocity*
machine   *viscosity*
mechanical advantage   *wedge*
mechanism   *wheel and axle*
moment   *work*

After reading this chapter, you will be able to:

- Identify the six simple machines.
- Describe types of gears and list uses for each type.
- Explain methods of applying pressure to increase or decrease effort applied to an object.
- Design and make a product that incorporates one or more mechanisms.
- Explain how friction can be reduced or overcome in a system.

Useful Web sites:
www.alifesdesign.com/about.asp
www.nationaldesignawards.org/2008/honoree/charles-harrison/?p=109

Machines help us to produce food on the land and make the electronics you play with. They help us dig trenches and bore tunnels. Around the home, the automatic washing machine, lawn mower, and vacuum cleaner lighten our workload. The dentist's drill removes decay from our teeth. The jet plane speeds us to our vacation site.

A *machine* is a device that uses energy to do some kind of work. For example, a lawn mower uses gasoline to trim the lawn. Machines do not need to be large or complicated. A knife, a bottle opener, and a claw hammer are also machines.

Machines have been used for thousands of years to make work easier. For example, people discovered that they could lift heavy objects more easily by throwing a rope over a tree branch. This allowed them to pull down instead of lifting to move the objects. This idea was later refined to use a wheel with a groove for a rope, chain, or belt. See **Figure 8-1**.

Simple Machines

Over the years, six basic types of machines were developed. They include the lever, pulley, wheel and axle, inclined plane, wedge, and screw. We refer to these as *simple machines*. These six simple machines can be divided into two groups. One is based on the lever, and the other is based on the inclined plane. See **Figure 8-2**.

Figure 8-1. Changing the direction of applied force makes it easier to lift heavy objects. As the woman pulls down on the rope, the pail moves upward.

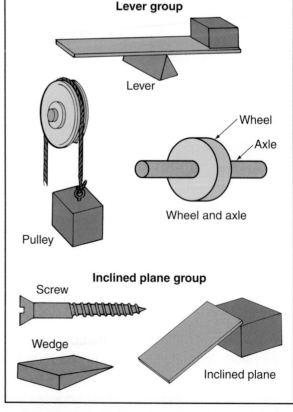

Figure 8-2. The six simple machines continue to be important inventions.

The Lever

You have probably played or seen a game of baseball. See **Figure 8-3**. A baseball bat is a lever. A *lever* has a fulcrum, effort, and resistance. The fulcrum is the point where the bat is held. The batter's muscles supply the effort, and the resistance is the ball. See **Figure 8-4**.

To understand the principle of the lever, look at the boy in **Figure 8-5**. He is using a branch to move a heavy rock. The branch is the lever. The mass of the larger rock is the resistance (R). The boy's muscle power pushing down on the lever provides the effort (E). The smaller rock on which the lever is pivoting is the fulcrum (F). These three elements—resistance, effort, and fulcrum—are always present in a lever. However, they can be arranged in different ways to create three different classes of levers.

In a Class 1 lever, the fulcrum is placed between the effort and the resistance, as shown in **Figure 8-6**. Some applications of Class 1 levers are illustrated in **Figure 8-7**.

In Class 2 levers, the resistance is placed between the effort and the fulcrum, as shown in **Figure 8-8**. Some applications of Class 2 levers are illustrated in **Figure 8-9**.

Figure 8-3. Baseball players use a bat as a lever to strike the baseball with greater speed.

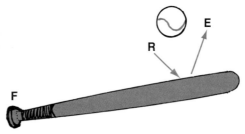

Figure 8-4. The baseball bat is an example of a lever: F–Fulcrum; R–Resistance; E–Effort.

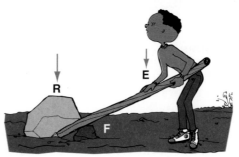

Figure 8-5. A lever being used to move a heavy load.

Figure 8-6. In a Class 1 lever, the fulcrum is between the effort and resistance.

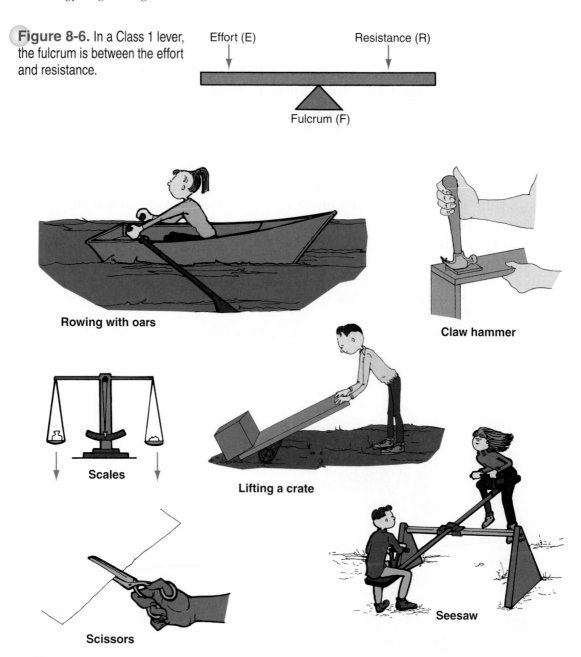

Effort (E)   Resistance (R)

Fulcrum (F)

Rowing with oars

Claw hammer

Scales

Lifting a crate

Scissors

Seesaw

Figure 8-7. Everyday uses of Class 1 levers.

In Class 3 levers, the effort is applied between the resistance and the fulcrum, as shown in **Figure 8-10**. Some applications of Class 3 levers are illustrated in **Figure 8-11**.

From the many examples shown, you can see that some levers are designed to increase the force available. Examples are a wheelbarrow, a bar used to move a crate, and a garden spade. Other levers are designed to increase the distance a force moves or the speed at which it moves. Examples are a fishing rod and a human arm.

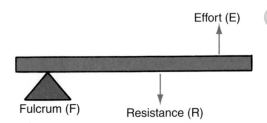

Figure 8-8. In a Class 2 lever, the resistance is between the effort and fulcrum.

Figure 8-9. Everyday uses of Class 2 levers.

Mechanical Advantage

Using a smaller effort to move a large resistance creates an advantage. This is called the *mechanical advantage* of the lever.

Mechanical advantage is equal to the resistance divided by the effort. The greater the resistance that can be moved for a given effort, the greater the mechanical advantage is. The formula is:

$$\text{Mechanical Advantage (M.A.)} = \frac{\text{Resistance}}{\text{Effort}}$$

For example, if a lever can make it possible to overcome a resistance of 90 newtons (N) when an effort of 30 N is applied, the mechanical advantage is 3. The newton (N) is the metric unit of force or effort.

$$\text{(M.A.)} = \frac{90\text{N}}{30\text{N}} = 3$$

Figure 8-10. In a Class 3 lever, the effort is between the resistance and fulcrum.

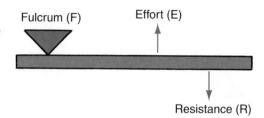

Figure 8-11. Everyday uses of Class 3 levers.

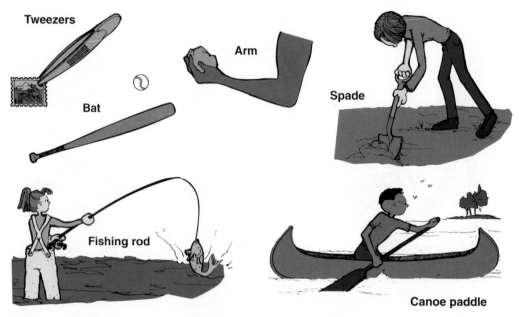

In other words, the human effort applied is being multiplied by the machine. In this case, the machine is a lever. The effort required becomes less as the fulcrum and resistance are brought closer together. However, as the fulcrum and resistance are moved closer, the load moves a shorter distance. See **Figure 8-12**.

Figure 8-12. Moving the fulcrum of a Class 1 lever affects both the effort required and the distance the object is moved.

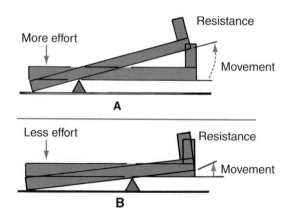

Science Application

Sports and Levers

Without realizing it, people use the principle of leverage in many sports, including tennis, baseball, and golf. For example, when a tennis player serves a ball, the player's arm and tennis racket work together as a lever.

A pitcher on the mound winds up, draws the right side of his body back and his left foot forward. Suddenly, he throws his weight forward, flings his arm forward, and lets fly with a fast ball. His arm has been used as a lever.

A golfer tries to keep his golf club on the correct swing path. As his arm and club come back down, he reaches a point where the elbow and right hip are close together. This position creates a lot of leverage.

Science Activity

Look at the three diagrams labeled A, B, and C. In each diagram, state where the *resistance*, *force*, and *fulcrum* are found. Also state whether the player's movement is an example of a first, second, or third class lever.

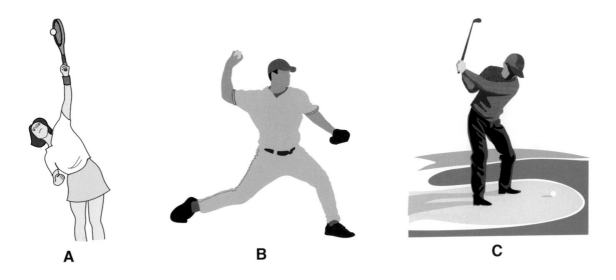

A   B   C

Moments and Levers

Imagine a lever with a fulcrum in the middle. On one side is an effort and on the other a resistance. When at rest, this lever is said to be balanced. If the effort is increased, the lever turns in a counterclockwise direction. If the resistance is increased, the lever turns in a clockwise direction, as shown in **Figure 8-13**. The turning force is called a *moment*.

The moment depends on two things: the effort and the distance of the effort from the fulcrum.

$$\text{Moment} = \text{Effort} \times \text{Distance}$$

If a beam is in balance, the clockwise moments are equal to the counterclockwise moments.

$$4 \times 50 = 8 \times 25$$

The levers discussed so far have been used to increase force, distance moved, or speed. Levers can also be used to reverse the direction of motion.

Think of a lever with a fulcrum in the center. If it pivots about its fulcrum, the ends move in opposite directions. One end moves down, and the other end moves up. See **Figure 8-14**. A single lever with a pivot in the center reverses an input motion.

This idea is used in linkages. A *linkage* is a system of levers used to transmit motion. **Figure 8-15** illustrates a reverse motion linkage. The input force and output force are equal.

If the pivot is not at the center, the input force is increased or decreased at the output. This is shown in **Figure 8-16**.

The Pulley

The *pulley* is a special kind of Class 1 lever. See **Figure 8-17**. Its action is continuous. The resistance arm is the same length as the effort arm. The length of each arm is the radius of the pulley. See **Figure 8-18**.

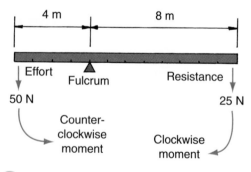

Figure 8-13. Forces acting on a lever are called *moments*.

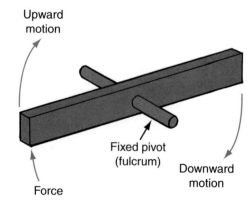

Figure 8-14. Some levers are designed to change motion.

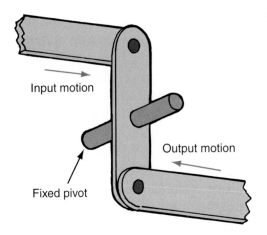

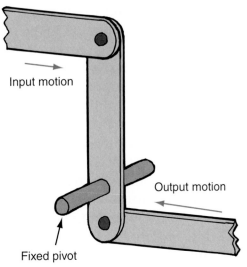

Figure 8-15. In a reverse motion linkage, the pivot is fixed at the center of one lever. Input force equals output force.

Figure 8-16. In this example, the pivot has been moved closer to the output motion. Is the input force increased or decreased at the output?

Figure 8-17. A simple pulley changes direction once.

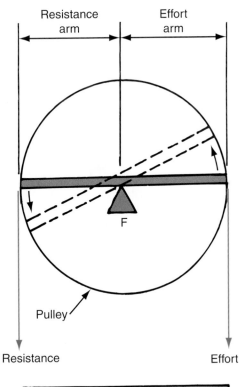

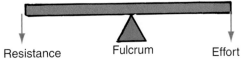

Figure 8-18. Look at this drawing and explain why a pulley is a special type of Class 1 lever.

Pulleys are used to lift heavy objects. See **Figure 8-19**. A bale of hay can be lifted into a hayloft using a single pulley suspended from a beam. A car engine hoist enables one person to lift a car engine having a mass of over 450 lb. (200 kg). Cranes use pulleys to lift enormous loads.

Two types of pulleys are used to lift heavy objects: fixed and movable. In a single, fixed pulley system, **Figure 8-20**, the effort is equal to the resistance. There is no mechanical advantage. It is easier, however, for the operator to pull down instead of up. There has been a change in direction of force. The distance moved by the effort (effort distance) is equal to the distance moved by the resistance (resistance distance).

A single, movable pulley system has a mechanical advantage of two. Both ropes support the resistance equally. See **Figure 8-21**. The amount of effort required is half of the resistance. The disadvantage is that the operator must pull upward. Also, the effort distance is two times the resistance distance. In all pulley systems, as the effort decreases, the effort distance increases.

To have the advantages of change of direction and decreased effort, movable and fixed pulleys can be combined as shown in **Figure 8-22**. In both examples, the mechanical advantage is two. However, the effort must be exerted over twice the distance.

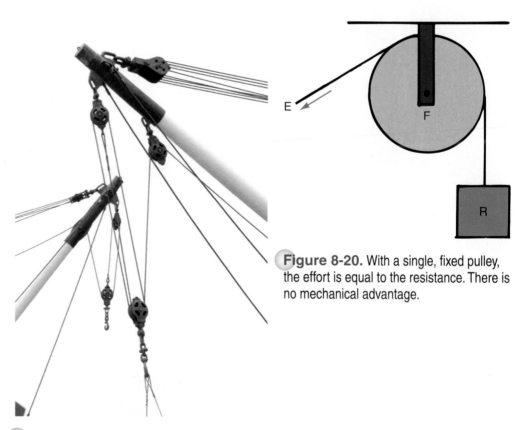

Figure 8-20. With a single, fixed pulley, the effort is equal to the resistance. There is no mechanical advantage.

Figure 8-19. Pulleys can be used alone or together in different ways to lift and move large objects easily.

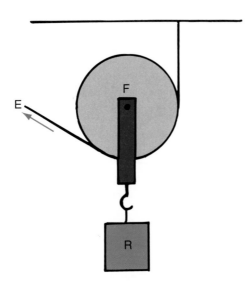

Figure 8-21. With a single, movable pulley, effort equals half the resistance. The effort moves twice the distance of the resistance.

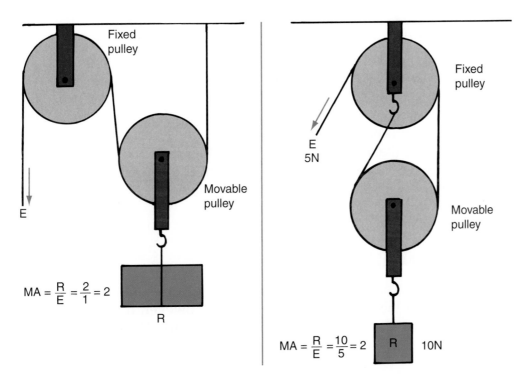

$$MA = \frac{R}{E} = \frac{2}{1} = 2$$

$$MA = \frac{R}{E} = \frac{10}{5} = 2$$

Figure 8-22. The fixed pulley changes direction of the effort. The movable pulley decreases the amount of effort required.

Pulleys may also be used to transmit motion, increase or decrease speed, reverse the direction of motion, or change motion through 90°. See **Figure 8-23**. These types of pulley systems may be used in cars (fan belt), upright vacuum cleaners, washing machines, and electrical appliances.

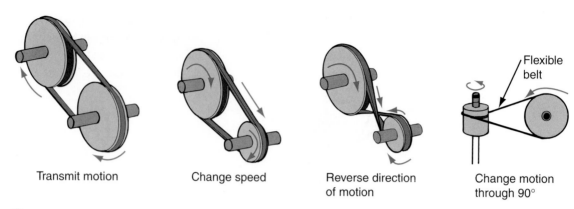

Transmit motion   Change speed   Reverse direction   Change motion
                                                of motion   through 90°

Figure 8-23. Pulleys can transmit motion from one point to another.

The Wheel and Axle

The *wheel and axle* is a simple machine that consists of a large diameter disk or wheel that is attached rigidly to a smaller diameter bar (axle). Effort applied to the outer edge of the wheel is transmitted through the axle. Think of it as a special kind of Class 1 lever. See **Figure 8-24**.

Like levers, the wheel and axle contains three elements: effort, fulcrum, and resistance. In the case of the doorknob, the effort is applied to the rim of the wheel (knob). The knob multiplies the effort and transmits it through the axle (bar). The resistance is the door latch.

Even the simple pizza cutter uses a wheel and axle. The effort is supplied by the person pressing on the handle and the resistance is the force of the wheel on the pizza. See **Figure 8-25**.

The Inclined Plane

There are at least two ways of loading the furniture onto the truck in **Figure 8-26**. One way is simply to lift it. However, it is much easier and safer to push the motorcycle up a sloping plank. The sloping surface formed by the plank is called an *inclined plane* or ramp.

A   B

Figure 8-24. A—A doorknob is a common example of the wheel and axle. B—Where are the effort, fulcrum, and resistance?

Figure 8-25. A pizza cutter is a form of wheel and axle. Where are the effort, fulcrum, and resistance located?

Figure 8-26. It would be much more difficult to lift furniture in and out of the truck without a ramp. Would a shorter ramp make the task harder or easier?

Uses for Inclined Planes

Common uses for inclined planes include:

- Loading and unloading a car transporter.
- Replacing steps so people with strollers or wheelchairs have access.

Figure 8-27 shows other common uses for inclined planes.

Calculating Inclined Planes

How do people who build ramps decide about their length? Guessing about it is too costly. They might have to tear down the ramp and start over. Therefore, they use mathematics.

Figure 8-27. We use inclined planes for many purposes. A—Skilled skateboarders use ramps to perform various feats. B—Ramps provide access to multistory parking garages.

In general, heavier objects need longer ramps or more effort to move them upward: the more gradual the rise, the less effort required. We can say the length of the inclined plane is directly related to the mass of the object. This means that as mass increases, ramp length increases. **Figure 8-28** shows this relationship. So does the following formula:

Effort (E) × Effort Distance (ED) = Resistance (R) × Resistance Distance (RD)

Now, let us see how we might use the formula to work out a problem. Suppose, for example, that a mass of 450 lb. (200 kg) needs to be raised 6′ (2 m). Then suppose that the most force that can be exerted is 100 lb. (50 kg). How long must the ramp be so the force is able to move the object?

Formula:

$$ED = R \times \frac{RD}{E}$$

U.S. Customary calculation:

$$450 \times \frac{6}{100} = \frac{2700}{100} = 27′$$

Metric calculation:

$$200 \times \frac{2}{50} = \frac{400}{50} = 8 \text{ m}$$

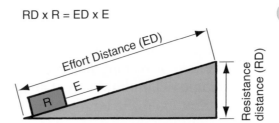

RD x R = ED x E

Effort Distance (ED)

E

R

Resistance distance (RD)

Figure 8-28. The mass of an object and the distance it is raised are related to the length of the inclined plane.

The Wedge

The *wedge* is a special version of the inclined plane, as shown in **Figure 8-29**. It is two inclined planes back-to-back.

The shape is effective because the force exerted pushes out in two directions as it enters the object. Do you see the difference between the wedge in **Figure 8-29** and the inclined plane in **Figure 8-28**? Other applications of the wedge include the plow, a doorstop, the blade of a knife, and the prow of a boat. See **Figure 8-30**.

Figure 8-29. Because of its shape, an axe enters material easily and can split it apart.

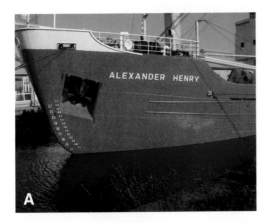

A

B

Figure 8-30. We use the wedge in many products. A—The wedge-shaped prow (front) of a ship cuts easily through the water. B—The teeth on the bucket of a backhoe are wedge-shaped to enter the soil more easily.

The Screw

A *screw* is an inclined plane wrapped in the form of a cylinder. To illustrate how this works, take a rectangular piece of paper and cut it along a diagonal. This triangle will remind you of an inclined plane. Now wrap it around a pencil. Roll from the edge of the triangle toward the point to shape it like a screw. See **Figure 8-31**.

When we examine the inclined plane, we find that the time taken for the load to be pushed up a longer inclined plane increases, but the effort decreases. The same is true in the case of the screw threads on a nut and bolt. The greater the number of threads, the shallower the slope, and the longer it takes to move the nut to the head of the bolt. However, a greater number of threads makes it easier to move the nut against a resistance.

The wedge-shaped section of a tapering wood screw reveals another application of the wedge. See **Figure 8-32**. It allows the screw to force its way into the wood.

Screw threads may be used in two quite different ways. They may be used to fasten, as with wood screws, machine screws, and light bulbs. They may also transmit motion and force. Examples are C-clamps, vises, and car jacks. See **Figure 8-33**. Note also that a screw converts rotary motion into straight-line motion. See **Figure 8-34**.

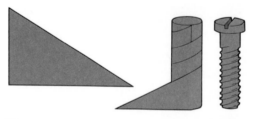

Figure 8-31. A screw is an inclined plane wrapped into a cylinder.

Figure 8-32. A wood screw also has a wedge shape to push aside the wood fibers as it enters the wood.

A

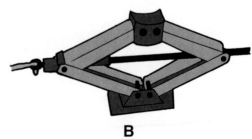

B

Figure 8-33. A—A scissor jack lifts heavy loads with less effort applied. B—As the screw of the jack is turned, the two ends of the jack are forced to move toward each other, and the car is raised.

Figure 8-34. Screw threads are used on vises and C-clamps. They convert rotary motion to straight-line motion.

Gears

Gears are *not* classified as one of the six simple machines. They are similar to pulleys in that their motion is usually circular and continuous. However, gears have an advantage over pulleys. They cannot slip.

Many bicycles have gears to help make pedaling as effortless as possible. See **Figure 8-35.** When climbing hills, the cyclist selects a low gear to make pedaling easier. When descending a hill, the cyclist uses a higher gear that provides a high speed in return for slower pedaling.

A mechanical clock contains many different sized gear wheels. They are arranged so that they rotate the clock hands at different speeds.

Figure 8-35. Notice how gears are used in this bicycle. What is the advantage of using gears instead of pulleys?

Gears, like pulleys, are modified levers. They transmit rotary motion. They increase or decrease speed, change the direction of motion, or transmit a force. See **Figure 8-36.** This force, known as *torque*, acts at a distance from the center of rotation. To understand this concept more easily, think of torque as a measure of turning effort. It is similar to using a wrench to tighten a bolt. See **Figure 8-37** and **Figure 8-38.**

An effort (E) is applied at a distance (R) from the center of the nut. The torque on the nut is calculated by multiplying the effort (E) by the distance (R). The applied force is measured in pounds (newtons) and the distance from the center of rotation is measured in feet. Therefore, torque is measured in ft.-lb. In the metric system, the applied force is measured in newtons, the distance from the center of rotation is measured in meters, and torque is measured in newton-meters (N•m).

As you look at a gear, you can consider the center of the gear to be like the nut. Consider the end of the gear tooth to be like the end of the wrench.

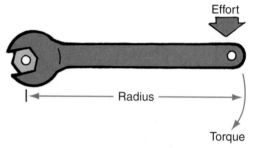

Figure 8-37. Torque is force applied to a radius.

Figure 8-38. A gear can be compared to a wrench turning a nut. How is the gear similar to the wrench?

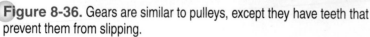

Figure 8-36. Gears are similar to pulleys, except they have teeth that prevent them from slipping.

Gears are used in groups of two or more. A group of gears is called a *gear train*. The gears in a train are arranged so that their teeth closely interlock, or mesh.

When two gears are of the same size, they act as a simple torque transmitter. They both turn at the same speed, but in opposite directions, as shown in **Figure 8-39**. The input motion and force of the drive gear are applied to the driven gear. The driven gear transmits the output motion and forces.

When two gears are of different sizes, as shown in **Figure 8-40**, they act as torque converters. The larger gear is called a *wheel*. The smaller gear is called a *pinion*. The pinion gear revolves faster, but the wheel delivers more force.

Gear Calculations

The amount of torque delivered by a gear is described as a ratio. For example, suppose a gear that has 10 teeth meshes with a gear that has 30 teeth. The smaller gear will make three revolutions for each revolution of the larger gear. As the small gear makes one revolution, its 10 teeth will have meshed with 10 teeth on the larger gear. The large gear will have turned through ten-thirtieths or one-third of a revolution. The small gear has to make three revolutions to turn the large gear through one full revolution. The gear ratio is therefore 3:1. See **Figure 8-41**.

Gear trains are either simple or compound. In a simple gear train, there is only one meshed gear on each shaft. **Figure 8-42** shows an idler gear placed between the drive gear and the driven gear. The driver gear and the driven gear now rotate in the same direction. The idler gear does not change the gear ratio between the driver gear and the driven gear.

A compound gear train also has a drive gear and a driven gear. However, the intermediate gears are fixed together on one common shaft, as shown in **Figure 8-43**. The gear wheels on the intermediate shaft are not idlers, because one is a driven gear and the other is a drive gear. They do affect the ratio of the gear train.

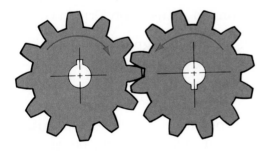

Figure 8-39. On a gear train, as one gear moves, its torque is transmitted to another gear.

Figure 8-40. Two gears act like levers to convert torque.

Figure 8-41. When a small gear drives a larger gear, the input torque is multiplied.

Figure 8-42. A simple gear train.

Figure 8-43. A compound gear train. Two gears are attached to the same shaft.

Just like the six simple machines, gears provide mechanical advantage. This advantage is calculated as follows:

$$\text{Mechanical Advantage (M.A.)} = \frac{\text{Number of Teeth on Driven Gear}}{\text{Number of Teeth on Driver Gear}}$$

The *velocity*, or speed, of the driven gear is calculated as follows:

Velocity of Driven Gear

$$= \text{Number of Teeth on Driver Gear} \times \frac{\text{Number of Teeth on Driver Gear}}{\text{Number of Teeth on Driven Gear}}$$

An example of these gear calculations is shown in **Figure 8-44**.

Types of Gears

Gears are designed in a variety of types for a variety of purposes. The five most common gear types are spur, helical, worm, bevel, and rack-and-pinion.

The spur gear is the simplest and most fundamental gear design. Its teeth are cut parallel to the center axis of the gear, as shown in **Figure 8-45**. The strength of spur gears is no greater than the strength of an individual tooth. Only one tooth is in mesh at any given time.

$$MA = \frac{96}{32} = \frac{3}{1} = 3$$

$$Velocity = \frac{32 \times 288}{96} = 96 \text{ rpm}$$

Figure 8-44. The mechanical advantage and the velocity of compound gears can be calculated as shown here.

Figure 8-45. On spur gears, the teeth are cut straight across the width of the gears.

To overcome this weakness, helical gears are sometimes used, as shown in **Figure 8-46**. Since the teeth on helical gears are cut at an angle, more than one tooth is in contact at a time. The increased contact allows more force to be transmitted.

A worm is a gear with only one tooth. The tooth is shaped like a screw thread. A wormwheel meshes with the worm. See **Figure 8-47**. The wormwheel is a helical gear with teeth inclined so that they can engage with the threadlike worm. This system changes the direction of motion through 90°. It also has the ability to make major changes in mechanical advantage and speed. Input into the worm gear system is usually through the worm gear. A high mechanical advantage is possible because the helical gear advances only one tooth for each complete revolution of the worm gear. The worm gear in **Figure 8-47** will rotate 40 times to turn the helical gear only once. This is a mechanical advantage of 40:1. Worm gear mechanisms are very quiet running.

Figure 8-46. Helical gears, cut at an angle, allow several teeth to engage at one time.

Figure 8-47. In a worm and worm wheel, the worm (bottom) has only one tooth. It spirals like a screw thread.

Bevel gears, **Figure 8-48**, change the direction in which the force is applied. This type of gear can be straight-cut like spur gears, or they may be cut at an angle, like helical gears.

Rack-and-pinion gears, **Figure 8-49**, use a round spur gear (the pinion). It meshes with a spur gear that has teeth cut in a straight line (the rack). The rack-and-pinion transforms rotary motion into linear (straight-line) motion and vice versa. **Figure 8-50** shows two uses for rack-and-pinion gears.

Figure 8-48. A—Bevel gears change the direction of the applied force. B—A hand drill uses bevel gears.

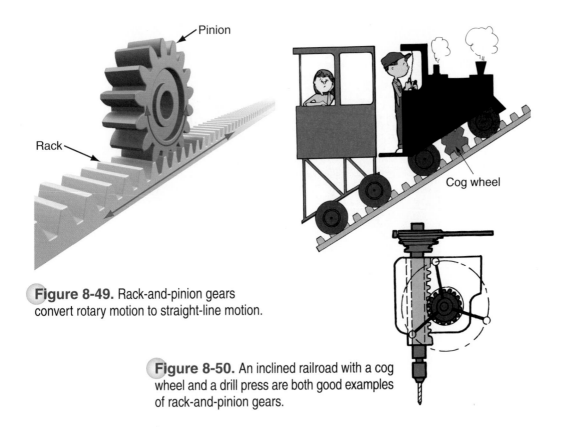

Figure 8-49. Rack-and-pinion gears convert rotary motion to straight-line motion.

Figure 8-50. An inclined railroad with a cog wheel and a drill press are both good examples of rack-and-pinion gears.

Transmissions

A transmission provides gear shifting to allow an increase or decrease in speed without overworking either an engine in a vehicle or a human pedalling a bicycle. See **Figure 8-51**. If you are a cyclist, you know that it is very difficult for you to start in the highest gear. You must start out in a low gear and gradually change to higher gears so that the speed of pedalling remains relatively constant. The same is true for a car. Without a transmission, a car would not have enough power to accelerate from a standstill. When a transmission is in low gear, the engine has to turn several times to make the drive shaft and the wheels turn once. As the transmission moves through the gears, from low to high, the drive shaft and the engine turn at approximately the same speed. The vehicle speed increases and the engine speed drops.

An automatic transmission performs the same function as a standard transmission, except that it shifts gears automatically by using internal oil pressure. A third type of transmission is a continuously variable transmission (CVT). This type uses belt systems to provide an infinite number of gear ratios. It improves engine efficiency by allowing the engine to match its speed to the load more efficiently. Currently there are also many other choices that are partly manual and partly automatic (automated manuals).

Figure 8-51. The gears on a multispeed bicycle allow you to decrease effort or increase speed. How is this like the transmission on an automobile?

Pressure

You have seen how simple machines and gears are able to move a greater resistance with a smaller effort. Now you will see that pressure can increase the effort applied. *Pressure* is the effort applied to a given area. **Figure 8-52A** shows a diagram of a simple system for multiplying force through use of pressure. **Figure 8-52B** shows a simple pressure device used to lift an automobile.

To understand how pressure can be increased, think about the area on which the effort presses. Would you rather a woman step on your foot with a small, pointed heel or with a larger, flat heel? See **Figure 8-53.** When a surface area is small, a little effort produces a large pressure.

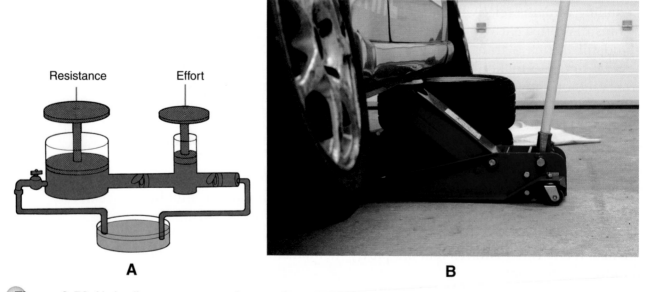

Resistance   Effort

A   B

Figure 8-52. Hydraulics use pressure to increase force. A—In a hydraulic system, a fluid transmits effort from where it is applied to where it is used. B—A hydraulic lift is used to support the mass of an automobile.

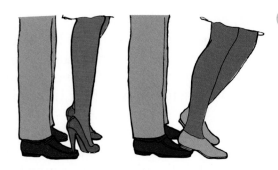

Calculating Pressure

To calculate pressure, divide effort by area. The formula is:

$$\text{Pressure} = \frac{\text{Effort}}{\text{Area}}$$

For example, if a 120-lb. woman rests her mass on a 4-in² heel, the pressure is 30 psi (pounds per square inch). On the other hand, if she rests her mass on a 1/4-in² heel, the pressure increases to 480 psi.

In the metric system, effort is measured in newtons. Area is measured in meters squared (m²). Pressure is calculated in newtons per meter squared (N/m²). The metric unit of measure for pressure is the pascal (Pa). Since a pascal is small, kilopascals are generally used (1 kPa = 1000 Pa).

Consider another example of how area affects pressure. A knife has a sharp edge. Pressed against a surface, it takes up a very small area. That is why it cuts: the material offers little resistance to such a tiny surface. A dinner fork works in a similar way. The narrow prongs place enough pressure on the food to pierce it easily.

Hydraulics and Pneumatics

The study and technology of the characteristics of liquids at rest and in motion is called *hydraulics. Pneumatics* is the study and technology of the characteristics of gases. Unlike solids, liquids and gases flow freely in all directions. Pressure, therefore, can be transmitted in all directions. For example, water will flow in a garden hose even when the hose is curved in many directions.

Figure 8-54 shows a model of a hydraulic lift. The effort is being applied to a piston. Pressure produced by the effort is being transmitted by the liquid to a second piston. This piston moves the resistance. The second piston has a larger area and so the pressure presses on a larger area. This produces a larger effort. If the resistance piston has four times the area of the effort piston, the effort on it will be four times greater.

From this example, you can see that the effort acting on a piston from a liquid under pressure depends on the area of the piston. The larger the area is, the larger the effort will be. However, the distance moved by the larger piston will be less than the distance moved by the smaller piston. In **Figure 8-54**, the smaller piston moves four times the distance of the larger piston.

The hydraulic brake system on passenger cars operates using the same principles as the hydraulic lift. See **Figure 8-55**. Using the brake pedal, a driver applies a small effort to the piston. Hydraulic fluid is transmitted through the brake lines to a larger piston. The larger piston forces the brake pads and shoes against the discs and drums. The brake systems of large trucks, buses, and trains are often pneumatically operated.

Among the many common applications of hydraulic power are dentist and barber chairs, door closers, and power steering. Common applications of pneumatic power include a variety of tools such as air drills, screwdrivers, and jackhammers. Sometimes hydraulic and pneumatic systems are combined. For example, air pressure forces hydraulic fluid to raise the lift in a garage, as shown in **Figure 8-52B**.

Because of their many advantages, most industries use hydraulic and pneumatic systems. These advantages include the ability to:

- Multiply a force using minimal space
- Transmit power to wherever pipe, hose, or tubing can be located
- Transmit motion rapidly and smoothly
- Operate with less breakage than occurs with mechanical parts
- Transmit effort over considerable distance with relatively small loss

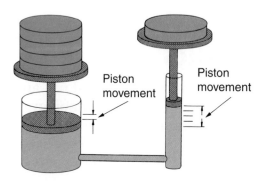

Figure 8-54. In this hydraulic lift, the large movement in the piston on the right causes a smaller movement in the piston on the left. Why is this an advantage?

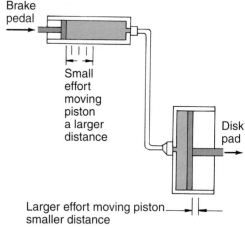

Figure 8-55. An automobile braking system is an example of a hydraulic system.

Mechanism

A *mechanism* is a way of changing one kind of effort into another kind of effort. For example, a C-clamp holds two pieces of wood together while glue sets. Rotary motion in the screw is changed to linear motion to apply pressure. See **Figure 8-56**.

Mechanisms can be combined to form machines. Their advantages include:

- Changing the direction of an effort
- Increasing the amount of effort applied
- Decreasing the amount of effort applied
- Applying an effort to a place that is otherwise hard to reach
- Increasing or decreasing the speed of an operation

Machines change one kind of energy into another and do work. The amount of *work* done is approximately equal to the amount of energy changed.

$$\text{Work} = (\text{Energy Change})$$
$$= \text{Effort} \times \text{Distance Moved in}$$
$$\text{Direction of Effort}$$

For example, how much work is done to move a 50 lb. resistance a distance of 5'?

$$\text{Work} = 50 \times 5$$
$$= 250 \text{ ft.-lb.}$$

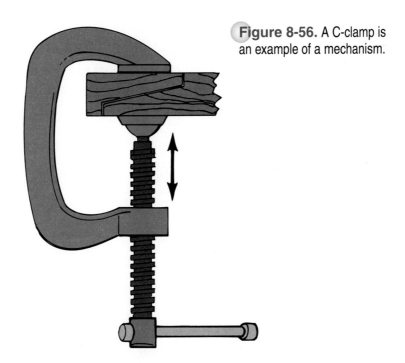

Figure 8-56. A C-clamp is an example of a mechanism.

In metric, how much work would be done to move a 50 N resistance through 4 meters?

$$\text{Work} = 50 \times 4 = 200 \text{ joules}$$

Machines make it easier to do work. However, no machine does as much work as the energy put into it. If a machine did the same amount of work as the energy supplied, it would be 100% efficient. Most machines lose energy as heat or light. Approximate efficiencies of some common machines are listed in **Figure 8-57**.

Another important term associated with work is power. *Power* is the rate at which work is done or the rate at which energy is converted from one form to another. It can also be the rate at which energy is transferred from one place to another. Power can be expressed in this formula:

$$\text{Power} = \frac{\text{Work}}{\text{Time}} \qquad P = \frac{W}{T}$$

Mechanisms use or create motion. See **Figure 8-58**. The four basic kinds of motion are:

- Linear—straight-line motion
- Rotary—motion in a circle
- Reciprocating—backward and forward motion in a straight line
- Oscillating—backward and forward arc motion, like a pendulum

Mechanisms are often used to change one kind of motion into another kind. Some examples are shown in **Figure 8-59**.

Friction

Friction is a force that acts like a brake on moving objects. Your finger will slide without much effort on a pane of glass. But if you do the same thing on sandpaper, you can feel a resistance slowing your movement.

Figure 8-57. Scientists, engineers, and manufacturers are always trying to design more efficient engines, motors, and machines.

Watt's Steam Engine	3%
Modern Steam Engine	10%
Gasoline Engine	30%
Nuclear Power Plant	30%
Aircraft Gas Turbine	36%
Diesel Engine	37%
Rocket Engine	48%
Electric Motor	80%

Motion	Application	Symbol
Linear		→
Rotary		 ↻
Reciprocating		↕
Oscillating		 ↝

Figure 8-58. Mechanisms are used to create different kinds of motion.

Friction helps us in many ways. Without friction, cars would skid off the road and nails would pop out of walls. Friction also enables us to walk on most surfaces without sliding. But it can also slow us down and produce unwanted heat. When you pedal your bicycle, you are working against friction.

The moving parts of mechanisms do not have perfectly smooth surfaces. The tiny projections on the surfaces rub on one another. This creates friction and results in heat. The friction between the moving parts must be minimized so that:

- Less energy will be needed to work the machine
- Wear and tear will be reduced
- Moving parts will stay cooler

Figure 8-59. Mechanisms change motion from one form to another.

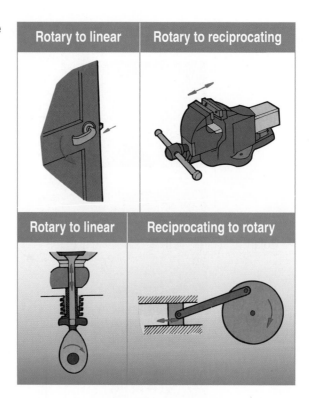

Reducing Friction

Friction may be reduced in several ways, as shown in **Figure 8-60** through **Figure 8-62**.

- Ball bearings—steel balls enable surfaces to roll over one another instead of sliding.
- Air or water cushions—compressed air or water separates moving parts.
- Streamlining—the shape of a fast-moving object can be changed to reduce its resistance to air or water.

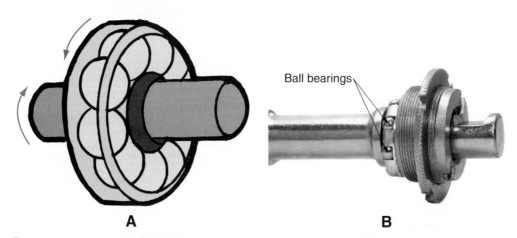

Figure 8-60. A—Ball bearings allow parts to roll over one another, reducing friction. B—The bearings in this shaft are separated to provide more structural stability to the product.

Figure 8-61. Hovercraft use a cushion of air to reduce friction.

Figure 8-62. One of the effects of streamlining is to reduce air drag, which is a form of friction.

Friction may also be reduced in other ways. In the game of curling, for example, "sweeping" the ice in front of the rock makes the rock travel further and curl less. The heat from the sweeping action melts the ice, creating a micro-thin layer of water on which the rock can ride. Sweeping can increase the distance a rock travels by several feet. See **Figure 8-63**.

Lubrication

If you rub your hands together quickly, you will feel heat build up. That heat is caused by friction. If your two hands were coated with oil, friction would be reduced so they could more easily slide against one another. *Lubrication* is the application of a smooth or slippery substance between two objects to reduce friction.

Figure 8-63. Sweeping melts the ice in front of the rock. Why does this make the rock travel further?

Oil separates two surfaces that would otherwise touch, rub, and wear each other away. See **Figure 8-64**. For good lubrication, oil must have the right *viscosity* (thickness). Viscosity is a kind of friction in liquids. It is caused by the molecules of a liquid rubbing together. Water has a low viscosity and pours easily. Honey and molasses have a high viscosity, so they ooze out of containers. Heating honey and some oils will cause them to flow more easily.

Motor oil becomes thinner (less viscous) as its temperature increases. In the warm summer months, a thicker oil is needed. However, in the icy cold of a winter's morning, a thinner oil is needed for easier starting. Therefore, multi-grade oils have been developed to meet the needs of automobile engines in a range of temperatures.

Figure 8-64. Oil lubricates parts to reduce friction.

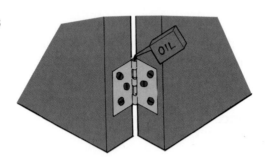

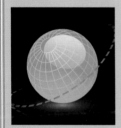

Think **Green**

Green Lubrication

Lubricants such as the motor oil used in vehicles are necessary, but they are not environmentally friendly. Motor oil is not biodegradable, and in some states, it is considered a hazardous waste. Automobile technicians are required to recycle used oil and to document the process. Individuals who change the oil in their vehicles should take the used oil to collection sites for recycling.

New alternatives to traditional motor oil are becoming available, however. Argonne National Laboratory has used nanotechnology to improve traditional motor oils. They combine extremely small particles of boric acid with motor oils. Each particle of boric acid is only 50 nanometers in diameter—less than one-thousandth the width of a human hair.

The resulting product is biodegradable, is not a health or environmental hazard, and does a better job of reducing friction than traditional motor oil. Engines that have less friction to overcome are more efficient and use less gasoline. Also, lowering friction can extend the life of the lubricated parts. Finally, these lubricants only have to be changed every 100,000 miles, so the vehicle owner spends less time and money changing the oil.

Over the centuries, humans have created many kinds of machines to make their work easier. From the six simple machines to today's complex computer systems, machines continue to help us in both work and leisure activities.

In the past few decades, computers have become much smaller and "smarter." At the same time, we have learned more about how humans think and reason. This information has been combined to make "intelligent" machines such as a robot that can vacuum the floor without guidance. Intelligent machines are just beginning to be widely used, but their possible uses are almost endless. Computer algorithms (sets of programming rules) are allowing computers to make more and more complex decisions.

What machines might people create in the next 50 years? How will they compare to the machines and mechanisms included in this chapter? You will have a part in deciding the answers to these questions. As you do, be sure to keep in mind the responsibilities that go along with invention. The machines created in this century and their effects on humans and the environment will help shape the world you live in.

- The six simple machines are the lever, pulley, wheel and axle, inclined plane, wedge, and screw.
- Gears are modified levers used in the transmission of rotary motion. They can increase or decrease speed, change the direction of motion, or transmit a force.
- Pressure can be used to increase an effort applied. Pressure in liquids is called *hydraulics*, and pressure in gases is called *pneumatics*.
- Mechanisms change one kind of effort into another kind of effort. They can be combined to form a machine. Machines change one kind of energy into another kind and do work.
- Friction results from resistance between the surfaces of moving parts. Lubrication and other methods can be used to reduce friction and increase the efficiency of machines.

Finding the Meaning of Unknown Words

Copy the following graphic organizer onto a separate sheet of paper. In the left column, record words from the chapter that you do not understand. As you read the chapter, try to guess their meanings and record your guesses in your chart. After you have read the chapter, look up each word in a dictionary. How close were your guesses? After you look up each word, go back and reread that portion of the chapter. Do you understand the chapter better?

Unknown Word	What I Think the Word Means	Dictionary Definition

Test Your Knowledge

Write your answers to these review questions on a separate sheet of paper.

1. When you hit a ball with a baseball bat, the bat is an example of a lever. Explain why.

2. Explain which of the following is an example of a Class 1 lever, and why: nutcracker, wheelbarrow, hockey stick, scissors.

3. A lever is used to move a load of 1500 newtons with a force of 300 newtons. What is the mechanical advantage of the lever?

4. A laborer using a Class 1 lever places the load the same distance from the fulcrum as the effort. If the fulcrum is moved closer to the load, what happens to the mechanical advantage?

5. What force would be required to raise a load of 15 newtons using a single, movable pulley?

6. List three examples of a wheel and axle.

7. List three practical applications of an inclined plane.

8. A 100 N load has to be moved to the top of an inclined plane that is 10 m long and 2 m high. What effort is required?

9. Explain the difference between a wedge and an inclined plane.

10. What is the connection between an inclined plane and a screw thread?

11. A gear is a modified form of which simple machine?

12. Describe three common uses for gears.

13. The wheel in a two-gear train has 60 teeth, and the pinion has 15 teeth. What is the ratio of the gear train?

14. Sketch a simple gear train in which the first and last gears are rotating in the same direction. Use arrows to show the direction of rotation of each gear.

15. How is pressure calculated?

16. What is the difference between hydraulics and pneumatics?

17. What is the difference between work and power?

18. Name the four basic kinds of motion and give one practical example of each.

19. List two benefits and two disadvantages of friction.

20. List three ways in which the friction between two objects can be reduced.

Critical Thinking

1. Gears are considered an improvement over pulleys because the teeth on a gear prevent a belt or chain from slipping. Why, then, are pulleys still in use today?

2. Which would work better for lubricating a rusty lock so that a key turns in it more easily: an oil with high, low, or medium viscosity? Why?

Apply Your Knowledge

1. Design and build a robotic arm that will move an AA dry cell from one location to another. Use syringes and tubing in your design.

2. Design and build a mechanism that will make a loud noise. Your solution must contain at least two simple machines.

3. Design and build a method of measuring the mass of a series of weights from 1 oz (25 g) to 16 oz (500 g).

4. Consult your library or the Internet (www.macchinedilenoardo.com) to learn about the mechanical inventions of Leonardo da Vinci. Which of his mechanical inventions are somewhat similar to the machines and vehicles we see today? Write a technical report explaining how innovations have been made to his machines over time to arrive at the products we use today. Evaluate the innovations that have been made. How did each innovation meet the needs of people at the time?

5. Research green alternatives to lubricants that are available in your area. Analyze each option to determine its advantages and disadvantages. Prepare a slide show or multimedia presentation and present it to the class to share the information you have found.

6. Research the principles of an atomic clock. Use library and Internet sources to find manuals that explain how an atomic clock works, or talk with people who are experienced in working with or repairing these clocks. Write a technical report explaining how atomic clocks work and describing any shortcomings or features you think can be improved.

7. Research one career related to the information you have studied in this chapter. Create a report that states the following:

   - The occupation you selected
   - The education requirements to enter this occupation
   - The possibilities for promotion to a higher level
   - What someone with this career does on a daily basis
   - The earning potential for someone with this career

You might find this information on the Internet or in your library. If possible, interview a person who already works in this field to answer the five points. Finally, state why you might or might not be interested in pursuing this occupation when you finish school.

Apply Your Knowledge *(Continued)*

8. Changing technology often results in changing career opportunities. Think about the machines described in this chapter and research how they have been modified, improved, or built upon to create new machines through the last 200 years. Write an essay explaining how these modifications have affected career opportunities in technology.

9. Search the Internet to find the lubrication and maintenance specifications for several large machines. Write a report to explain your findings. Include answers to the following questions: What tasks are necessary to keep the machines running smoothly? What might happen if the maintenance schedule is not followed?

STEM Applications

1. **ENGINEERING** Go to www.rubegoldberg.com and look at the examples of some of his complicated mechanisms that do simple tasks. Then create your own "Rube Goldberg" mechanism that will rake leaves from a lawn or park area and place them in a bag. Your mechanism must use all six simple machines at least once.

2. **ENGINEERING** The Americans with Disabilities Act (ADA) provides guidelines for wheelchair ramps in public buildings. According to the ADA, ramp slopes should be between 1:16 and 1:20. The ramp should be at least 36 inches wide, and level landings must be provided at both ends. The landings must be at least as wide as the ramp and a minimum of 60 inches long. If two or more ramps are used together, the landings between the ramps must be at least 60 inches × 60 inches. Using these guidelines, design a ramp to allow wheelchair access to a theater stage that is 8 feet high. Create dimensioned working drawings to show the specifics of your design.

Simple machines help people relax and stay fit. How many simple machines can you identify in this exercise equipment?

CHAPTER 9

Transportation

The Bikedispenser allows one-way trips to other rental stations.

Hans Schreuder designed the Bikedispenser®

Hans Schreuder, a Dutch industrial designer, wants to encourage more people to use public transportation. Hans and his team noted that a lot of people use a bicycle to travel between their home and a train station. But when they arrive at their destination, they have to use a bus or taxi, both of which are expensive and add to air pollution, to complete their commute. The Bikedispenser is a fully automated public bicycle rental system. It can store 50–100 bicycles in a compact and safe environment at a train or bus station. You rent a bicycle, cycle to work and leave the bicycle in the company parking lot. At the end of the day, you cycle back to the station and return the bicycle to the Bikedispenser. Not only does this reduce the use of polluting buses, but cycling is healthier and often quicker than using public transportation.

Bicycle rental systems can have a positive effect on the environment. What are the environmental advantages and disadvantages?

"We must create better alternatives to those imperfect solutions currently in use, such as cars."

Reading Target

Finding the Main Idea

As you read this chapter, look for the key points, or main ideas, in each part of the chapter. Then look for important details that support each main idea. After you have read the entire chapter, use the Reading Target graphic organizer at the end of the chapter to organize your thoughts about what you have read.

Key Terms

automated guided vehicle (AGV)
diesel engine
electric motor
electric vehicle
engine
external combustion engine
fuel cell vehicle
gasoline engine
human-powered vehicle (HPV)
intermodal transport

internal combustion engine
jet engine
mass transit
on-site transportation
recumbent
steam turbine
thrust
transportation
transportation system
turbine
turbofan
turboshaft

Objectives

After reading this chapter, you will be able to:
- State the advantages and disadvantages of various modes of transportation.
- Explain the principles of various types of engines and motors.
- Describe how industries rely on transportation systems.
- Identify the processes that enable a transportation system to work.
- Explain how transportation technology influences everyday life.
- Discuss the environmental impact of transportation systems.

Useful Web sites:
www.bikedispenser.com/home-english.html
www.youtube.com/watch?v=pcwll9nzEvc

Transportation forms a vital part of our lives, giving us access to education, recreation, jobs, goods, services, and other people. Different transportation systems enable people, packages, and commodities to travel from house to house, from town to town, from country to country, and even from earth to the moon.

Modes of Transportation

Suppose you want to send a package to a friend who lives in a distant city. You might transport the parcel to the post office using your bicycle. From the post office, a truck might take the parcel to the airport. An airplane delivers it to a city across the continent. There, another truck takes it to a central depot for sorting. Finally, a mail carrier might use a bus or mail truck to deliver the parcel to your friend's house. The bicycle, truck, airplane, and bus are individual modes of transportation. Together, they form a transportation system.

You are probably familiar with the millions of miles of public roads, but you might not know about other transportation networks, including the commercially navigable waterways and hundreds of ports on our inland waterways, the Great Lakes, and our coastal regions. You have probably never thought about the millions of miles of pipelines in North America, but they too are part of our transportation system. They transport oil and natural gas.

Land Transportation

Transportation burns much of the world's supply of petroleum. Also, hydrocarbon fuels produce carbon dioxide, a greenhouse gas thought to be the main cause of global climate change. Although regulations for our vehicles have ensured that the vehicles pollute less than they used to, the number of vehicles has increased, and we use the vehicles more often. For example, over 90 percent of workers in North America travel to work by car.

By contrast, most people in the rest of the world use other forms of transportation. See **Figure 9-1**. While we are used to traveling by car, most people in the world cannot afford a car. Walking, cycling, and taking minibuses or trains are their means of getting around.

Even in North America, public transportation can often be the fastest way to travel. When city streets are congested, buses traveling in dedicated bus lanes, metro cars moving underground, and high-speed trains going between major urban centers can all travel at much higher speeds than private vehicles. See **Figure 9-2**. If transportation were public, we would have more people in each vehicle, fewer cars would be on the roads, and traffic jams might be eliminated.

Figure 9-1. Trolley buses are an option in many urban areas. What are the advantages and disadvantages of trolley buses?

Figure 9-2. Many major cities have dedicated bus lanes. How can this help improve traffic flow?

When large numbers of passengers are transported, we call it *mass transit*. Mass transit is a good description because underground systems, such as those in London and Hong Kong, carry millions of passengers each day. Underground systems solve some of the worst traffic problems, especially at rush hour, because weather and street-level congestion do not affect them. They are, however, extremely costly to build. Many mass transit railways include not only underground tracks, but surface and elevated tracks. San Francisco's Bay Area Rapid Transit (BART) includes 19 miles (30 km) of tunnels, 25 miles (40 km) of surface track, and 31 miles (50 km) of elevated track.

Road Transportation

Roads carry an extraordinary variety of motor vehicles. In addition to cars, buses, and many sizes of delivery trucks, they handle fire engines, ambulances, police vehicles, mixer trucks, crane carriers, and garbage trucks, to name only a few types of vehicles.

Tractor-trailers are the most common vehicles for long-distance hauling. See **Figure 9-3**. The tractor part has a cab, an engine, and a transmission with a turntable (or fifth wheel) on top of the rear axle, where it hooks to the trailer. Rules mandate the number of hours tractor-trailer drivers can drive and how much rest or sleep time is necessary. Long-distance trucks are, therefore, often equipped with sleeper cabs, where the drivers can rest while not driving. These are equipped like small apartments, with good sound systems, stoves, refrigerators, storage places, and wash basins. They are air-conditioned and insulated against noise and vibration, so the drivers stay fresh and alert.

Figure 9-3. Many tractor-trailers have a cab equipped like a small apartment.

Rail Transportation

Railroads are part of North America's history and folklore. Today, the Trans-Siberian railroad in Russia is the longest in the world, stretching almost 5780 miles (9300 km), much of that over frozen wasteland. The Orient Express in Europe has been the setting of several films and novels. There are also fictional trains, such as the Hogwarts Express, which transports Harry Potter to Hogwarts Academy.

The diesel-electric system is a common type of fueling system for locomotives, **Figure 9-4.** However, more systems are starting to use electric trains, because these trains create less air pollution. Electric locomotives are also generally faster than diesels. For example, the Spanish Alta Velocidad Española runs between Madrid and Seville at speeds of up to 300 km per hour. See **Figure 9-5.** Electric trains pick up their power from overhead power lines or a third rail alongside the ordinary track. Trains make efficient use of fuel because there is not a lot of friction between the steel wheels of the train and the steel rails on which they run. Some special kinds of trains run on unique guides, such as monorails, maglevs, rubber-tired subways, funiculars (cable railways), and cog trains.

Figure 9-4. Diesel-electric locomotives are used in many rail systems. What is the disadvantage of this type of locomotive?

Figure 9-5. The Spanish Alta Velocidad Española is an example of a high-speed electric train. What are the advantages of this type of train?

Figure 9-6. Merchant ships are commonly used to transport cargo over long distances. What are the advantages and disadvantages of this form of transportation?

Water Transportation

For more than 5000 years, people have used ships to travel, trade, and fight in far-off lands. See **Figure 9-6**. Today, planes have largely replaced ships as people carriers, but every day, thousands of merchant ships ply the seas. These ships vary in size from simple cargo carriers to giant oil tankers. Although they differ in size, all of them have double hulls, which are double steel plates with a space in between, for safety reasons. Most ships use diesel power, although a few use nuclear reactors to make steam to drive turbines. While a ship's advantage lies in the amount of goods it can transport economically, its disadvantage is its slow speed, due to the resistance, or drag, of water on the hull.

Hydrofoils and hovercraft overcome this disadvantage by reducing the drag of water on the hull. When at rest or moving slowly, a hydrofoil looks like any other boat, but when it increases its speed, the entire hull lifts out of the water. The only parts that remain submerged are the rudder and part of its wings, or foils. This lift is made possible by the shape of the foils.

A hovercraft flies a few centimeters above the surface on a cushion of air. This means it is free of friction from the ground or water underneath. In the early days of hovercraft, transportation regulators could not decide whether hovercraft were boats or planes. Hovercraft are used as passenger and cargo ferries and by the military as patrol craft and troop carriers. See **Figure 9-7**.

The most popular method of transporting goods by sea is container ships. See **Figure 9-8**. Modern container ships are relatively fast. They can cross the Atlantic Ocean in seven days. Other ships transport loose cargo, such as coal, grain, and mineral ores in holds below deck. Others specialize in transporting mail, cargo, and passengers to smaller communities, which sometimes are not even connected by roads. Cruise ships create a "fun ship" image by transporting vacationers to islands in the Caribbean, the Mediterranean, and the South Seas.

Figure 9-7. Hovercraft fly just above the water's surface on a cushion of air.

Figure 9-8. Containers allow the maximum amount of goods to be stored in the minimum amount of space.

Air Transportation

A huge variety of transportation vehicles fly through the air. We are most familiar with planes. Regularly scheduled passenger planes operate according to timetables, and some companies hire charter planes to fly passengers. See **Figure 9-9**.

Lift

Why doesn't a heavy airplane drop out of the sky? The principle discovered by Daniel Bernoulli explains how the shape of a plane's wing produces lift and keeps the airplane in the sky. Airplane wings, as shown in **Figure 9-10**, are designed to be more curved on the top surface. Thus the air travels a larger distance over the wing than below it. The speed of the air above the wing is increased, but the pressure is reduced. Therefore, pressure on the underside of the wing is greater. This pressure difference gives the wing lift and allows the plane to fly.

Figure 9-9. Small passenger planes can move people between both small and large airports.

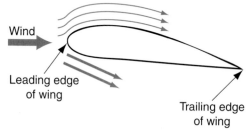

Figure 9-10. Air moving above the wing travels faster than air traveling below the wing. Why does this create lift?

Types of Aircraft

Although airplanes are the most common type of aircraft, many other types are also commonly used. For example, helicopters are often called on to help in rescue operations, such as lifting a pet to safety from the side of a cliff, helping people who are lost in remote areas, or rescuing earthquake survivors. See **Figure 9-11**.

Sports enthusiasts use many other means of air travel, including gliders, hang gliders, parachutes, and ultralight aircraft. See **Figure 9-12**. Gliding is a sport that enables people to experience the exhilaration of flying in a vehicle that looks like a small plane but has no engine. The most common way to launch a glider is to pull it into the air by means of a steel cable connected to a powerful winch, or to be towed into the sky behind another airplane.

Hang gliders are quite different; they do not even need a tow. The riders launch themselves from high places and return to the ground using movements of their bodies to turn and land. Ultralight aircraft are like hang gliders with very small engines. They operate at a maximum speed of 55 knots per hour and fly only between sunrise and sunset.

Other flying vehicles include hot air balloons and blimps. In a hot air balloon, passengers ride in a basket suspended under the balloon. See **Figure 9-13**. A propane heater warms the air inside the balloon, and because air expands as it is warmed, it rises and lifts the balloon up through the cooler outside air.

Blimps are often seen at major sporting events. They are usually used for photography or advertising. Blimps work in a manner similar to a balloon, except that instead of hot air, they are filled with helium, which is lighter than the surrounding air. They also usually have a motor and rudder, which make them more maneuverable than hot air balloons. The strangest aircraft is the ornithopter, an engine-powered vehicle in which the mechanical flapping of its wings creates the thrust and nearly all the lift!

Figure 9-11. Helicopters use rotating blades instead of a fixed wing to give them lift. Why are helicopters suited to rescue missions?

Figure 9-12. This glider has a retractable engine that allows the aircraft to be self-launching. The engine also provides a back-up source of power for those times when it flies too low.

Intermodal Transportation

Airplanes, ships, trains, and trucks all have advantages, but no single system is better than the others. Ships can move the largest loads from country to country. Trains carry large loads efficiently over long distances on land, but they cannot collect and deliver goods to your door or to the store around the corner. The amount of cargo that trucks carry is limited. While planes are the fastest means of travel, the cost of air freight is the highest of any method.

The most cost effective and efficient transportation system is *intermodal transport*, which integrates separate transportation systems through an intermodal chain. The most important development for intermodal transport was a simple one: the container. A container is a box, usually measuring 20' long, 8' high, and 8' wide. Its size has been standardized worldwide, so lifting equipment is available in all ports and rail terminals. Cranes can handle the containers quickly.

Goods are shipped in bulk within the containers, so costs are reduced. The contents are protected from damage and theft, and containers can be stacked to minimize storage space. Freight might be packed in containers and transported by ship, for example, to San Francisco from China.

Each container can be loaded onto a flatbed railcar when it arrives at a port. See **Figure 9-14.** When the railcar arrives at a large city, each container might be loaded onto a tractor-trailer and taken to a warehouse. There the containers are opened and the contents removed. Finally, a forklift truck loads items onto smaller trucks that deliver the items to local stores.

On-Site Transportation

On-site transportation moves people and materials from one place to another in a defined location, such as a warehouse or factory. See **Figure 9-15.** Other examples of sites include gravel pits, shopping malls, airports, and mines. Elevators, escalators, and conveyors move people or materials in a building. Robots in a factory move materials cheaply and

Figure 9-14. Containers are often transported first by boat (A) and then by train (B). What happens to the containers when they reach the train terminal?

quickly from one spot to another. Some are stationary and move parts short distances. Others are *automated guided vehicles (AGVs)*. They follow a set path and transport parts and materials over longer distances. Conveyors, trucks, and pipelines move materials at many sites, including mines, refineries, wells, and construction sites.

Figure 9-15. A forklift truck can move heavy pallets around a warehouse or can load them onto a truck.

Human-Powered Vehicles

A *human-powered vehicle (HPV)* is any vehicle powered solely by one or more humans. Rowboats, canoes, and bicycles are common examples. See **Figure 9-16**. The shape of the bikes you see every day has changed very little in the past 100 years. See **Figure 9-17**. However, there is now a greater choice of materials, including steel, aluminum, carbon fiber composite, and titanium. Top-of-the-line off-road bikes now include features such as hydraulic disk brakes and dual suspension for riding on off-road terrain that has drops, big roots, logs, and other obstacles.

Conventional bicycles and other pedaled vehicles have several disadvantages. A limited amount of downward pressure can be placed on the pedals. Also, as much as 80% of a rider's energy is used just to push away the air in front of the vehicle.

Wind resistance can be reduced in several ways. Making the tubing oval-shaped and hiding parts such as cables within the frame makes all surfaces smoother. A second way is to lower the rider into a *recumbent* position, lying on his or her back or stomach. A third method is to cover the machine and rider by a streamlined shape that resembles an aircraft wing or teardrop. See **Figure 9-18**.

Several cities have bike sharing systems. Bikes can be taken from one location and dropped off at another. See **Figure 9-19**. Many countries, and many cities within the United States, also have dedicated bike paths.

Figure 9-16. Any vehicle that is powered by humans is considered an HPV.

Figure 9-17. Bicycles are among the most common human-powered vehicles. What are the advantages and disadvantages of using a bicycle for transportation?

Math Application

Division of Whole Numbers

In the transportation industry, two important practical concepts are:

- How long will it take the product being shipped to reach its destination?
- How much fuel will be needed, and how much will it cost?

The answers to these questions can be found by division. Division is the reverse of multiplication. It determines how many times one quantity is contained in another. For example, if you ride your bike at 10 miles per hour (mph), how long will it take you to travel 70 miles? To find the answer, divide 70 miles by 10 miles per hour:

$$\frac{70 \text{ miles}}{10 \text{ mph}} = 7 \text{ hours}$$

Math Activity

Calculate the answers to the following questions about practical concerns in transportation and shipping.

1. If a car travels 320 miles on a tank of gasoline and the tank holds 20 gallons, how many miles can the car travel on a gallon of gas?

2. If the total load carried by a container ship is 50,000,000 lbs., and there are 5,000 containers on board, how many pounds does the average container hold?

3. If a space shuttle travels at 18,000 mph and the distance to the moon is 234,000 miles, how long would it take to travel to the moon?

Figure 9-18. Wind resistance is a major force that must be overcome by a cyclist. What three things have been done to these bicycles to lower wind resistance?

Figure 9-19. Bike sharing is a common system in some cities and countries. What are the advantages and disadvantages of bike sharing?

Engines and Motors

Most forms of transportation need engines to make them move. *Engines* are machines made up of many mechanisms. They convert energy into useful work. An engine needs a constant supply of energy to keep it working. The energy may come from burning fuel such as gasoline, diesel, or kerosene. Engines that burn this fuel inside them, such as those used in cars, are called *internal combustion engines*. Internal combustion engines depend on hot, expanding gases for power.

An *external combustion engine* burns its fuel outside the engine. A steam engine is an example of this type of engine. The fuels used as energy sources include coal and oil. External combustion engines heat water or gases to produce power.

Four-Stroke Gasoline Engine

Imagine trying to pedal the family car at 55 mph (90 km/h). The driver and one passenger would have to pump their legs up and down about 60 times a second. Although it is impossible for legs to move that fast, pistons in a gasoline engine slide up and down at this speed. The pistons are like short metal drums moving inside cylindrical holes in a metal block. The power to move the pistons comes from a mixture of air and gasoline. This mixture is ignited by an electric spark. Expanding gases push the pistons down to rotate the crankshaft. The crankshaft then transmits turning power to the drive train. The four-stroke *gasoline engine* is an internal combustion engine. See **Figure 9-20.**

Figure 9-20. The four-stroke gasoline engine is a popular power source for transportation.

In a four-stroke cycle, a piston travels down and up the cylinder four times (down twice and up twice) to produce a complete sequence of events, or cycle. See **Figure 9-21**. The following is what happens during one complete cycle of the engine:

1. **First Stroke—Intake**

   The intake valve opens. The rotating crankshaft pulls the piston downward. The piston sucks fuel and air into the cylinder.

2. **Second Stroke—Compression**

   The intake valve closes. The crankshaft pushes the piston upward. The air-fuel mixture is squeezed into a small space to make it burn with an explosion. Just before the piston reaches the top of the stroke, the spark plug ignites the mixture.

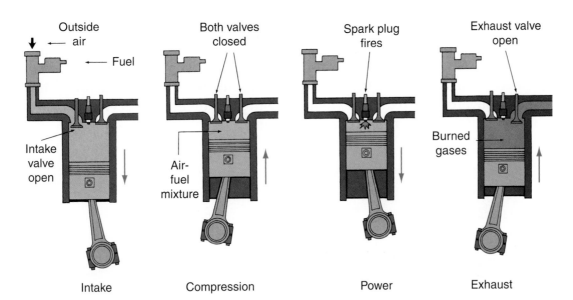

Figure 9-21. The four strokes of a four-stroke gasoline engine.

3. **Third Stroke—Power**

   Both valves are closed, so expanding gases cannot escape. The hot, expanding gases force the piston downward with great force. The piston pushes hard on the crankshaft, making it rotate.

4. **Fourth Stroke—Exhaust**

   At the bottom of the power stroke, the exhaust valve opens. The piston comes up again and pushes the gases from the burned fuel out of the cylinder. The cycle starts over again.

The four-stroke gasoline engine can be used in many different types of vehicles, as shown in **Figure 9-22**. It has the following advantages:

- It adapts easily to speed changes.
- It provides good acceleration.
- It has sufficient power for medium size machines.

However, the gasoline engine also has some disadvantages:

- It is not strong enough for very heavy work.
- It pollutes the air.
- It burns a relatively expensive fuel.

The Four-Stroke Diesel Engine

Diesel fuel is different from gasoline. It contains larger hydrocarbon molecules. This makes it heavier and oilier than gasoline, so it evaporates more slowly. The result is that the fuel can be ignited by pressure alone. So, unlike the gasoline engine, the *diesel engine* does not need a spark to ignite its fuel. It uses the heat from compressed air to ignite the fuel. Pure air is drawn into the cylinders. This air is squeezed to a much higher pressure than the air-fuel mixture in a gasoline engine. The pressure raises the air temperature to 1300–1500° F (700–900° C). At the top of the compression stroke, a fine spray of diesel fuel is injected into the cylinder at high pressure. When the fuel spray meets the hot, compressed air, the fuel ignites and burns very rapidly. See **Figure 9-23**. This forces the piston downward. Because of the high pressures involved, diesel engines have to be built stronger than gasoline engines.

Figure 9-22. Gasoline engines power all kinds of vehicles.

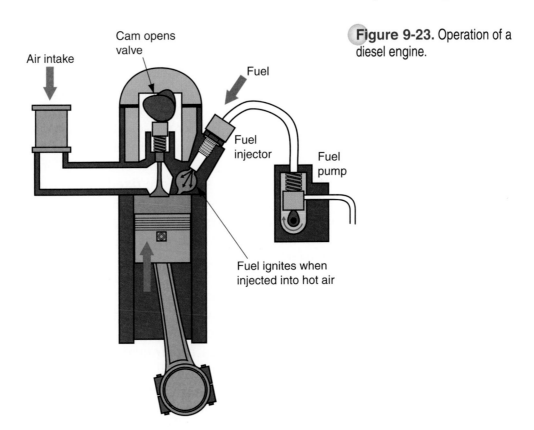

Figure 9-23. Operation of a diesel engine.

The four-stroke diesel engine has the following advantages:

- It uses less fuel than a gasoline engine.
- It lasts longer than a gasoline engine.
- It stands up well to long, hard work.

Early diesel engines were noisy, had poor acceleration, and created more pollution than gasoline engines. Most of these problems have now been reduced. Diesel engines are frequently used in passenger cars in Europe, and cars with diesel engines are becoming more popular in North America. They are commonly used everywhere to power large machines and vehicles, including tractors and buses. See **Figure 9-24.**

Figure 9-24. Diesel engines power: A—Tractors; B—Buses.

The Steam Turbine

The term *turbine* originally described machines driven by falling water, such as the waterwheels described in the Engineering Application. Later, the term *steam turbine* was also given to heat engines powered by steam. Engineers discovered that a strong jet of steam could turn a large wheel with blades on it. A steam turbine's hundreds of blades are set on a long shaft enclosed in a strong metal case. The blades on the rotor spin to drive the shaft. Blades on the stator are fixed to the outer casing and cannot spin, as shown in **Figure 9-25**. Each stator fan guides the steam flow so that as it moves along the turbine it has plenty of thrust to move the next rotor fan in its path. The spinning fans turn the drive shaft. The shaft, in turn, drives a propeller.

After its journey through the turbine, the steam cools and turns into water. This water then flows back to the boiler. Heated to steam again, it returns to work the turbine. The water is heated either by burning fuel or by a nuclear reactor outside the turbine.

The steam turbine has several advantages. It runs smoothly, has a long life, and is powerful and suited to slow, large machines that require an engine bigger than a diesel. This is shown in **Figure 9-26**.

One disadvantage is that it needs a lot of room, because space is needed for a boiler. Another is that it does not adapt as easily to changes in speed as piston engines.

Jet Engines

You have learned how Bernoulli's principle keeps an airplane in the sky. To get an airplane into the air, though, you need pushing power or *thrust*. What is thrust? Think of stepping forward off a skateboard, as shown in **Figure 9-27**. As you go forward, the skateboard moves backward.

turning shaft

Figure 9-25. This cutaway of a steam turbine shows how it is similar to a waterwheel. The stator blades (colored blue) do not move. They guide the steam so that it passes over and turns the rotor blades.

Engineering Application

Testing Waterwheels

The original, water-driven turbines were also known as waterwheels. There are two types. The overshot wheel is driven with water from above. The wheel rotates clockwise. An undershot wheel is driven with water from below so that the wheel turns counterclockwise.

Overshot

Undershot

Which type of wheel do you think was more powerful? What are the advantages and disadvantages of each type? To find out, perform the activity described below.

Engineering Activity

Design and make a waterwheel to test the advantages and disadvantages of overshot and undershot waterwheels. Materials you may need include:

- Corrugated cardboard or foam board (to make the wheel)
- Hot glue gun (to glue the blades to the sides of the wheel)
- Metal rod cut from coat hangers or wooden barbecue skewers (to make the axle)
- Pencil, ruler, compass, and protractor (to mark out and draw the parts for the wheels)
- Box cutter (to cut the cardboard or foam board)

One example of a waterwheel model is shown at the right.

When the wheel is assembled, place it in a sink and allow a small flow of water to spin the wheel. Position the wheel as necessary to use your model as an undershot waterwheel and an overshot waterwheel. Use the same flow of water for both tests. Does the wheel spin faster when it is used as an overshot or undershot waterwheel? Which type is affected by backwater pressure?

Write up your design, experiments, and results in a formal technical report. Be sure to evaluate your design and include ideas about how to improve the waterwheel and your testing process. Add the report to your portfolio.

Figure 9-26. Steam turbines power large seagoing vessels.

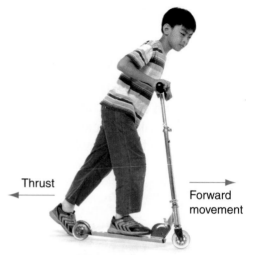

Thrust

Forward movement

Figure 9-27. Pushing backward against the ground with one foot creates thrust. This pushes the scooter and the boy forward.

Think about what happens when you turn on a garden hose. Suddenly, a jet of water bursts out, and the hose jumps backward. Firemen are sometimes pushed over by this backward force. Blow up a balloon and let it go without tying the neck. The balloon is driven in much the same way as the hose and skateboard.

These are just three examples of Isaac Newton's third law of motion. This law states that for every force in one direction, there is always an equal force in the opposite direction.

All jets work on this principle—for every action there is an equal and opposite reaction. The reaction to the rush of gases out of a *jet engine* is a thrust that drives the airplane forward. The engine sucks in air at the front, squeezes it, and mixes it with fuel. The mixture ignites and burns quickly. This creates a strong blast of gases. These hot gases expand and rush out of the back of the engine at great speed. Many people believe that the gases "push" against the air to propel the plane forward. This is not true. As the gases shoot out backward, the jet goes forward.

The *turbofan* is one kind of jet engine that powers aircraft. See **Figure 9-28**. A jet engine produces a very noisy gas stream. In a turbofan engine, however, the gas stream drives a large fan located at the front of the engine. This creates a slower blast of air. Thrust is as great as a simple jet, but the engine is quieter.

Air is drawn into the engine by the compressor fans. As pressure increases, the compressed air mixes with fuel. Ignition takes place, and temperature and pressure increase even more. The burned mixture leaves the engine through the turbine, which drives the compressor and the fan at the front of the engine. Pressure thrusts the engine forward, while the exhaust gases rush out of the back in a jet stream. The turbofan is an internal combustion engine.

The turbofan engine has several advantages. It is relatively lightweight, very powerful, and uses less fuel than other jet engines.

Science Application

Newton's Third Law of Motion

We may think that the forces we experience are one-sided. However, when a bird's wings push the air downward, the air reacts by pushing the wings, and the bird, upward. When playing ice hockey, a player pushes backward with one skate, and the ice pushes the player forward. When the player stops on ice, the skate's sharp blade cuts into the ice deep enough to create a barrier. As the player pushes against the barrier, it pushes in the opposite direction and the skater stops. When a rocket is fired, it pushes hard on the ground and the ground pushes upward with an equal force. All of these situations are examples of Newton's third law of motion: For every action, there is an equal and opposite reaction.

Science Activity

Design and make a dragster to demonstrate Newton's third law of motion. The diagram shows one simple shape for the dragster, but you should modify the shape to improve its efficiency. After you have built the model, test it on a 33 ft. (30 m) track to see how far it can travel.

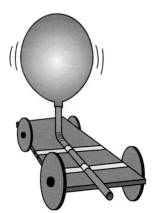

Materials you may need include cardboard, masking tape, bendable straw, box cutter, compass, pencil, ruler, party balloon, and pins to fasten the wheels in place. Follow these steps:

1. Assemble the parts according to your design.
2. Inflate the balloon to stretch it and then let the air out.
3. Place it over the short end of the bendable straw.
4. Inflate the balloon again by blowing into the long end of the straw.
5. Place the dragster on the floor and let go of the straw.
6. Measure the distance the dragster travels.

Write up your design, experiments, and results in a formal technical report. Include a chart to compare your results with those of several classmates. Explain how Newton's third law of motion is demonstrated in this experiment, and include labeled sketches to illustrate the principle. Add the report to your portfolio.

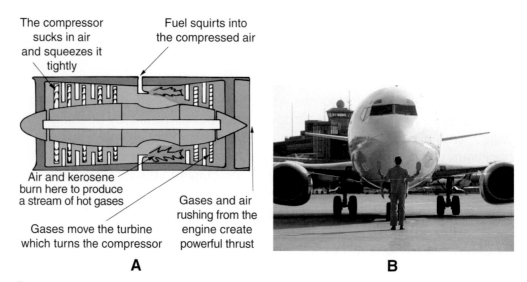

The compressor sucks in air and squeezes it tightly

Fuel squirts into the compressed air

Air and kerosene burn here to produce a stream of hot gases

Gases move the turbine which turns the compressor

Gases and air rushing from the engine create powerful thrust

A

B

Figure 9-28. A—Operation of a turbofan engine. B—Most passenger jets today use turbofan engines.

The Turboshaft Engine

Like the turbofan engine, the *turboshaft* engine uses a stream of gases to drive turbine blades. The blades turn a shaft, as shown in **Figure 9-29**. However, the turboshaft differs from the turbofan in that this shaft is connected to rotors or propellers. The spinning propeller blades are angled to push air backward similar to the jet engine thrust, and the vehicle moves forward.

The Rocket Engine

As described in the section on jet engines, when you release an inflated balloon, it zooms around the room. The compressed air that was forced into the balloon rushes out of its nozzle. This creates thrust, resulting in forward motion. Inside a rocket, a fuel is burnt to produce hot, high-pressure gas. This escapes from the rocket to provide the thrust.

Figure 9-29. Turboshaft engines are used on all but the smallest helicopters.

What pushes rockets forward? Imagine that the fuel is burning and the rocket's exhaust is closed. As the high-pressure gas burns, it pushes out in all directions against the inside of the rocket. See **Figure 9-30A**. The rocket does not move because the force is equal in all directions. Now imagine that the exhaust is opened. Hot gas will rush through the opening. There is little or no downward force on the bottom of the combustion chamber, but there is upward force on the top. The rocket is pushed up. See **Figure 9-30B**.

The difference between a jet engine and a rocket engine is how each obtains oxygen to burn its fuel. The jet uses the oxygen in the surrounding air. A rocket must carry its own oxygen if it operates outside the earth's atmosphere, where there is no oxygen-containing air.

Rocket engines burn a variety of fuels called *propellants*. Some propellants are solid, and others are liquid. Nearly all space rockets use liquid propellants, such as kerosene or liquid hydrogen, plus liquid oxygen or some substance that can provide the oxygen for combustion. See **Figure 9-31**.

Alternative Motors and Engines

For the last several decades, automobile companies and other inventors have experimented with other types of motors and engines for use in passenger vehicles. Their purpose is to reduce pollutants as well as reduce dependence on petroleum-based products.

Electric and Hybrid Vehicles

Electric vehicles and small delivery vehicles have been in use for 100 years. *Electric motors* change electrical energy into mechanical energy. They provide smooth turning power to drive a shaft. They have many advantages, including little maintenance and no oil changes.

A **B**

Figure 9-30. A—This rocket will not move because the pressure inside is pushing equally in all directions. B—When the exhaust valve at the rear of the rocket is opened, the rocket moves forward because the pushing force is greatest in the forward direction.

Figure 9-31. Because space has no oxygen, rockets that travel in space must carry both fuel and oxygen for combustion.

In the past, electric vehicles have had several disadvantages. Most use batteries that provide a driving range of only 100 to 200 miles (160 to 320 km). In contrast, gasoline-powered vehicles usually have a driving range of more than 300 miles. Also, recharging the batteries takes several hours. Finally, the batteries are large and heavy, and they take up a large amount of space in the vehicle.

These limitations mean that, until recently, forklift trucks, local delivery vehicles, and golf carts have been the most common electric vehicles. However, this is starting to change as new, smaller batteries that last longer are being developed. Several types of gasoline-electric hybrid vehicles have recently become popular. They combine a gasoline engine with electric motors. See **Figure 9-32**.

The Tesla Roadster, produced in California, is 100% electric. It goes from zero to 60 mph (100 km/h) in four seconds. It can go 250 miles (400 km) on a charge, and then its lithium-ion batteries can be recharged in only three-and-a-half hours. See **Figure 9-33**.

Some vehicles, such as trains and trolleys, are connected directly to a source of electricity by overhead cables. Others, such as subway trains, are attached to electrified rails. See **Figure 9-34**.

Solar Vehicles

Universities and technical colleges have been making experimental solar cars for many years. Usually, the solar cells that convert sunlight to electricity totally cover the upper surfaces of these cars. Each solar cell produces only about 1/2 volt of electricity, so hundreds of them are needed to power the car.

The sun's energy not only powers the car's motor, but it also charges a battery that can supply the energy needed when the sun is hidden behind clouds. This method of powering a car is very interesting because the energy source is inexhaustible and no pollution results from its use. The 2009 North American Solar Challenge Race was the eighth such race. The 24 teams racing from Dallas to Calgary were mostly from American

Figure 9-32. Gas-electric hybrid cars such as this Toyota Prius employ a regenerative braking process, which can provide power to the battery or directly to the electric motors.

Figure 9-33. The all-electric Tesla provides high performance and an attractive sports-car design.

Figure 9-34. Electric motors power several types of transportation vehicles. A—Some trains and most trolleys are connected directly to a source of electricity by overhead cables. B—Other electric trains and subway trains use electrified rails.

universities and colleges, but they also included Canadian entries such as the Éclipse VI from École de technologie supérieure (ETS, Montreal), **Figure 9-35**, and Schulich 1 from the University of Calgary.

Fuel Cell Vehicles

Fuel cell cars may solve the problem of distance, while also having the same low emissions level of electric vehicles. The *fuel cell vehicle* creates power by combining oxygen from the air with hydrogen from an on-board tank. See **Figure 9-36**. The power is used to turn electric motors that drive the wheels. The only material leaving the exhaust pipe is water vapor. There are several different designs for fuel cell vehicles. **Figure 9-37** hydrogen-powered vehicles at a hydrogen filling station.

Figure 9-35. Eclipse VI was a competitor in the 2009 North American Solar Challenge. The solar cells, which cover most of the upper body surface, convert the energy in light into electrical energy.

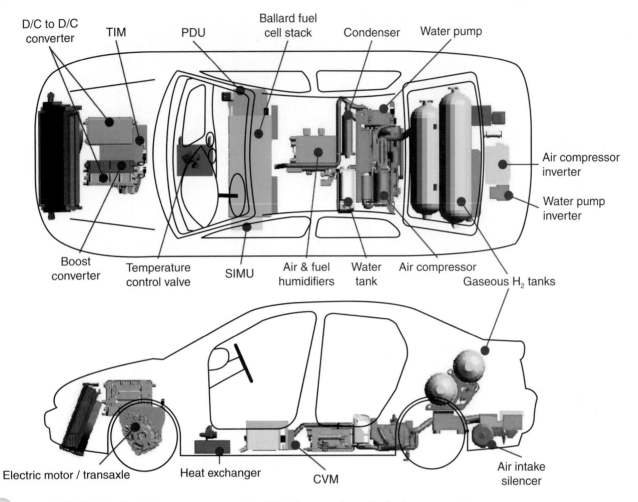

Figure 9-36. Fuel cell vehicles are very complex. This diagram shows typical components.

Figure 9-37. As more hydrogen-powered cars become available, hydrogen filling stations like this one will become more common.

Think **Green**

Transportation Alternatives

This chapter describes several technologies that may provide good alternatives to today's pollution-rich vehicles. Some of them are in limited use today. Many of these solutions are still in development, however, and it may be years before they become feasible for most people.

Other alternatives that provide at least partial solutions are available today. Ride-sharing, for example, cuts fuel emissions by reducing the number of vehicles on the road. If three people share a ride to work or school instead of taking separate cars, they reduce their contribution to vehicle-related pollution by two-thirds. Using public transportation when possible is another way to "go green." For short distances, you may consider walking or riding a bicycle.

Most of these alternatives save money as well as lower pollution levels. Why don't more people use them? There are several reasons. For example, bus stops may not be conveniently located for some people. Others may just enjoy the "quiet time" alone in their cars while they commute.

In many cases, however, our means of transportation is purely a habit. We hop into a car without thinking about possible alternatives. Think about your transportation needs. Discuss options with your friends. Try to develop an alternate means of transportation for getting to ball games and other activities. What are your alternatives?

Transportation Systems

Modern vehicles, such as cars and buses, are complex machines composed of a number of subsystems. Automobiles have systems that control the emissions, lights, speed, and many other functions.

Systems in Vehicles

The modern car is composed of the following subsystems: electrical, emission control, computer, fuel, axles and drive train, steering, suspension, brake, climate control, navigation, and engine. If one of these systems is not working correctly, the car's performance and efficiency will suffer. For example, power brakes use a vacuum to help you push the pedal. A leak in its vacuum lines is dangerous because it leads to a loss of braking power.

Systems to Coordinate Transportation

Transportation systems that coordinate modes of travel within a geographic area are usually run by the local government. For example, in Chicago, the Chicago Transit Authority (CTA) has schedules for all three of its different modes of transportation—bus, subway, and "L" train. All three of these parts of the CTA must run on time. To achieve this, both the vehicles and the people in the systems must be reliable. People depend on these forms of transportation for timely arrivals to their destinations.

Imagine that you are meeting friends at a movie theater across town. Your journey might start with a bus ride from your local bus stop at 6:30 to the central bus station. From there, you could take another bus at 6:50 in order to catch the 7:10 subway train. Upon arrival at the second stop, you walk two blocks to the theater to meet your friends. But what if the 6:50 bus were running late? You would miss the 7:10 train and your friends might think you weren't coming. If one part of the CTA is late, people pay the price of being late for their daily activities.

Transportation systems are not limited to those at the local level, such as the CTA. They also include national and international systems that utilize planes, helicopters, boats, barges, trucks, and trains to deliver passengers, foods, and other goods. Schedules are made for each of these systems. Within each of these systems, processes may include receiving, holding, storing, loading, moving, unloading, evaluating, marketing, managing, and communicating. These are all necessary for the successful day-to-day operation of our modern transportation systems. Like regional systems, these systems rely on both people and vehicles to transport people and goods. However, the design and operation of national and international transportation systems must be guided by government regulations.

Parts of a Transportation System

Like all other technological systems, transportation systems include inputs, processes, outputs, and feedback. Inputs include the people and materials that are transported; the people who develop, operate, and maintain the system; and the machines and structures within the system. Processes include managing and organizing the system, along with the actual transport of materials or people from one location to another. Outputs include the successful transportation of materials or people, the impacts on society, and the impacts on the environment. Feedback involves periodic checks comparing the location of cargo within the system to the transportation schedule. Another example of feedback is the evaluation of the completed transportation process. Did the cargo arrive at the intended location at the intended time without unintended incidents?

Transportation systems play a vital role in today's economy. Their influence reaches nearly all today's industries:

- Agriculture—Trucks, trains, and planes deliver fruit, vegetables, and other agricultural goods to vendors and grocery stores.

- Manufacturing—Coal, iron ore, steel, and lumber are taken to mills and plants by boats, trains, and tractor-trailers.
- Construction—Trucks take lumber, bricks, mortar, and concrete to job sites.
- Communication—News vans, trucks, and helicopters go on-location with camera operators and news anchors to broadcast stories. Helicopters also fly over the city and report the traffic to TV and radio stations.
- Health—Vans, ambulances, and helicopters take organs from donors to recipients and take injured people to the hospital.
- Safety—Police and government agencies use squad cars, vans, buses, planes, and helicopters to enforce laws and respond to accidents.

Exploring Space

People have always wondered about what lies beyond Earth. In the 1960s, we reached a milestone when the first human set foot on the moon. More recently, scientists have sent probes to the far reaches of our solar system. As transportation technology improves, we can send more sophisticated equipment over longer distances to find out more about our solar system and beyond.

Space Shuttle

The space shuttles were the first reusable space vehicles. They enabled people to fly into Earth's orbit and remain for several days before returning to Earth.

Imagine being one of seven people strapped to rockets the height of a 20-story building. When the solid rocket boosters (SRBs) are ignited there is no going back. The fuel cannot be extinguished. The booster rockets propel the orbiter to a speed of 3,500 miles per hour (5,700 km per hour). Once the vehicle has left earth's atmosphere, large thrusts of energy are not needed. The spacecraft is no longer trying to overcome the earth's gravity. Therefore, after two minutes of flight, the SRBs separate from the orbiter at an altitude of 30 miles (50 km). They fall to the earth with the aid of parachutes and are picked up by retrieval ships.

The orbiter's main propulsion unit then takes over. It consists of three engines burning liquid hydrogen and oxygen. After about eight minutes of flight, the orbiter's main engines shut down and the small orbiting maneuvering system (OMS) engines take over. To return to earth, the orbiter turns around and fires its OMS engines. After descending through the atmosphere, it lands like a glider. On-board computers control all the functions.

The Constellation Program

The space shuttle orbiters were retired in 2011. The new successor, Orion, will be a key component of the Constellation program. It will be able to accommodate larger crews than Apollo plus an emergency escape rocket. Testing of the launch vehicles, including the Ares I rocket shown in **Figure 9-38**, began in 2009.

International Space Station

There was a time when the destination of flights into space was the moon. Today, flights with humans aboard are destined for the $100 billion International Space Station (ISS), in order to deliver component parts to build or maintain the space station. See **Figure 9-39**. The ISS is a sprawling assembly of laboratories, living space, service areas, and solar panels the size of two football fields. Sixteen countries are cooperating in its construction.

No large cranes are available in space to lift the pieces of the ISS into place. In their place, astronauts use a "space arm" known as the Space Station Remote Manipulator System (SSRMS). See **Figure 9-40**. The SSRMS not only assembles the station but also maintains the station and maneuvers equipment and payloads. Its two hollow booms, which are the diameter of telephone poles, are made of 19 layers of carbon fiber. They are joined at an aluminum elbow and end in a "hand" that is similar to sockets from a mechanic's ratchet set.

Space station technology involves programs to develop and transfer technology to other sectors of our economy. Space arm mechanisms are being modified to control prosthetic hands for children. Space arm hardware and software are used in robotic devices.

Figure 9-38. New, improved rocket boosters are being designed for the Constellation program. Testing of the Ares 1 rocket began in 2009.

Figure 9-39. The International Space Station (ISS) is in use as an engineering and science laboratory.

Figure 9-40. The Space Station Remote Manipulation System (SSRMS) contributed to the International Space Station by the Canadian Space Agency.

Living on the ISS is, in one way, like going to the Arctic or to a drilling rig in the Atlantic Ocean. People stay there for one to three months, then return to Earth for a long rest. Since the ISS is expected to have a 15-year life span, there could be a hundred visits from space shuttles over this period in order to deliver personnel, supplies, and experimental cargo.

Future Objectives

Future space exploration has three objectives. First, scientists want to prepare a defense against asteroids. While very large asteroids collide with Earth on average once every 100 million years, there is a much greater risk of smaller asteroids striking at any time. The aim is to evaluate the potential of one landing and make a trial test to deflect an asteroid in space.

Another objective is to look for life in our solar system. Some planets appear to have underground seas and other raw materials for life. *Phoenix*, a recent American mission to Mars, landed near one of Mars' poles where ice and water may be present. The current goal is to analyze rocks and dirt from Mars and to get ready for exploration of two of Jupiter's moons: Europa and Titan.

Finally, scientists want to explore the outer boundaries of our solar system. An interstellar probe could use a solar sail, like a big mirror, to capture sunlight to provide power on its way to Mercury and Jupiter.

The Impact of Transportation

Transportation systems make our lives much more convenient, but they do have negative impacts. See **Figure 9-41**. Several modes of transportation pollute the environment. The pollutants in exhaust emissions from cars, trucks, buses, and airplanes cause many problems.

Figure 9-41. Private transportation has both positive and negative impacts. What are the impacts of today's passenger vehicles?

Air pollution causes a bad smell in the air and may contain disease-causing particles. The pollution level may be so high that people have difficulty breathing. Their eyes water, and they experience a burning sensation in their lungs. Air pollution can also corrode the surface of buildings. Noise pollution is another problem.

Technology can provide the answers to some of these problems. In the next few decades, you and your friends may be able to help. You may find new methods of transportation that reduce the negative impacts of current transportation technology. You may also find ways to help fix problems, such as pollution, that occurred as a result of earlier technology.

Transportation systems move people or freight from a point of origin to a point of destination. These systems rely on both the vehicles and the people that operate, maintain, and repair the vehicles. A system may use one or more modes of transportation including bicycles, cars, trucks, buses, trains, subway cars, ships, airplane, or a space shuttle.

Most modes of transportation need an engine to make them move, and most of today's vehicles have internal combustion engines that use gasoline and diesel as fuels. As environmental and climate concerns become greater, scientists and inventors have begun experimenting with alternatives to gasoline and diesel engines.

Electric and gas-electric hybrid vehicles are current alternatives to gasoline engines. Car manufacturers are also experimenting with fuel cells, compressed natural gas, and biofuels.

The transition from gasoline powered to electrically powered cars will depend largely on the development of suitable batteries such as lithium ion and nickel-metal hydride batteries. Public transit, intercity trains, cars and trucks can all be powered by electricity or hybrid systems. Electric buses could even be run by capacitors, without batteries. A bus could go about five stops, then recharge in less than a minute, at a quick-charging station, as passengers get on and off the bus.

- Modes of transportation include land, water, and air transportation. These are often combined to form intermodal means of transportation.

- The most common types of internal combustion engines are gasoline and diesel engines, but other engines are being developed to address environmental and economic concerns.

- Like other types of systems, transportation systems include inputs, processes, outputs, and feedback. Transportation systems play a large role in all of today's industries.

- National and international efforts are ongoing to explore Earth's moon, other planets in our solar system, and even far beyond solar system.

- Transportation technology has both positive and negative impacts on our lives.

Finding the Main Idea

Create a bubble graph for each main idea in this chapter. Place the main idea in a central circle or "bubble." Then place the supporting details in smaller bubbles surrounding the main idea. A bubble graph for the first part of the chapter is shown here as an example, but your bubble graph may look different.

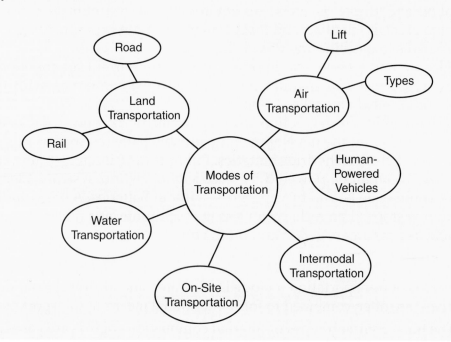

Test Your Knowledge

Write your answers to these review questions on a separate sheet of paper.

1. Describe the ways in which the modes of transportation affect the way you live.

2. State the best mode of transportation to move:

   A. A student traveling five blocks to the sports field

   B. A huge load of grain from one continent to another

   C. A prize bull from a farm in Texas to a farm in Alberta

   D. 10,000 commuters from a suburb to a city center

   E. A transplant organ from one city to another city 500 miles away

   F. A large supply of fresh fruit from South America to North America

3. Name, in the correct sequence, the four strokes of an internal combustion engine.

4. What is the major difference between a gasoline engine and a diesel engine?

5. Make a sketch to show how a jet of steam can be used to turn a shaft.

6. What is the basic principle of a jet engine?

7. Explain why a rocket engine must carry its own oxygen.

Test Your Knowledge *(Continued)*

8. In what ways do industries rely on transportation systems?

9. An electric motor changes electrical energy into what kind of energy?

10. Several current modes of transportation pollute the environment. State one way in which this pollution can be minimized.

Critical Thinking

1. Recreate the chart shown below using a word processing program. Do not write in this textbook. Fill in the Advantages and Disadvantages columns for each mode of transportation. Then use critical thinking skills to fill in the Ways to Minimize Disadvantages column.

2. Identify one problem in your town related to a mode of transportation. Research the problem and determine specific causes. Then suggest specific ways to solve the problem that would work in your town.

Mode of Transportation	Advantages	Disadvantages	Ways to Minimize Disadvantages
Bicycle			
Car			
Truck			
Bus			
Train			
Subway train			
Ship			
Airplane			
Space vehicle			

Apply Your Knowledge

1. Form a group with five classmates. Brainstorm future possibilities for powering spacecraft. Compile a list of at least 10 ideas. Then agree on one idea that seems the most feasible. As you decide, be courteous and show appreciation for the ideas of all team members. Draw a picture of your chosen method, describe its operation, and comment on its advantages and disadvantages.

2. Design a poster to show the relative distances of Earth, Mars, Venus, and Jupiter. Add interesting information about each of the planets to your poster.

3. If we were able to make contact with a civilization that is 250 light-years away and ask a question, it would take 500 light-years to get a reply. What question would you ask, and why?

4. Maintenance is the process of inspecting and servicing a product or system on a regular basis. Maintenance is done to extend the life of an item, help it continue to function properly, or upgrade its capability. Research the maintenance needs of a four-cylinder gasoline engine and those of a rocket engine. Write a report comparing and contrasting the needs of the two types of engine. What are the major concerns for each? Are similar processes used? Why or why not?

5. Research one career related to the information you have studied in this chapter. Create a report that states the following:

   - The occupation you selected
   - The education requirements to enter this occupation
   - The possibilities for promotion to a higher level
   - What someone with this career does on a daily basis
   - The earning potential for someone with this career

   You might find this information on the Internet or in your library. If possible, interview a person who already works in this field to answer the five points. Finally, state why you might or might not be interested in pursuing this occupation when you finish school.

STEM Applications

1. **SCIENCE** The human body is adapted specifically for life in the conditions on Earth. When we send people into space, the conditions are much different. Find out what negative effects an astronaut may experience by being in space for prolonged periods. What happens to the person's muscles, ears, sense of touch, heart, spine, bones, lungs, stomach, and eyes? Why?

2. **TECHNOLOGY** Design an extraterrestrial character, either as a drawing or in 3D form, for use in a new movie. Specify where the character came from (a specific planet, another solar system, or other place). Research the conditions in that place and give the character the features it would need to live there. Create a basic 3D model of your character using computer software.

3. **ENGINEERING** Use a formal design process to design a vehicle to transport eggs up and down a flight of stairs without breaking. The vehicle must be able to carry at least 6 eggs in each trip. Build a prototype and test it thoroughly. Be sure to test it first without eggs to make sure the vehicle is stable. Make any design modifications necessary and retest the model. When you are sure the eggs will not break, add the eggs for your final tests. When you are satisfied with the result, document the design, your test procedures, and the results. Place the documentation in your portfolio.

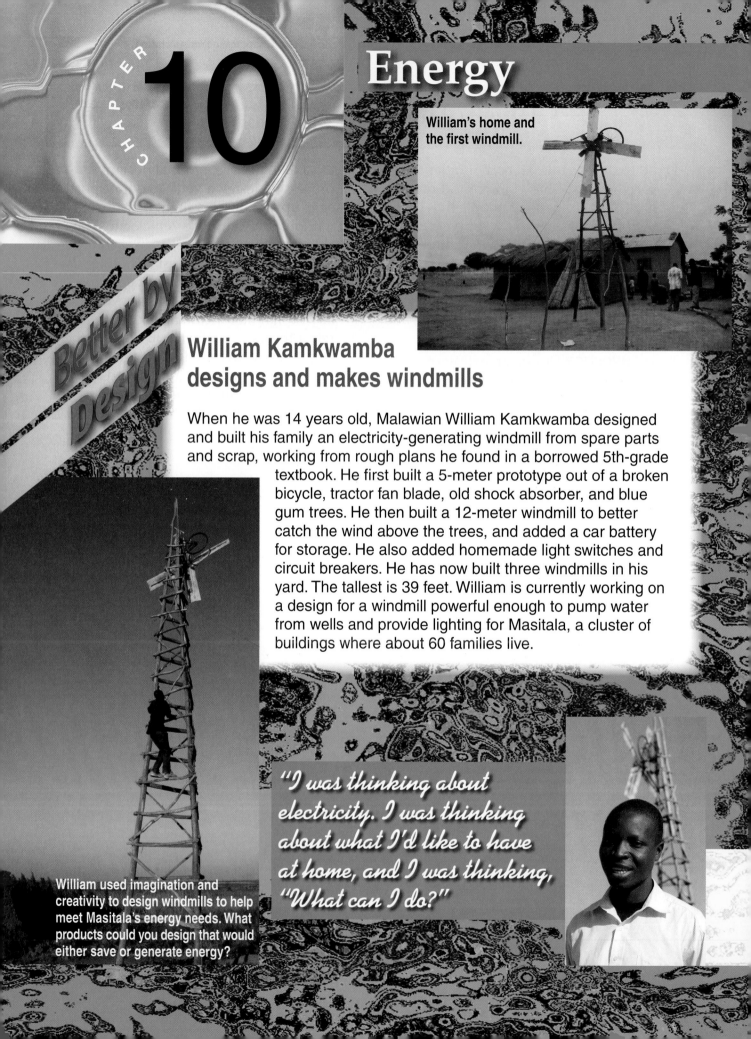

10

Energy

William's home and the first windmill.

Better by Design

William Kamkwamba designs and makes windmills

When he was 14 years old, Malawian William Kamkwamba designed and built his family an electricity-generating windmill from spare parts and scrap, working from rough plans he found in a borrowed 5th-grade textbook. He first built a 5-meter prototype out of a broken bicycle, tractor fan blade, old shock absorber, and blue gum trees. He then built a 12-meter windmill to better catch the wind above the trees, and added a car battery for storage. He also added homemade light switches and circuit breakers. He has now built three windmills in his yard. The tallest is 39 feet. William is currently working on a design for a windmill powerful enough to pump water from wells and provide lighting for Masitala, a cluster of buildings where about 60 families live.

"I was thinking about electricity. I was thinking about what I'd like to have at home, and I was thinking, "What can I do?"

William used imagination and creativity to design windmills to help meet Masitala's energy needs. What products could you design that would either save or generate energy?

Summarizing Information

A *summary* is a short paragraph that describes the main idea of a selection of text. Making a summary can help you remember what you read. As you read each section of this chapter, think about the main ideas presented. Then use the Reading Target graphic organizer at the end of the chapter to summarize the chapter content.

biofuels
biomass energy
chemical energy
conduction
convection
convection currents
elastic materials
electrical energy
electromagnetic waves
energy
frequency
fuel cell
geothermal energy
gravitational energy
hydroelectricity

kinetic energy
mechanical energy
nonrenewable energy
nuclear energy
nuclear fission
nuclear fusion
photovoltaic cells
potential energy
radiation
renewable energy
solar energy
sound energy
strain energy
thermal energy
wavelength

After reading this chapter, you will be able to:

- ◯ Explain society's dependence on energy.
- ◯ Describe the difference between potential and kinetic energy.
- ◯ Identify the various forms of energy and their applications
- ◯ Describe how energy can be changed from one form to another.
- ◯ Distinguish between nonrenewable and renewable sources of energy.
- ◯ List the advantages and disadvantages of each source of energy.
- ◯ Describe the components of energy and power systems.

Useful Web sites:
williamkamkwamba.typepad.com/

Every day we use energy in one form or another. When you ride your bicycle to meet friends, you are using your own physical energy to turn the pedals. When you fly a kite, wind provides energy to keep the kite in the air. The family car uses the energy in gasoline.

The sun provides heat energy when it shines through your windows and warms your house. In winter, the energy stored in wood, oil, or other fuels heats your home. When you turn on a light, you are using electrical energy.

Energy Basics

What is energy? *Energy* is the capacity to do work. To understand it more fully, we need to look at two categories: potential energy and kinetic energy.

Think about lifting a sledgehammer to drive a post into the ground. As you hold the hammer in the air, it has *potential energy*. Potential energy is also called "stored energy." Energy is stored in the hammer until it is dropped and the energy is released to hit the post. A stretched elastic band also has potential energy. When you wind up a spring in a toy, you are giving the spring potential energy. Once again, energy is being stored.

While working with springs, Robert Hooke discovered Hooke's Law, which says that the amount a spring extends is proportional to the force with which you pull it. By pulling twice as hard, you stretch it twice as far. Materials that obey Hooke's Law are known as *elastic materials*.

When the head of the hammer is dropped to hit the post or the spring is released to drive the toy, both the hammer and the toy gain kinetic energy. *Kinetic energy* is the energy an object has because it is moving. See **Figure 10-1**.

Forms of Energy

Energy is available in many different forms. **Figure 10-2** through **Figure 10-7** describe different forms of energy and the way energy gets things moving so that work is done.

Chemical Energy

A great deal of *chemical energy* is locked away in different kinds of substances. Chemical energy is found in the molecules that make up food, wood, gasoline, and oil. The energy is often released by burning the chemical. Burning rearranges the substance's molecules and releases heat.

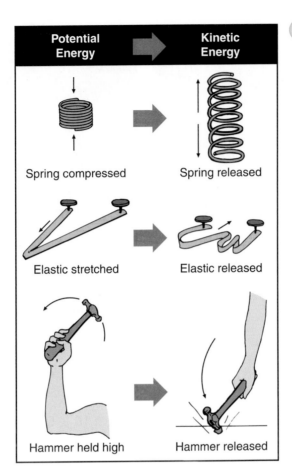

Figure 10-1. Examples of kinetic and potential energy.

Chemical energy is used in both natural and artificial systems. For example, the human body converts food into chemicals that provide the muscles with the energy to do work. Gasoline provides the chemical energy needed to keep a motorcycle moving. See **Figure 10-2**.

In artificial systems, chemical energy is often stored in batteries for future use. In the human body, it is stored in body cells in the form of a molecule called *ATP*. The body breaks down the ATP molecules as required to supply the energy we need.

Gravitational Energy

Objects always tend to move toward the lowest possible level. This is due to the *gravitational energy* (attraction or pull) of the Earth. It causes objects to fall. It is why water runs or objects roll downhill. For example, a skateboard at the top of a hill has gravitational energy. See **Figure 10-3**. This energy is available because of the pull of gravity.

While the skateboard is standing still at the top of the hill, the gravitational energy is potential energy. It changes to kinetic energy when the skateboard starts rolling down the hill.

Figure 10-2. Chemical energy is used in natural systems, such as the human body, and in technological systems, such as engines.

Figure 10-3. Objects have gravitational energy because of their position. When is gravity useful to you?

Mechanical Energy

Energy of motion is *mechanical energy*. Mechanical energy is often associated with or caused by a machine. However, it is not always caused by a machine. Two good examples of mechanical energy are a waterfall (natural mechanical energy) and a hydroelectric power plant (machine-related mechanical energy). See **Figure 10-4**. The power plant harnesses the falling water's mechanical energy by using it to turn turbines. These turbines are connected to generators, which produce electricity.

Strain Energy

Certain materials that can be stretched or compressed have a tendency to return to their original shape. This is known as *strain energy* or the energy of deformation. It is the kind of energy most easily seen in a rubber band, a bungee cord, or a bow used to shoot arrows. See **Figure 10-5**. However, many materials, including steel and carbon fiber, have a constant elasticity, and so follow Hooke's Law. When constructing buildings, architects and engineers take these properties into account so that the structure does not buckle when heavy loads are applied.

Figure 10-4. Mechanical energy is demonstrated in two ways: naturally by the falling water and machine-related by the movement of the turbines in a hydroelectric dam.

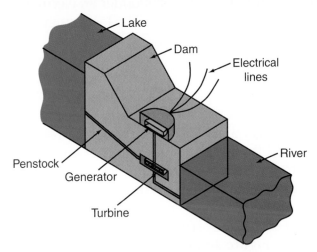

Figure 10-5. When an arrow is fired from a bow, the potential (strain) energy of the bow and string are converted to kinetic energy in the arrow.

Electrical Energy

Electrical energy is the movement of electrons from one atom to another in a conductor. This process is described in Chapter 11. Electrical energy provides the power to operate electrical devices such as motors and heaters. See **Figure 10-6**. It can move from place to place and readily changes into other forms such as heat, light, and sound. Electrical energy can be stored in batteries or produced in a generating station.

Thermal Energy

Thermal energy, or heat energy, occurs as the atoms of a material become more active. If you could look at atoms under an electron microscope, you would see that the atoms move about. The faster they move, the warmer the material.

Heat energy travels through matter in three ways: convection, conduction, and radiation. *Convection* occurs when expanded warm liquid or gas rises above a cooler liquid or gas. See **Figure 10-7A**. When liquid is heated, it expands and its volume increases. The amount of material (its mass) does not change. Since the mass is more spread out, hot liquid is less dense than cold liquid. In a mixture of hot and cold liquid, the cold liquid will sink to the bottom and the hot liquid will rise to the top. This creates a current in the liquid that is known as a convection current.

Conduction occurs when heat energy passes from molecule to molecule in a solid. The heat energy moves even though there is no obvious movement of the material. See **Figure 10-7B**.

Radiation occurs when heat energy is moving in the form of electromagnetic waves. For example, when you stand in the sunshine or in front of an electric heater, heat is transmitted through the air. No material is required between you and the source of heat for this type of heat energy. See **Figure 10-7C**.

Figure 10-6. Electrical energy provides the power to operate electrical devices. How many of the devices you use every day need electrical energy?

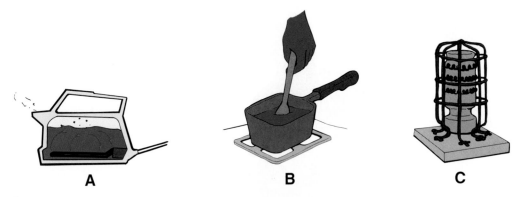

Figure 10-7. Heat energy can move through matter by convection, conduction, and radiation.

Solar Energy

Solar energy is related to heat energy. Another name for it is *radiant energy*. This type of energy comes to us from the sun. It travels in straight lines as a wave motion. See **Figure 10-8.** Objects such as TV pictures, lamps, and the sun are seen because of the light they send out. Most other objects can be seen because they reflect light.

Light travels at approximately 186,000 miles (300,000 km) per second. The speed of light does not change. Albert Einstein was the first to explain that a beam of light from the headlamp of a speeding train does not move faster than a beam of light from a stationary train.

Sound Energy

Sound energy is a form of kinetic energy. It is produced when matter, such as a tuning fork or human vocal cords, vibrates. The vibrating object has kinetic energy due to movement. The string on a guitar has potential energy when it is pulled back. When released, it has kinetic energy. See **Figure 10-9.** Sound energy moves at about 1100' (331 m) per second. This is much slower than light energy.

Figure 10-8. A flashlight is an example of an object that sends out light in the form of light waves.

Figure 10-9. Sound waves carry vibrations from a source to our ears and make our eardrums vibrate.

Nuclear Energy

Nuclear energy occurs as atoms of certain material are split or are forced together. This action, called *nuclear fission*, creates huge amounts of energy. Most of it is in the form of heat. Nuclear plants therefore need large amounts of water for cooling. See **Figure 10-10**.

Figure 10-10. Uranium is used in nuclear power stations to produce heat to turn water into steam.

Electromagnetic Waves

The air around us is full of *electromagnetic waves* (waves of energy that have both electric and magnetic properties). An example is the gamma rays given off by nuclear reactions. Waves with medium wavelengths are X-rays that are useful in X-raying your teeth, but over-exposure can damage living cells. Ultraviolet rays come primarily from the sun and can cause sunburn. Infrared rays include the heat waves you feel when sitting in front of a radiant heater. Microwaves are in common use in microwave ovens and in satellite communications. Radio waves are used for communication and are the largest of all electrical waves.

All of these waves travel at the speed of light, but they differ from one another in several ways. They have different wavelengths and frequency, and in the amount of energy they can carry. *Wavelength* is the length of one wave cycle, as shown in **Figure 10-11.** *Frequency* is the number of cycles that occur in a second. Waves with the shortest wavelength have the highest frequency and carry the most energy.

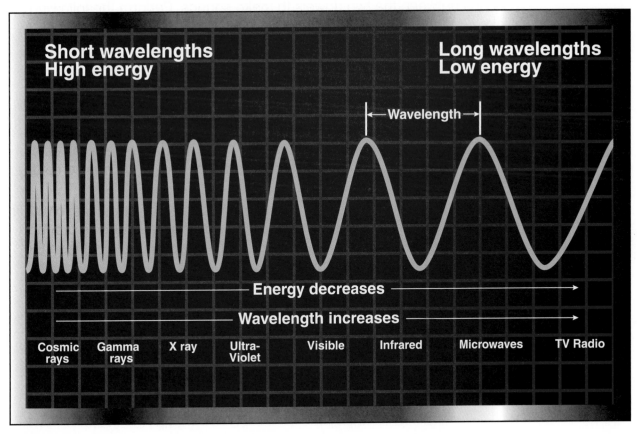

Figure 10-11. As the wavelength of an electromagnetic wave increases, its frequency, or cycles per second, decreases.

Science Application

Vibrations and Sound

Music can be made by causing part of an instrument, or the air inside the instrument, to vibrate. This vibration produces sound waves in the air that we hear as musical sounds. The frequency of the sound—the number of vibrations per second—is measured in hertz (Hz). String instruments have strings that are stretched tightly on the instrument, but are left free to vibrate. How fast they vibrate depends on how tight the strings are.

Science Activity

To demonstrate the principle of sound energy, make a simple string instrument. You will need four rubber bands, three empty boxes (such as a cereal box) of different sizes, and two 1/2″ dowel rods. Wrap the rubber bands around one of the boxes. Then place the dowels under the bands, 1″ from each end of the box, to lift them off the box. Pluck the rubber bands between the two dowels to make sounds. Perform the following experiments:

1. Move the dowels to different positions, such as 2″, 3″, and 4″ from the edges of the box. What difference does the position of the dowels make in the sound? Record your findings.

2. With the dowels 2″ from each edge of the box, tighten the rubber bands. What happens to the sound? Record your findings.

3. Repeat steps 1 and 2 for each box. What difference does the size of the box make? Is this any different from changing the position of the dowels? Record your findings.

Document your experiments and results. Record your ideas for other tests you could do to find out more about how string vibrations produce sound. What changes could you make to the instrument to improve the quality of sound? Place your report in your portfolio.

Energy Conversion

It is important to realize that energy can be neither created nor destroyed. We simply change its form. The first law of thermodynamics states that the total amount of energy in the universe remains constant. More energy cannot be created, and existing energy cannot be destroyed. It can only be converted from one form to another. Energy conservation is a rule of physics. It states that the total amount of energy is unchanged even though it may change from one type to another.

For example, the energy you use to pedal your bicycle comes from the food you eat. Your body has made a change in the form the energy takes. The chemical energy in the muscles is changed to kinetic energy of the bicycle. When the brakes are applied, this kinetic energy is changed into heat energy as a result of friction between the brake shoes and the wheel.

When a flashlight is switched on, the chemical energy in the battery is changed to electrical energy. Electrical energy is changed to light energy when the bulb is lit. A bungee jumper has gravitational energy because of the height above the ground. This energy is changed to kinetic energy during the dive. If the horn of a car blows, electrical energy becomes sound energy.

Losses during Conversion

When switching on a light bulb, you may expect to change all of the electrical energy to light energy. Not so! Only a portion of the electrical energy is converted into light energy. The rest is converted into heat energy. You can feel the heat produced by holding your hand near the light bulb. See **Figure 10-12**. In all energy changes, some of the energy is used as intended, but some is wasted. However, remember that although there has been a change in the form of energy, the total amount of energy remains the same: energy is neither created nor lost.

Figure 10-12. A light bulb is one example of energy conversion. Electrical energy converts into light energy and wasted heat energy.

Reducing Energy Waste

Partly because incandescent light bulbs waste so much energy, they will be banned in Canada by 2012 and in the United States by 2014. In the United States, they will start being phased out in 2012, beginning with the 100-watt bulb. In most of Europe, the ban is now being phased in, and incandescent bulbs are already banned in Australia and Cuba.

Where Does Energy Come From?

Much of the energy we use comes from the sun. The sun's heat keeps us warm. Heat from the sun also causes wind and rain. Most plants need light energy from the sun for growth. These plants provide humans and animals with the energy they need to do work. Over millions of years, some of these plants have been changed into petroleum and coal. These fossil fuels may be used to provide energy for machines.

All sources of energy make up two groups: nonrenewable and renewable. *Nonrenewable energy* sources will eventually be used up and cannot be replaced. They include coal, oil, natural gas, and nuclear energy. *Renewable energy* sources will always be available. They include the sun, wind, water, and geothermal energy. These two major groups are summarized in **Figure 10-13**.

Nonrenewable Sources of Energy

Most nonrenewable sources of energy were formed from the remains of living matter. Although other sources of energy are being explored, coal and petroleum are currently the most important nonrenewable sources of energy.

Coal

Coal developed from the remains of plants that died millions of years ago. For this reason, it is often referred to as a fossil fuel. The coal-forming plants probably grew in swamps. As the plants died, they gradually formed a thick layer of vegetable material. Sometimes, ancient seas covered this layer. Sediments (fine particles of sand or gravel) settled to the bottom to form layers of sandstone or shale. As this process was repeated, the layers of vegetable material became squeezed under great pressure and heat for a long time. The result was coal. See **Figure 10-14**.

Removing coal from the ground is called *mining*. Coal mines are of two types: surface mines and underground mines. Surface mining involves stripping away the soil and rock that lie over a coal deposit. The coal can then be dug up and hauled away. See **Figure 10-15**.

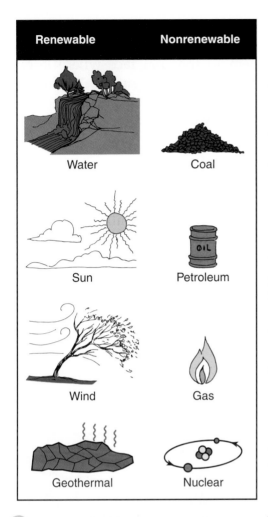

Renewable	Nonrenewable
Water	Coal
Sun	Petroleum
Wind	Gas
Geothermal	Nuclear

Figure 10-13. All energy comes from a source that is either renewable or nonrenewable.

Water

Organic material from forests becomes covered by water.

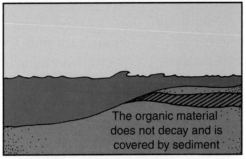

The organic material does not decay and is covered by sediment

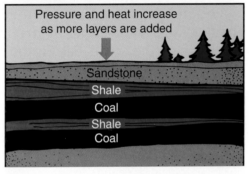

Pressure and heat increase as more layers are added

Sandstone
Shale
Coal
Shale
Coal

Figure 10-14. After millions of years of heat and pressure, organic material becomes coal.

Figure 10-15. In open pit or surface mining, a dragline removes soil and rock to expose a coal deposit.

Underground mining involves digging tunnels to reach the coal deposit. Miners go down a shaft in a large elevator and then ride through the tunnels in cars. The cars take them to the coal face where large machines rip coal from its million-year-old home. See **Figure 10-16**. Coal is primarily used as a fuel for electric power generating stations. It is also used to power industrial processes, particularly those manufacturing steel.

Petroleum and Natural Gas

Petroleum was formed from the bodies of countless billions of microscopic plants and animals that lived in the seas millions of years ago. As these plants and animals died, they sank to the bottom to mix with mud and sand on the sea floor. Fossil fuels formed as the result of millions of years of heat and pressure on their remains. As the deposits became buried deeper, the pressure and temperature increased. Over millions of years, the material was slowly changed into complex hydrocarbons that we call coal and gas. See **Figure 10-17**.

Oil and gas are removed from the ground by drilling deep holes. The holes are made either by drilling rigs located on land or by drilling platforms on the ocean. See **Figure 10-18**.

Figure 10-16. In an underground coal mine, coal cars carry coal and miners to and from the coal face.

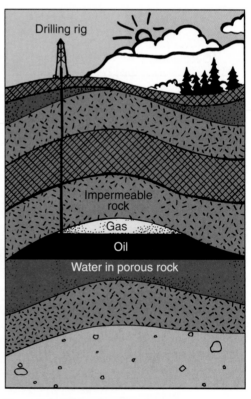

Figure 10-17. When an oil well is drilled, the oil may come out in a gusher. This is due to underground pressure. In other cases, the oil must be pumped to the surface.

Figure 10-18. An oil drilling rig removes oil from below the ground.

In its natural or crude form, oil removed from the ground is useless. Crude oil is processed in an oil refinery. The process involves heating the crude oil to approximately 350° C in a building known as a *fractionating tower*. As the crude oil heats, different compounds are separated. The lighter components rise to the top and the heavier components remain at the bottom. Light hydrocarbons such as ethane and propane come off the top of the distillation tower. They contain between one and four carbon atoms. See **Figure 10-19.** The next lightest are used for gasoline; they contain between five and ten carbon atoms. Diesel fuel contains between twelve and sixteen carbon atoms and is distilled off next. Finally, the heavy residues such as bitumen that contain more than twenty-five carbon atoms are distilled.

The supply of fossil fuels is limited. With our known reserves, and at our present rate of use, we probably have enough coal to last about 200 more years, natural gas for about 75 years, and oil for 60 years or less.

Nuclear Energy

Nuclear power currently provides 15% of the world's electricity. There are currently close to 500 nuclear power plants operating around the world. Nuclear energy is created using one of two processes: nuclear fission and nuclear fusion.

Nuclear Fusion

Nuclear fusion is the same process that powers our sun and the stars. It requires enormous temperatures and pressures. Nuclear fusion has been achieved in research fusion reactors on a limited scale. To date, however, continuous nuclear fusion has not been possible. If we could harness this power on Earth, fusion could be the key to unlimited clean energy.

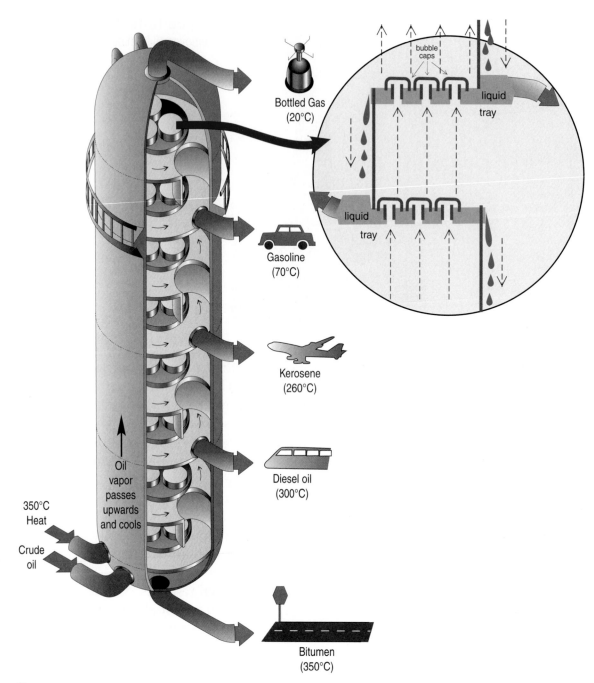

Bottled Gas
(20°C)

bubble
caps

liquid
tray

liquid
tray

Gasoline
(70°C)

Kerosene
(260°C)

Oil
vapor
passes
upwards
and cools

Diesel oil
(300°C)

350°C
Heat

Crude
oil

Bitumen
(350°C)

Figure 10-19. By distillation, or "cracking," crude oil can be converted into more than 800 products.

Nuclear Fission

At present, nuclear power stations use only the nuclear fission process. What is nuclear fission? Remember that all solids, liquids, and gases are composed of chemical elements. The smallest unit of each element that still retains the properties of that element is an atom. Although atoms are very small, they are made of even smaller subatomic particles called *protons*, *neutrons*, and *electrons*. At the center of each atom is a nucleus made up of protons and neutrons. See **Figure 10-20**.

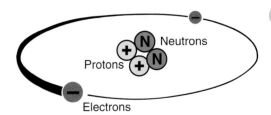

Figure 10-20. The nucleus of an atom contains protons (positively charged) and neutrons (no charge). Electrons, (negatively charged) orbit around the outside the nucleus in a "cloud."

Most atoms have a stable nucleus, which means they do not change. In a few atoms, the nucleus is unstable. These unstable nuclei try to become stable. They throw off particles or rays. These rays are called *radiation*. The atoms are radioactive.

Uranium is a metal. Its atoms have very large nuclei. Very large nuclei are often particularly unstable. When a neutron hits an atom of uranium, the nucleus of the atom splits. See **Figure 10-21**.

The atom splits into two parts, called *fission fragments*. Together, the fragments weigh slightly less than the original atom. The loss in mass turns into energy. On average, an atom that undergoes fission produces one, two, or three neutrons. These neutrons may hit other uranium nuclei. When they do, these nuclei may also split. This throws out more neutrons, and a chain reaction occurs.

To understand a chain reaction, imagine 200 marbles lying on a flat surface and arranged in a circle. What would happen if another marble was thrown at them? They would fly all around in different directions, and each marble would probably hit two or more other marbles. The single marble caused a chain reaction. This is what happens in nuclear fission. *Nuclear fission* is the energy produced by splitting atomic nuclei.

The heat produced by nuclear fission is used to heat water, which turns into steam. This steam is used to drive turbines. Generators attached to the turbines produce electricity.

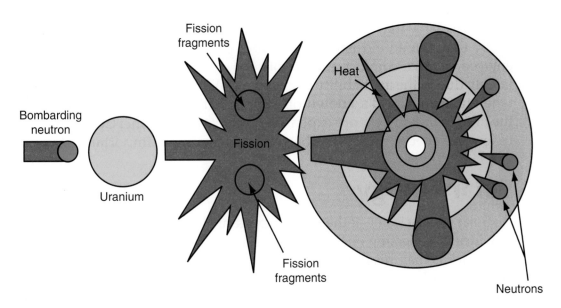

Figure 10-21. Nuclear fission occurs when an atom of uranium splits.

The fission process is noted for the large amount of heat energy it releases. The fissioning of 2.2 lb. (1 kg) of uranium produces about the same amount of heat as burning 2.9 *million* lb. (1.3 million kg) of coal!

Nuclear power has a number of disadvantages. First, some nuclear generating stations have been shut down because of technical and other problems. Second, the general public is concerned about the possibility of major disasters, such as the ones that happened at Three Mile Island (United States) and Chernobyl (Russia). Third, costs per kilowatt-hour are at least twice that of other conventional sources. Finally, safe disposal of nuclear waste is vital. Just a few minutes of exposure to a single bundle of spent fuel one year after it is removed from the reactor, at a distance of 12" (30 cm), would be fatal. Nuclear waste remains dangerously radioactive for thousands of years. How to best dispose of these wastes is a question that is still being debated worldwide.

Nevertheless, people who favor nuclear power point out that climate change and the cost and dangers of coal and petroleum production change our priorities. For example, the gas, oil, and coal-fired plants that produce most of the world's electricity cause more deaths in a year from mining accidents alone than can be traced to nuclear power plants in the past 50 years. Furthermore, fail-safe measures are built into new plants which "kick in" automatically in the event of an emergency. Today 70 percent of France's power is nuclear, compared to approximately 20 percent in the United States and United Kingdom.

Renewable Sources of Energy

Renewable energy is energy that can be replaced rapidly with natural processes. The energy provided by the source can be renewed as it is used. In the past, most energy has been obtained from burning nonrenewable fossil fuels. But as the nonrenewable sources of energy become scarce, alternative sources (solar, wind, tidal, geothermal, and biomass) are being developed.

By living in an industrialized world, people in many countries have enjoyed relative comfort and wealth for over 100 years. This has been largely due to the availability of cheap oil. We are now reaching the point when half the world's known oil supply will have been extracted, and the other half will be more expensive to extract and refine. Consequently, renewable energy sources are becoming increasingly important.

Solar Energy

Solar energy is one of the most important alternative sources of energy. The idea of collecting energy from the sun is a very good one. The main drawback is that this type of energy is not always available. In the winter and on cloudy days, there may be too little. At night, there is none. Yet these may be the times when energy is most needed. However, we can use solar panels to collect and store energy from the sun for later use.

The simplest type of solar panel collects heat directly from the sun's rays. The heat is carried away to provide hot water or to heat buildings. In one kind of solar panel, water flows through pipes or channels under a plate of glass. These pipes or channels are painted black to absorb heat better. This heat transfers to the water. Pipes carry it to the hot water system, where the heat is released. Solar panels are usually placed on the roof of a building. See **Figure 10-22**.

Photovoltaic Cells

Most solar energy is generated by two other methods. The first uses solar cells called *photovoltaic cells* to convert the sun's energy directly into electrical energy. The cells are silicon wafers. Photovoltaic cells work because visible and ultraviolet light are powerful enough to knock electrons free from atoms. The loose electrons move through conductors. This creates an electric current.

Photovoltaic solar power is best known for its use in space. The International Space Station (ISS) has eight solar arrays. Together, the arrays contain a total of 262,400 solar panels. See **Figure 10-23**. During its 90-minute orbit, approximately 40 minutes is without sunlight. An electrical power storage system with rechargeable batteries is therefore necessary.

Uses of solar power on Earth vary from portable devices, such as pocket calculators, to recreational use in distant locations. A remote home can be virtually self-sufficient with solar power. An inverter that converts direct current (DC) to alternating current (AC) can be used to run most domestic appliances.

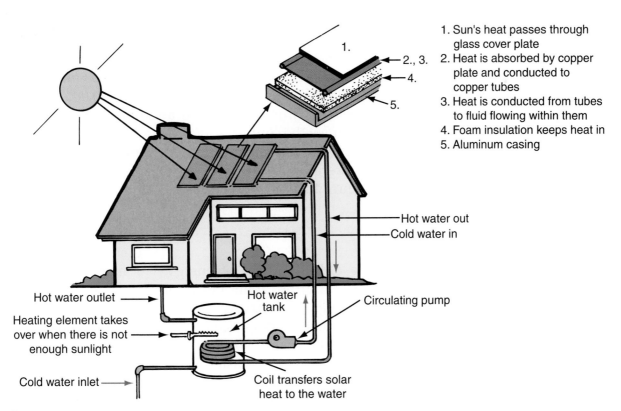

1. Sun's heat passes through glass cover plate
2. Heat is absorbed by copper plate and conducted to copper tubes
3. Heat is conducted from tubes to fluid flowing within them
4. Foam insulation keeps heat in
5. Aluminum casing

Hot water out
Cold water in

Hot water outlet
Heating element takes over when there is not enough sunlight
Cold water inlet
Hot water tank
Coil transfers solar heat to the water
Circulating pump

Figure 10-22. Solar panels collect heat from the sun. This heat is used to provide hot water.

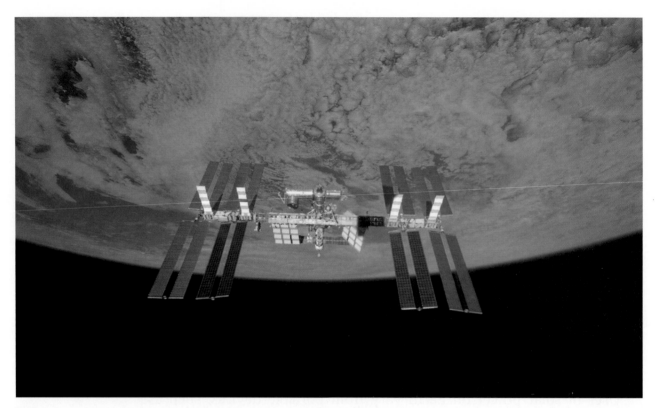

Figure 10-23. The International Space Station has solar panels that produce approximately 110 kW of power.

Concentrating Solar Power Systems

Solar power can also be produced by concentrating solar power (CSP) systems. CSP systems use mirrors to capture and focus the sun's rays on a single point. A fluid, such as water, is heated to high temperatures, which then drives a turbine. A large solar farm in the Mojave Desert, California, currently covers four square miles (10.3 sq.km) and uses 400,000 mirrors to capture the sun's energy. It produces enough electricity for 900,000 homes. Deserts in the southwestern United States are an abundant source of sunshine that could help meet the country's increasing demand for power without releasing any CO_2. This system differs from photovoltaic solar systems in which light interacts directly with semiconductor materials to generate power. See **Figure 10-24**.

Wind Energy

Wind is one of the oldest sources of energy. For many centuries, wind has turned wheels to grind grain and pump water. Today, wind is increasingly used to generate electricity. Wind power costs about half as much as power from a dam, and a large windmill can be erected and running in one week. Wind can spin wind turbines that are situated on top of high towers, where wind blows faster. The carbon fiber-reinforced blades can be up to 100 ft. (30 m) long, and there may be hundreds of turbines in one wind farm. See **Figure 10-25**.

Figure 10-24. A concentrating solar power (CSP) system.

Figure 10-25. Wind turbines on a wind farm capture the kinetic energy in surface winds and convert it into electrical energy in the form of electricity.

As long as the sun continues to shine, the wind will continue to blow. Wind currents occur because of differences in temperature between different parts of the Earth. More heat from the sun reaches the equator than the poles. See **Figure 10-26**. This is because the sun's rays hit the Earth directly at the equator (A). Near the poles (B), they hit at an angle. The heat is spread over a wider area.

The air above the equator expands most and rises by convection. When it reaches about 30° latitude, the warm air cools and falls. At about 60° latitude, it meets cold air from the poles and rises over it. This air movement creates more convection currents. See **Figure 10-27**.

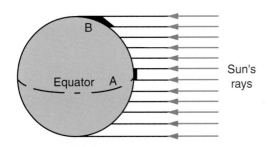

Figure 10-26. More solar heat is delivered at the equator than at the poles. Can you explain why?

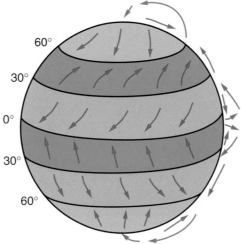

Figure 10-27. Air currents follow a certain pattern over the entire Earth.

Convection currents are also created as a result of the difference in temperature between the land and the sea. During the day, the land heats up more quickly than the sea. As the air warms up, it expands and becomes less dense. This causes it to rise. Cooler, denser air from the sea moves in to take its place. See **Figure 10-28**. At night, the land cools more quickly than the sea. The process reverses, and warm air rises from the sea. Cooler air from the land moves in to replace it, **Figure 10-29**.

The wind is free, but it is also unreliable. Meteorologists can determine where the most powerful winds blow. However, no one can predict when it will blow. Many turbines need a breeze of 8 mph (14 km/h) to start them rotating. At 40 mph (60 km/h) they are working at full capacity. At wind speeds above 55 mph (90 km/h), the turbine may have to be shut down to prevent damage to the equipment. Newer turbines have variable geometry blades that can flex, reducing their speed if the wind is too strong.

Currently, wind turbines provide only about 1% of electricity use in the United States. However, it is predicted that this amount may increase to 15% by 2020. Farmers who allow wind turbines to be constructed on their land could earn far more than they would earn by planting grain to produce ethanol.

Energy from Moving Water

The energy of moving water is found in rivers, estuaries, and oceans. Why do rivers flow? As shown in **Figure 10-30**, water uses gravitational energy to flow from higher ground to lower ground.

The sun evaporates water mainly from the sea and also from rivers, lakes, and plants. This water vapor rises to form clouds. The clouds move with the land breezes. When they reach high ground, they are forced to rise. This causes them to cool, and they cannot hold as much water. Water falls as rain on high ground and forms rivers.

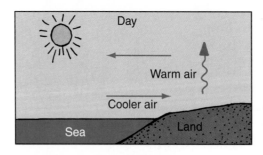

Figure 10-28. Sea breezes blow on shore as warmer air over land rises and cooler sea air moves in to replace it.

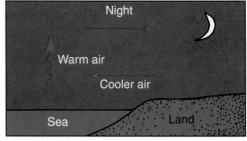

Figure 10-29. At night, land masses cool faster than oceans, causing the breezes to blow from land to sea.

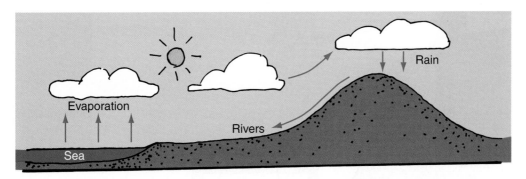

Figure 10-30. Rivers flow because of gravitational energy. Rains deliver the water to high ground. The water then flows to lower ground.

Hydroelectricity

Electricity generated from moving water over turbines is called *hydroelectricity*. Hydroelectric systems produce about 15% of the world's electricity, making hydroelectricity the most important renewable energy source. In Canada, about 60 percent of electrical power comes from hydroelectric systems.

In a hydroelectric system, water from rivers is generally stored behind a dam, as shown in **Figure 10-31**. To generate electricity, water flows through very large pipes called penstocks. The penstocks direct water onto turbine blades, spinning them. The turbines are connected to generators. See **Figure 10-32**.

Figure 10-31. Dams store water that will be used to produce electricity.

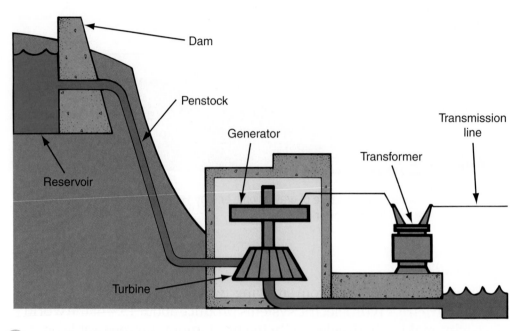

Figure 10-32. Stored water runs through the turbine with great force, causing it to spin rapidly. The turbine drives an electric generator, which produces electricity.

Hydroelectricity is economical compared to other electric power systems. After the initial expense of building the dam and generating station, the cost of producing electricity is small. No fuel is needed apart from the energy provided by the sun. However, hydroelectric power is not pollution-free. When a big dam is built, a large area of land is flooded. The vegetation and soils contain organic matter that rots underwater. This action creates carbon dioxide and methane gas. These gases enter the atmosphere and contribute to global climate change, especially in the tropics.

Wind and hydroelectric power systems complement each other. In winter, when rivers flow more slowly, wind power is at its strongest.

Tidal Energy

Tides from the sea are yet another alternate source of energy. Isaac Newton explained that tides occur because the moon pulls differently on oceans on the near and far sides of the earth. The different gravitational pull causes the surface water to bulge both toward and away from the moon, resulting in the rise and fall of tides every 12 hours.

The tides are regular and inexhaustible. The force of tidal currents can be used to produce electricity. The method is much the same as that used with waterfalls and rivers. The force of tidal water, however, can be captured when it is rising as well as when it is falling.

To understand this method, think of a dam-like structure being placed across the mouth of a bay. As the tide rises, the water flows through a tunnel in the dam. It turns a turbine inside the tunnel. As the tide falls, the water flows back toward the ocean. Once again, it turns the turbine. See **Figure 10-33.**

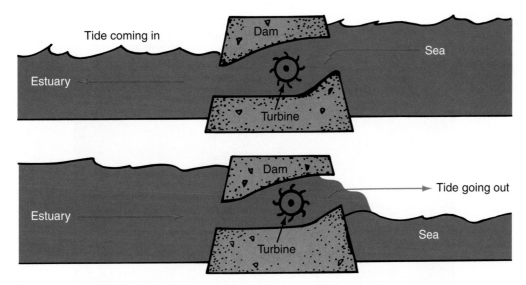

Figure 10-33. Tides can be used to produce electricity. Whether flowing into or out of the estuary, the water spins a turbine.

Newer technologies use large turbines lowered into deep ocean water to harness powerful tidal currents as water rushes past them. An average turbine can produce about two megawatts of power, similar to that of one wind turbine. A wave farm, to be built off the coast of California, is expected to be completed by 2012.

Geothermal Energy

Visitors to Yellowstone National Park will likely see Old Faithful erupting. A cone of boiling water shoots into the air at least 100 ft. (30 m) high at half-hour to two-hour intervals. This display is a demonstration of *geothermal energy*. It shows what happens when water is trapped underground, heated by hot rocks, and forced to the surface through cracks in the earth's crust.

Over 2,800 megawatts of electricity are generated from geothermal power plants in the United States. This electricity is used to supply four million homes with power. The total quantity of the earth's geothermal energy could satisfy the global consumption of energy for thousands of years. However, the challenge is to find a way to access the energy economically, because it mostly lies in depths of mile (1.6 km) or more below the surface.

To reach this energy source, a geothermal loop is made. See **Figure 10-34.** Two wells are drilled from the surface: an injection well and a production well. An explosive charge is placed at the bottom of the holes. The resulting fracture joins the holes. Cold water is pumped down the injection well. The hot water is forced to the surface through the production well. There, the mixture of steam and hot water powers a turbine generator to produce electricity. Next the steam is condensed by evaporation in the cooling tower. It is then pumped back down the injection well because it contains a high mineral content and cannot be allowed to pollute the surface.

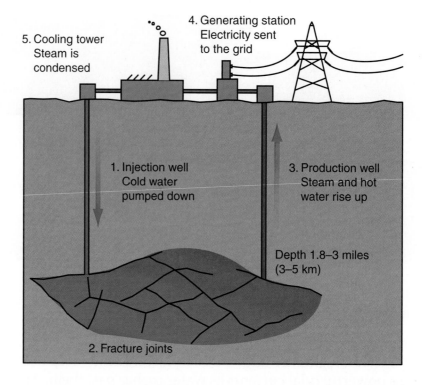

Figure 10-34. To produce geothermal energy, cold water is pumped down through one well to be heated by superheated rocks. Steam rises up a second well to the power plant.

5. Cooling tower
Steam is condensed

4. Generating station
Electricity sent to the grid

1. Injection well
Cold water pumped down

3. Production well
Steam and hot water rise up

Depth 1.8–3 miles (3–5 km)

2. Fracture joints

Other Renewable Sources of Energy

Other renewable sources of energy are being developed in different parts of the world. The most important are:

- Energy from plants
- Energy from decomposing matter
- Hydrogen

Together, these sources produce only a very small part of our energy needs.

Energy from Plants

Energy from plants is called *biomass energy*. Wood has been used as a fuel for thousands of years. It is still the most commonly used fuel in the developing world, where four out of five families depend on it as their main energy source. Some fast-growing plants can be burned as fuels, although currently they provide only 3 percent of the energy used in the United States.

When trees are harvested in North America, about 50% of the tree is converted into lumber or pulpwood. The remaining 50%, mainly branches, twigs, and bark, is often discarded, but can be used as fuel or made into other products. Many pulp and paper mills use wood-fired generators to make their own electricity. See **Figure 10-35**.

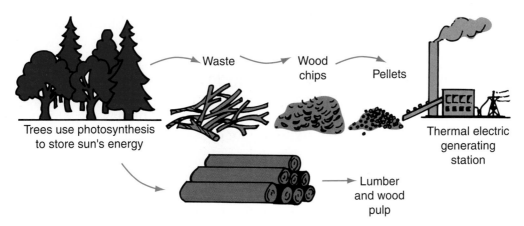

Figure 10-35. Waste wood products can provide energy for producing electricity. Wood is a biomass source of energy.

Biomass fuels, also known as *biofuels*, may also be in liquid form. Methanol and ethanol are alcohol fuels produced from biomass matter, such as corn, beets, sugar cane, wheat, and wood wastes. They are used as additives in gasoline and diesel fuel. Sugar produced from sugar cane can also be fermented to make alcohol, which can be mixed with gasoline to fuel vehicles. See **Figure 10-36**. In Brazil, 40 percent of the fuel used by cars is produced from sugar cane waste.

Because they are made from plants, biofuels are considered renewable. Unlike petroleum, the raw material for these fuels can be regenerated again and again. However, biofuels produce less heat than petroleum. Therefore, they are generally not used by themselves but are added in small amounts to gasoline.

Energy from Decomposing Matter

Another type of biofuel is decomposing matter. On farms, manure can be collected. Farms also have plant wastes. Pasture plants that have not been eaten, leftover feedstock, fruits, vegetables, and grains that are damaged or unsold are also sources for this type of energy.

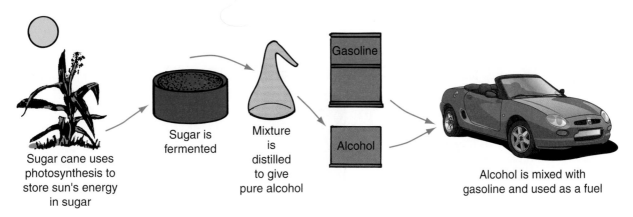

Figure 10-36. Sugar cane can be processed from cane and then used to produce alcohol. Blends of alcohol and gasoline are used to fuel automobiles.

When manure and organic wastes are put into closed tanks, bacteria will digest them. This produces methane gas, which scientists call *biogas*. Methane gas can be used for cooking, lighting, and running engines. It is a common method of producing energy in many parts of the world. In China, more than 24 million biogas digesters are in use, mostly in villages. Biogas is particularly easy to make on farms. A small biogas digester can produce enough gas for a family to use for cooking, heating, and lighting. See **Figure 10-37**.

Hydrogen

The word *hydrogen* comes from the Greek word meaning "water generator." The element was given this name because water is produced when hydrogen burns in the presence of oxygen. One of the first successful uses of hydrogen fuel was in the Saturn V rocket that took men to the moon. The space shuttle had a huge external tank filled with liquid hydrogen and liquid oxygen. This fuel not only lifted the shuttle into orbit but also produced the electricity needed during a mission.

The power to make the two elements combine comes from a fuel cell. A *fuel cell* is a device that allows hydrogen and oxygen to combine, without combustion, to generate electricity. A reaction occurs when electrons are released from the hydrogen and travel to the oxygen through an external circuit. As electrons travel through the circuit, they generate a current that can power electrical devices. When hydrogen fuel is used on space missions, the reaction also benefits astronauts in another way. The only by-product is water, which is the water that astronauts drink.

If fuel cells produce energy without any toxic by-products, why don't we use them in other vehicles? The main reason is cost. Fuel cells use platinum, a very expensive metal. Currently hydrogen, as a fuel, cannot compete economically with petroleum. Other issues also remain. The hydrogen fuel tank takes up most of the trunk space. Also, there are currently very few hydrogen fueling stations across North America. In addition, temperatures below freezing present a major problem when the by-product of the motor is water.

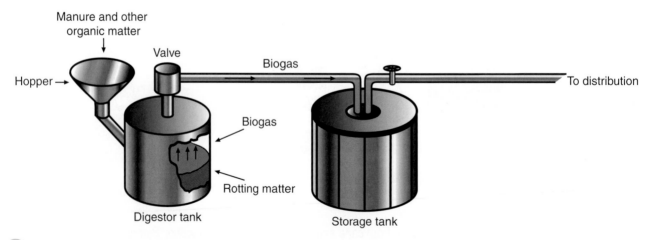

Figure 10-37. A biogas digester uses waste organic material to produce methane gas.

Environmentalists point out that, in order to be pollution-free, hydrogen must be made using renewable energy. If the hydrogen is manufactured by using fossil fuels, we are only shifting pollution from vehicle tailpipes to hydrogen production plants. We will have done nothing to reduce air pollution and greenhouse gases, if this happens. However, experimental vehicles are now on the road. Both Chicago and Vancouver have buses powered by hydrogen fuel cells.

Energy from Garbage: A Problem

Many people would like to believe that waste disappears when it is burned or that it can provide a nonpolluting source of renewable energy. Waste that is burned in a furnace or boiler generates heat, steam, and electricity, but the burnt waste does not disappear. It is transformed into ashes and gas. When this happens, chemical reactions in the atmosphere lead to the formation of new compounds. Some of these compounds are extremely toxic and carcinogenic (cancer-causing). These compounds include dioxins, acidic gases, particulates, and heavy metals.

Think **Green**

Renewable vs. Green Energy

Many people think that if a source of energy is renewable, it is automatically "green." Is this really true? As you can see in the "Energy from Garbage: A Problem" section, it is not.

Renewable energy sources *can* be green, or environmentally friendly. Solar energy, for example, is a green energy source unless the technology used to capture it contributes significantly to pollution or hazardous waste production. Currently, most solar energy systems are good sources of green energy.

Think about the other renewable sources of energy described in this chapter. What overall effects does each one have on the environment? Remember to include all elements of the environment: air, water, land, and human, wildlife, and plant populations. How environmentally friendly is each source of renewable energy?

Energy and Power Systems

Like all other technological systems, energy and power systems include inputs, processes, outputs, and feedback. Understanding these individual components is critical to understanding the system as a whole. (To refresh your memory on systems, refer to **Figure 7-29**.)

Energy and power systems have many types of inputs. Wind is an input for windmills. The sun is an input for solar power systems. Water falling by gravity is an input for waterwheels and turbines within dams. Other inputs include the people who develop, operate, and maintain the systems and the materials and machines that compose the systems.

Processes within energy and power systems are typically conversion processes. The conversion types include:

- Mechanical-to-electrical (electrical generating stations)
- Electrical-to-mechanical (electric motors)
- Chemical-to-mechanical (internal combustion engines)
- Chemical-to-electrical (batteries and fuel cells)

Other processes include the management and operation of the system.

Outputs of energy and power systems are typically the outputs of the conversion process. Power systems are often used to power other technological systems. For example, they can include the wheel rotation on an automobile, the thrust of a jet engine, the distribution of electricity to homes and businesses, the light from a flashlight, or the heating of water in a water heater. There are also societal outputs, such as the convenience of riding lawn mowers or readily available electrical power. Environmental outputs include the heat, smoke, and carbon dioxide produced by burning fossil fuels.

Feedback in energy and power systems is generally provided by system monitoring. This may include measuring and observing the speed of a motor, the amount of current flowing through an electrical wire, or the temperature of water in an aquarium.

Sources of energy can be divided into two groups: renewable and nonrenewable. Nonrenewable energy—coal, oil, and natural gas—cannot be replaced. Renewable energy—sun, wind, and water—will always be available.

Currently, most cars and trucks use nonrenewable petroleum products to power their engines. Research is underway to find a suitable replacement. Whatever fuel is used in the future should be a clean, sustainable energy source that will prevent climate change, air pollution, and further damage to the environment.

The technology exists to move toward a life without dependence on oil. The answer may be to think small, rather than big. Individuals should be encouraged to generate some of their own solar or wind power. In Germany, fifty percent of the owners of windmills are small farmers. If they generate more than is needed for their own use, they can return their surplus power to the power grid and receive credit for these surpluses. This system is called *net-metering*.

Chapter 10 Energy **331**

- The two basic categories of energy are potential energy and kinetic energy.
- Energy takes many different forms: chemical, gravitational, strain, electrical, heat, light, sound, and nuclear.
- Energy can be neither created nor destroyed. However, during energy conversions, some of the energy may be converted into an unwanted form.
- Some sources of energy are nonrenewable. When we use them up, they will no longer exist. These sources include coal, petroleum, natural gas, and nuclear energy.
- Some sources of energy are renewable and can be replaced as needed. These sources include energy from the sun, wind, water, and heat from deep within the Earth.
- The inputs into energy and power systems vary depending on the type of energy. Processes are usually conversion processes, and outputs are the results of the conversion processes. Feedback is provided by system monitoring.

Summarizing Information

Copy the following graphic organizer onto a separate sheet of paper. For each chapter section (topic) listed in the left column, write a short, one-paragraph summary of the topic in the right column. Do not write in this book.

Chapter Section (Topic)	Summary
Energy Basics	
Forms of Energy	
Energy Conversion	
Where Does Energy Come From?	
Nonrenewable Sources of Energy	
Renewable Sources of Energy	
Energy and Power Systems	

Test Your Knowledge

Write your answers to these review questions on a separate sheet of paper.

1. What category of energy does a wound spring have?

2. What category of energy does a falling boulder have?

3. List the nine forms of energy and give one example of each.

4. Give three examples to show that energy can neither be created nor destroyed.

5. List three examples of nonrenewable energy sources and three examples of renewable energy sources.

6. What are the major problems in using nuclear energy to produce electricity?

7. Describe two methods of collecting energy from the sun.

8. What is the major disadvantage of wind turbines as a source of electrical energy?

9. Describe how geothermal energy is used to produce electricity.

10. Describe the general inputs, processes, and outputs of a nuclear power system. What feedback mechanisms are needed?

Critical Thinking

1. Explain this statement: "Not all sources of clean energy are sustainable, and not all sustainable sources of energy are clean."

2. Nuclear energy is currently considered a nonrenewable source of energy. How might this type of energy be made renewable?

3. Think about society and the general attitude in the United States about conserving energy. How has this attitude changed during your lifetime? How might this affect the adoption or rejection of new technologies related to energy and fuel development?

Apply Your Knowledge

1. Think back to a time when a power failure occurred in your area. List the devices you were unable to use. Imagine that the power stayed off for 48 hours. What alternative sources of energy could you use?

2. List five devices in a typical home that use energy. Describe the energy change(s) that take place when each is used.

3. List the ways in which you use energy each day around the house, in traveling, at school, and for leisure. State whether the energy comes from a renewable or nonrenewable source.

4. Research the Internet, your library, or your public service utility to find ways to save energy. Create a poster that illustrates conservation ideas relating to your yard or garden, the vehicles your friends or parents drive, the types of food you buy, or the amount of recycling you do.

5. Research the number of households in the United States that had solar water heaters in the years 1990, 1995, 2000, 2005, and 2010. Also find out the average initial cost of a midsize solar water heater in each of those years. Plot your findings on a graph and study the result. Can you see a trend in either or both of these factors? What can you conclude from these trends, if anything? What effect might these trends have on future development of solar water systems? What does your data reveal about the acceptance and popularity of the use of solar energy in the home?

6. Research one career related to the information you have studied in this chapter. Create a report that states the following:

   - The occupation you selected
   - The education requirements to enter this occupation
   - The possibilities for promotion to a higher level
   - What someone with this career does on a daily basis
   - The earning potential for someone with this career

   You might find this information on the Internet or in your library. If possible, interview a person who already works in this field to answer the five points. Finally, state why you might or might not be interested in pursuing this occupation when you finish school.

STEM Applications

1. **ENGINEERING** Using popsicle sticks, rubber bands, or other commonly available items, design and make a working model of a wind turbine. Try to find a way to connect the turbine to a simple shaft in such a way that when wind turns the turbine, the shaft moves. Shaft movement can be either reciprocal (back and forth) or rotary. Test your model by taking it outside or by using a fan set on low speed. Keep in mind that the turbine must be strong enough to withstand the force of the wind.

2. **SCIENCE** When you place a bowl of water in direct sunlight, solar energy helps evaporate the water. This fact can be used to design a water purifier. For example, you could remove the salt from ocean water to make drinking water. Design a solar water purifier that removes the salt from a solution of saltwater. Use materials that are easily available. Test your design and document your results. Include your design in your portfolio.

3. **MATH** The Smith family is considering installing a new solar water heater. The new water heater will replace an older electric model. Before they can decide whether the solar water heater is a good investment, they want to find out how much money it will save them. In addition to the difference in actual operating costs, they want to know how long it will take them to achieve a return on their investment (ROI). Given the information below, calculate the ROI for the Smiths' new solar water heater. Show your work.

- Estimated annual cost to run the current electric water heater: $187.50
- Estimated annual cost to run the solar water heater: $148.64
- Startup costs for the solar water heater, including installation: $2,340.00

Several ROI calculators are also available on the Internet. Find one such site and input the same information you used in your calculations. Does the result match your calculation? If not, why?

The photos on this page show both old windmills, which produce mechanical power, and new wind turbines, which produce electrical power. Which pictures show windmills? Which show turbines? How is (or was) the power of each used?

Electricity and Magnetism

Better by Design

Christopher Horner designed the Solio™ solar charger

Using a solar panel allows you to generate your own free, clean electricity directly from the sun.

Christopher Horner and his design team want you to capture the sun's energy and convert it into electricity to power your portable electronics. When opened, a Solio hybrid charger captures light from the sun. When sunlight hits the solar cells, it creates an electric current. This current can either be used directly or stored in its battery. The stronger the sunlight, the more electricity is generated. According to Chris, using a solar-powered Solio is 100 times better for the environment than using a wall charger.

Solar energy is ideal for use when no electrical outlets are nearby, but it can also be used in other places. Where could you use a solar panel to support your daily activities and help the environment?

"We must make renewable energy tangible to the consumer who wants power on the go."

Reading Target

Key Terms

Preview and Prediction

Before you read this chapter, glance through it and read only the heads of each section. Based on this information, try to guess, or predict, what the chapter is about. Use the Reading Target graphic organizer at the end of the chapter to record your predictions.

alternating current (AC)
amperage
anode
battery
cathode

cell
circuit
commutator
conductor
cycle
direct current (DC)
distribution lines
dry cell
electric current
electrode
electrolyte
electromagnetism
electromotive force (EMF)
electron theory

energy density
frequency
generating station
generator
magnetism
primary cell
rotor
secondary cell
stator
step-down transformer
step-up transformer
transmission lines
voltage
voltaic cell

Objectives

After reading this chapter, you will be able to:
- Identify the different ways in which electricity can be produced.
- Describe the transmission and distribution of electricity.
- Explain the nature of electricity by referring to the movement of electrons.
- State the laws of magnetism.
- Explain the concept of electromagnetism.
- Describe how an electric motor operates.
- Explain the difference between cells and batteries.

Useful Web sites:
www.solio.com/
www.clean-energy-ideas.com/solar_panels.html
www.solarpanelinfo.com/

Imagine your town or city without electricity. It can happen. At approximately 4:15 in the afternoon in August 2003, the lives of 50 million people were suddenly interrupted. The power outage included 40 million people in eight states (New York, New Jersey, Vermont, Michigan, Ohio, Pennsylvania, Connecticut, and Massachusetts) and 10 million people in the Canadian province of Ontario. During the outage, 265 power plants shut down, including 22 nuclear power plants.

- Water systems in several cities lost pressure, forcing four million customers in the Detroit area to boil their water for four days.
- Passenger screening at affected airports was stopped, forcing the airports to shut down.
- Some television stations were knocked out of service.
- In New York City, 40,000 police and the entire fire department were on duty to maintain order.
- Commuters were stranded in New York City. Some found lodging, but many were forced to sleep in parks.
- A large number of factories closed in the affected area. Others outside the area were forced to close because of supply problems.
- Buildings became hot because air conditioning no longer functioned.
- Most of the interstate rail transport in the northeast corridor was interrupted.

In February 2004, a task force released a report explaining the main cause of the blackout: the failure to trim trees in parts of Ohio. High-voltage power lines shorted when they came into contact with overgrown trees. A domino effect resulted in the eventual shutdown of more than 100 power plants.

In our daily lives, we take electricity for granted. To most people in the United States, it is merely something that is always available. Only when it is gone do we realize how we depend on it. See **Figure 11-1**.

Figure 11-1. We depend on electrical energy more than we realize. How would a blackout this evening affect you?

Generating Electricity

The electrical energy supplied to your home comes from a generating station. A *generating station* uses energy from a source of power to turn turbines, which produce, or generate, electricity. The two principal types of generating stations are hydroelectric and thermal-electric.

Hydroelectric Generating Stations

Hydro is another word for water. Hydroelectric generating stations use the energy of flowing or falling water. The station is located at a waterfall or dam, as shown in **Figure 11-2**. As the water drops to a lower level, its mass spins a turbine. See **Figure 11-3**. A turbine is a finned wheel. When the falling water strikes the fins, the turbine turns rapidly. The turbines are connected to generators. A *generator* is a device that produces an electric current.

Thermal-Electric Generating Stations

Thermal-electric generating stations use steam to drive turbines. A heat source produces the steam. The steam is directed onto the blades of a turbine, spinning it rapidly. As in hydroelectric systems, the turbines drive generators to produce electricity.

Figure 11-2. A dam stores water and provides a powerful flow to drive turbines.

Figure 11-3. The turbine room of a hydroelectric generating station.

Heat for powering thermal-electric turbines comes from one of two sources: fossil fuels or nuclear fission. Refer to Chapter 10 for more information about these energy sources.

The process of burning fossil fuels to generate electricity is shown in **Figure 11-4.** In a nuclear station, the nuclear reactor does the same job as the furnace in fossil-fuel stations. See **Figure 11-5.** However, instead of combustion, the energy source used to heat the fluid is nuclear fission. Recall that nuclear fission is the process of splitting atoms. Atoms contain energy that binds them together. When the nucleus of an atom is split, some of the energy that was used to bind the atom together is released as heat. This heat is pumped to the heat exchanger, where it turns water to steam. The steam drives turbines connected to electrical generators. A thick concrete shield stops harmful radioactive substances from escaping.

Hydroelectric generating stations change the potential energy of water behind a dam. As it falls into the turbine, it becomes kinetic energy. Thermal-electric generating stations convert the energy stored in fossil fuels and uranium into kinetic energy. In both cases, the kinetic energy is converted to electrical energy by generators. See **Figure 11-6.**

Conventional Fossil Fuel Power Plant

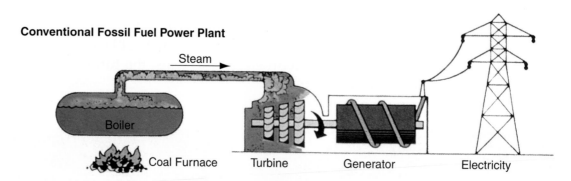

Figure 11-4. The parts of a coal-fired generating system.

Nuclear Power Plant

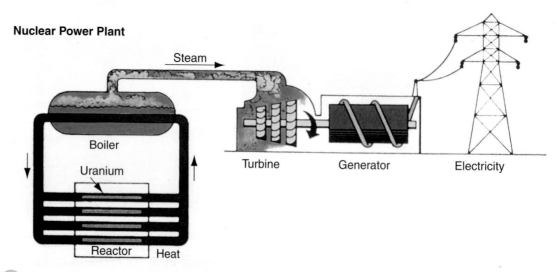

Figure 11-5. Splitting atoms, rather than burning coal, provides the heat in a nuclear power station.

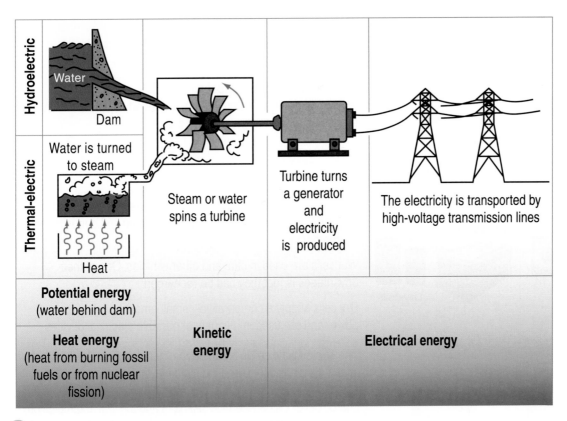

Figure 11-6. All generating stations are energy converters.

Most of the electricity used in homes and factories is produced either in hydroelectric or thermal-electric generating stations. There are, however, other methods. Friction, chemical action, light, heat, and pressure can also be used to generate electricity, but in much smaller amounts. See **Figure 11-7**.

Think Green

Green Power

Many power companies in the United States have begun offering a "green electricity" or "green power" option. This electricity is generated using "green" sources such as wind, hydroelectric, photovoltaic cells, landfill gas, or biogas.

In many areas, you can sign up to receive green power by calling your local utility company. The utility company charges a premium for the service. However, be aware that some systems are greener than others. For example, burning landfill gas produces greenhouse gases.

Therefore, before paying more for green electricity, find out exactly how the electricity is produced. Visit the Green Power Network Web site of the U.S. Department of Energy (eere.energy.gov/greenpower). This Web site contains links that show exactly what options are available in your area, along with the source of electricity for each.

Method	Application	Discussion
 Friction	 Person pulling off a sweater	Friction causes static electricity. After walking across a carpet on a dry day, you become electrically charged. If you touch a grounded object, the static electricity will discharge, creating a spark.
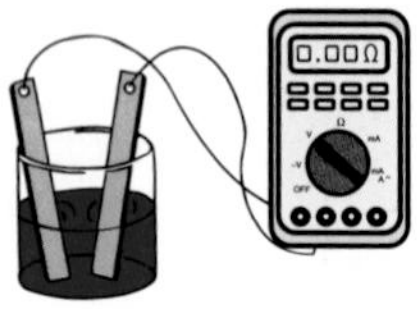 Chemical	Wet-cell battery 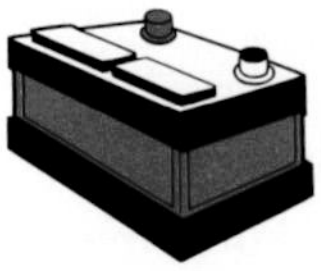  / Dry cells	An acid or salt solution, called an *electrolyte*, removes electrons by chemical action from one piece of material and deposits them on another. Wet cells are used in cars and other vehicles. One of their advantages is that they can be recharged. Dry cells supply a comparatively small amount of electrical power and are used in a variety of portable electrical devices.
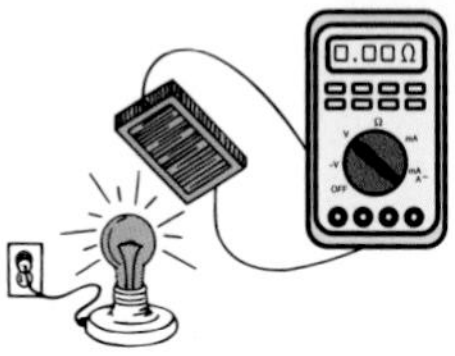 Light	 Solar powered calculator	The photovoltaic cell is a sandwich of three layers. The outside layers are translucent, and the inside layer is iron with a disk of selenium alloy. When light is focused on the selenium, an electric charge develops between the selenium and the iron. Examples of use are automatic headlight dimmers and portable solar-powered calculators. A second way of using light to produce electricity is photoconduction. A common application of this principle is the control of street lights that come on automatically when daylight fades. Light energy applied to a material that is normally a poor conductor causes free electrons to be released in the material so it becomes a better conductor.
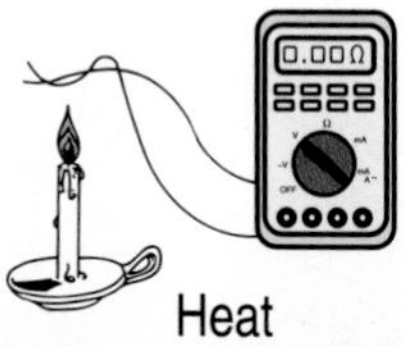 Heat	 Thermocouple	A small electric charge is generated if the ends of two wires are twisted together and heated. This is the principle of a thermocouple. Commercial thermocouples use unlike metals welded together. They do not supply a large amount of current and cannot be used to produce electric power. They are used as temperature sensors.
Pressure 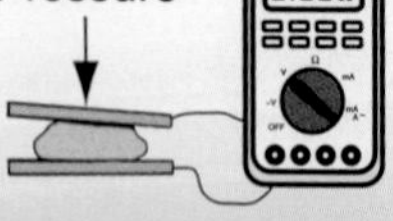	 Barbecue lighter	A small electric charge is generated if quartz is placed between two metal plates while pressure is applied. One application is a lighter of the type used for lighting gas grills.
Magnetism 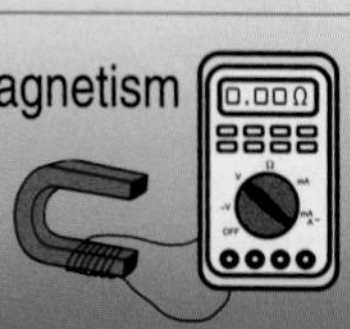	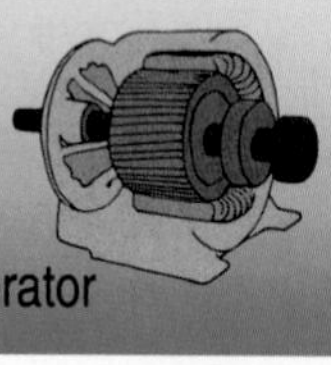 Generator	A generator uses magnetism to produce electricity. In an electric power generating station, generators are run by turbines.

Figure 11-7. Six ways to produce electricity.

Transmission and Distribution of Electricity

Generating stations are rarely found close to where the electrical energy is used. The electricity that comes to your home may have traveled a great distance.

After leaving the generating station, electricity is fed into a network of *transmission lines* and *distribution lines*, as shown in **Figure 11-8**.

Energy Conversions

Transmission lines, **Figure 11-9**, resist the flow of electrical energy, so some of the energy tends to be lost along the way. *Voltage* is a measure of electrical pressure, and *amperage* is a measure of the amount of current. Increasing voltage and reducing amperage greatly reduces the loss of energy from transmission lines. Therefore, when electricity is sent a long distance, it is more efficient and safer to send it as low-current, high-voltage electricity. A *step-up transformer* is used to increase the voltage for transmission. Each transformer consists of a pair of coils and a core.

Neither your home nor a factory can use electricity at the high voltage carried by transmission lines. It would destroy the wiring, appliances, and machines. The voltage must be reduced before current enters the distribution lines. *Step-down transformers* are used to reduce the voltage. The first drop in voltage occurs when electrical energy is transferred to

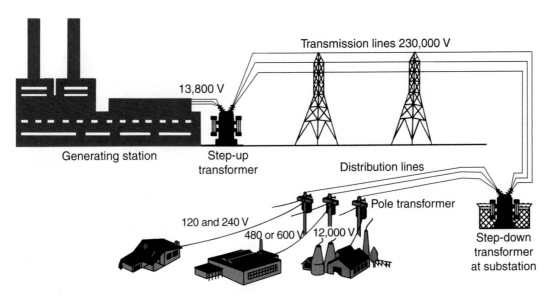

Figure 11-8. An electrical power transmission and distribution system.

Figure 11-9. Transmission lines are carried by towers. Insulators support the lines where they attach to the tower.

distribution lines. See **Figure 11-10**. Another step-down transformer reduces the voltage further when electrical energy is transferred from distribution lines to service lines. This transformer may be located on a pole or on the ground. See **Figure 11-11**. Electricity enters your home through a service line. It also passes through a meter and a main switch, as shown in **Figure 11-12**.

Figure 11-10. Distribution substations step down the voltage.

Figure 11-11. A service transformer further reduces the voltage in the distribution lines.

Figure 11-12. Electricity enters your home through a service line, meter, and main switch.

How Transformers Work

Basically, a transformer consists of two separated coils of wire around a magnetic core. **Figure 11-13** shows the construction of step-up and step-down transformers. One coil has more turns than the other. When the input voltage is connected to the coil with the least number of turns, the output voltage is increased (step-up). When the input voltage is connected to the coil with the greatest number of turns, the output voltage is decreased (step-down).

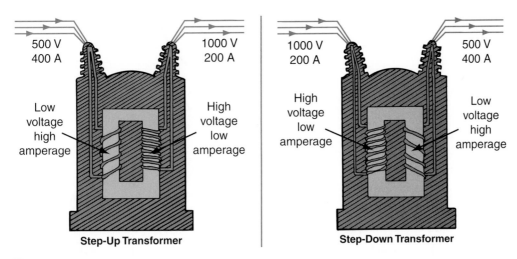

Figure 11-13. The number of turns on the coils in a transformer determine the amount of increase or decrease in the output voltage.

What Is Electricity?

When you flip a light switch, electricity flows through wires and lights a bulb. What exactly is it that flows through the wire when the switch is turned on? There is no perfect answer. A scientist would say that electric current consists of a flow of electrons. What does this mean?

One explanation is called the *electron theory*. As described in Chapter 10, everything around us is made from atoms. Atoms are made of even smaller particles called protons, electrons, and neutrons. The protons have a positive charge, and the electrons have a negative charge. Neutrons have no charge and play no role in electricity.

An atom normally has the same number of electrons as protons. The negative and positive charges cancel each other. Such atoms are electrically neutral. See **Figure 11-14**.

Metal wires conduct electricity because the outer electrons in each atom are not tightly bound. They can move from atom to atom. Metals "conduct" electricity as these electrons move. When a metal wire is connected in a circuit, the free electrons can all be pushed in the same direction. This flow of electrons is called an *electric current*. See **Figure 11-15**.

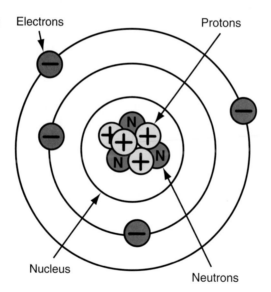

Figure 11-14. Protons and neutrons make up the nucleus of the atom. Electrons orbit around the nucleus.

Electrons   Protons

Nucleus   Neutrons

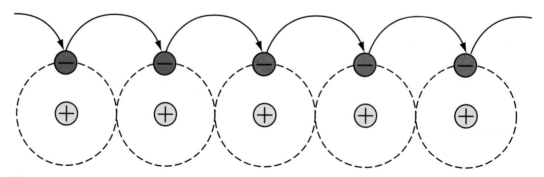

Figure 11-15. The movement of electrons occurs when a force is applied to one end of a wire. The free electrons move from one atom to the next, resulting in electric current.

A *conductor* is a material that allows an electric current to flow easily. Suppose that electrons are made to flow from one end of a metal wire to the other. Electrons can move through the wire, which acts as a conductor. The end that loses electrons becomes positively charged—the positive terminal. The end that gains electrons becomes negatively charged—the negative terminal. See **Figure 11-16**.

Why are electrons able to move through a wire? There are two reasons:

- A force pushes them along a path.
- There is a closed path, called a *circuit*, in which they can move.

In a circuit, an *electromotive force (EMF)* pushes electrons through a conductor. We also call this force *voltage*. The two most common sources of EMF are generators and chemical reactions. Generators use magnetism and mechanical energy to produce electricity. Dry cells and batteries use chemical reactions to produce electricity. These two sources provide most of the electricity that we use. Other sources of EMF are friction, light, heat, and pressure. (Look back at **Figure 11-7**.)

Magnets and Magnetism

The production of electricity depends on magnets and magnetism. *Magnetism* is the ability of a material to attract pieces of magnetic materials, such as iron or steel. Magnets do not attract nonmagnetic materials, such as aluminum, copper, glass, paper, and wood. See **Figure 11-17**. Magnets fall into three different groups: natural magnets, artificial magnets, and electromagnets.

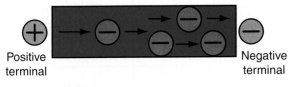

Figure 11-16. The terminal with a surplus of electrons is negative. The terminal with a shortage of electrons is positive.

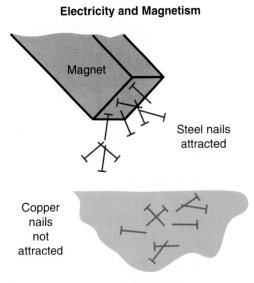

Figure 11-17. Magnets attract iron and steel, but not wood or other nonmagnetic items.

Math Application

Calculating the Cost of Electricity

Reading an electric bill is an excellent way to understand how much electricity your household uses each month. Electric utility companies usually have a key on the bill that helps customers understand what each entry means.

Normally, a bill has at least two readings: the previous reading and a current usage reading. These amounts have been read from the meter by an electric company employee. The readings are recorded in kilowatt-hours (kWh). The company bills the customer for the difference in the two readings at the rates marked on the bill.

An electric bill provides enough information to perform several calculations. To find the average kilowatt-hours used per day in any month, divide the number of kilowatt-hours used by the number of days. To find the cost per day, divide the amount of the bill by the number of days included in the bill. To find the cost per kilowatt-hour, divide the total amount of the bill by the number of kilowatt-hours.

Math Activity

The following information is from one family's bills for one year's supply of electricity. Use the information to practice calculating the cost of electricity by answering the questions below.

From	To	Days	kWh	$ Amount
08-07-2009	09-04-2009	28	360	34.22
09-05-2009	10-06-2009	31	420	35.17
10-07-2009	11-11-2009	35	680	56.91
11-12-2009	12-10-2009	28	710	59.18
12-11-2009	01-12-2010	32	990	82.04
01-13-2010	02-11-2010	29	920	77.58
02-12-2010	03-15-2010	31	1040	87.88
03-16-2010	04-13-2010	28	830	75.90
04-14-2010	05-11-2010	27	500	47.35
05-12-2010	06-09-2010	28	430	39.78
06-10-2010	07-07-2010	27	320	30.19
07-08-2010	08-06-2010	29	310	30.95

1. Find the average number of kilowatt-hours used per day from:
   (a) September 5 to October 6; (b) January 13 to February 11.
   Carry out your calculations to four decimal places.

2. Find the daily cost of electricity from December 11 to January 12.

3. Find the cost per kilowatt-hour from: (a) November 12 to December 10; (b) July 8 to August 6. Round the cost to the nearest one-hundredth (to the nearest penny).

4. Calculate the total number of kilowatt-hours used for the entire year and the total cost.

Natural Magnets

Natural magnets, such as lodestone, occur in nature. Lodestone is a blackish iron ore (magnetite). Its weak magnetic force varies greatly from stone to stone.

Artificial Magnets

Artificial magnets, also called *permanent magnets*, are made of hard and brittle alloys. Iron, nickel, cobalt, and other metals make up the alloys. The alloys are strongly magnetized during the manufacturing process.

Artificial magnets come in many shapes and sizes. The most common are horseshoe magnets, bar magnets, and magnets used in compasses. See **Figure 11-18**.

Natural and artificial magnets retain their magnetism indefinitely. A bike computer is a simple example of the use of magnets. The sensor attached to the bike's front fork can detect each time a magnet attached to a front wheel spoke passes the sensor. The computer uses this information to calculate the number of times the wheel turns. From this, it calculates the cyclist's speed, distance traveled, average speed, and fastest speed.

Electromagnets

Electromagnets are magnetized by an electric current. They consist of two main parts. One is a core of special steel. The other is a copper wire coil wound on the steel core. See **Figure 11-19**. Unlike permanent magnets, electromagnets can be turned on or off. Their magnetic force can be completely controlled.

For example, metal detectors work by creating magnetic fields. When a metal detector's electromagnetic field encounters a metal object, such as a lost coin, its own magnetic field generates a smaller magnetic field around the object. When this happens, it makes a sound to alert the user.

A   **B**

Figure 11-18. Three common types of artificial magnets. Which of these magnets is (a) a horseshoe magnet, (b) a bar magnet, and (c) a magnetic compass needle?

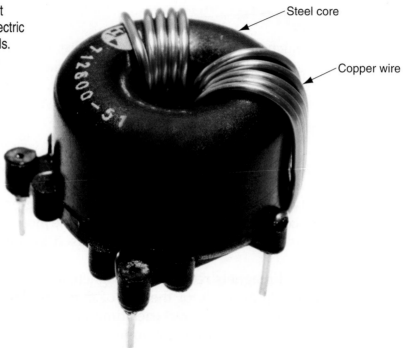

Figure 11-19. An electromagnet has a magnetic force only when electric current is sent through the wire coils.

Steel core

Copper wire

How Magnets Act

Figure 11-20 shows a bar magnet suspended from a loop of thread. Held this way, the magnet twists until it is lined up in a north-south direction. The end that points toward the north is called the north-seeking pole. The end that points toward the south is called the south-seeking pole.

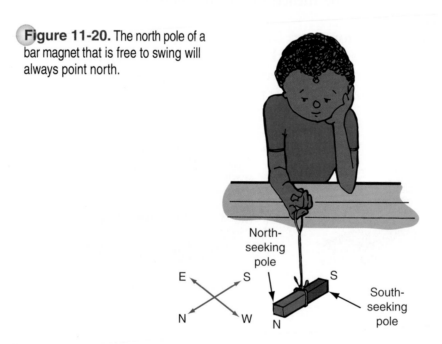

Figure 11-20. The north pole of a bar magnet that is free to swing will always point north.

North-seeking pole

South-seeking pole

If two magnets are suspended so that the north pole of one is close to the south pole of the other, the poles attract one another. See **Figure 11-21.** This is the first law of magnetism. It states that unlike magnetic poles attract each other.

If the north poles come close to one another, the magnets push away from each other. This is the second law of magnetism. It states that like magnetic poles repel one another. See **Figure 11-22.**

Lines of Force

Invisible lines of force surround a magnet. Although you cannot see them, you can prove they exist. Place a sheet of paper over a magnet and sprinkle iron filings on the paper. When you tap the paper gently, the small iron particles form a distinct pattern. See **Figure 11-23.** The lines of force shown by the iron filings take the shape shown in **Figure 11-24.**

Now place the two magnets end to end and repeat the experiment with iron filings. The lines of force demonstrate the two laws of magnetism. See **Figure 11-25.**

Magnetism and Electric Current

An electric current passed through a wire creates a magnetic field around the wire. Magnetism produced by this means is called *electromagnetism.* This principle is used to make the electromagnet in **Figure 11-26.**

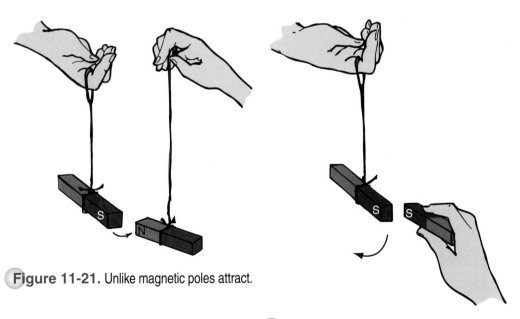

Figure 11-21. Unlike magnetic poles attract.

Figure 11-22. Like magnetic poles repel each other.

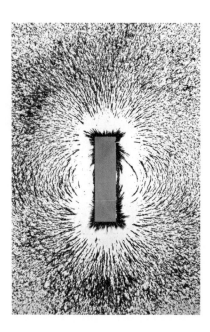

Figure 11-23. Iron filings show the lines of force around a magnet.

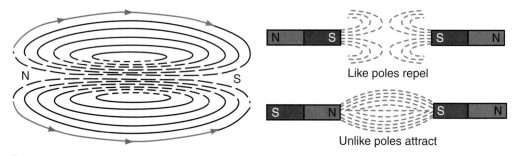

Figure 11-24. Note the pattern and direction of the lines of force.

Like poles repel

Unlike poles attract

Figure 11-25. Try this experiment with two magnets and iron filings.

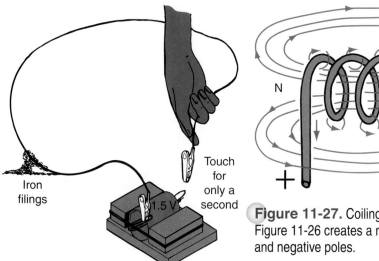

Iron filings

1.5 V

Touch for only a second

Figure 11-26. To produce a magnetic field, connect a wire to a dry-cell.

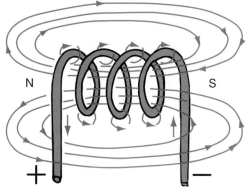

N   S

+   –

Figure 11-27. Coiling the wire shown in Figure 11-26 creates a magnet with positive and negative poles.

If the wire in **Figure 11-26** is wound to form a coil, it becomes a magnet, as shown in **Figure 11-27**. The magnetic strength of this coil depends on the strength of the current and the number of loops in the coil. By changing these factors, you can control the magnetic strength of the coil.

If a wire is wound around a core of magnetic material, such as a soft iron nail, the nail becomes an electromagnet. It remains strongly magnetic only as long as there is current in the wire. See **Figure 11-28**.

WARNING: The demonstrations shown in **Figure 11-26** and **Figure 11-28** should only be done by your teacher. A carbon-zinc cell should be used. **NEVER** use an alkaline cell. It may explode.

Generators

A generator is the most practical and economical method today of producing electricity on a large scale. It uses magnetism to cause electrons to flow.

To see this in action, connect a length of copper wire to a milliammeter. As shown in **Figure 11-29**, move part of the wire loop through a magnetic field. A small current flows while the wire is cutting across the magnetic field.

The strength of the current depends on two things. One is the strength of the magnetic field. The other is the rate at which the lines of force are cut. The stronger the magnetic field or the faster the rate at which the lines of force are cut, the greater the current.

The direction of electron flow depends on the direction in which the lines of force are cut. Look at **Figure 11-29** again. When the wire moves down through the lines of force of the magnet, electrons flow in one direction. When the wire moves up, electrons flow in the opposite direction. The end that loses electrons becomes positively charged. The end that gains electrons becomes negatively charged.

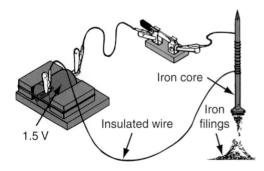

Figure 11-28. In this electromagnet, the nail remains magnetized as long as current flows through the coil.

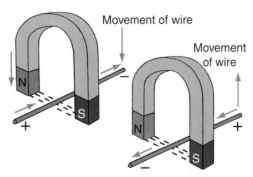

Figure 11-29. Moving a wire through a magnet creates a small current in the wire.

Alternating Current

Alternating current (AC) is electron flow that reverses direction on a regular basis. It is the type of current you use in your home. It is the type of current produced by power stations. How is it produced? This will become clear as the basic operation of a generator is explained.

Figure 11-30 shows a basic generator. It is no more than a loop of wire turning clockwise between the poles of a magnet. Remember that current is produced only when a wire cuts through lines of magnetic force.

With the loop (wire) in position A, no lines of force are cut. The generator produces no current. As the loop continues turning, it reaches position B. At this point, one side of the loop moves downward through the lines of force. At the same time, the other side of the loop is moving up through the lines of force. Because the wire is a closed loop, current travels through it in one direction.

As the loop reaches position C, half a revolution is completed. As in A, there is no current. Why? No lines of force are being cut.

The loop continues to turn until it reaches position D. The two sides once more cut lines of force. There is a difference, however. The side that moved downward before is now moving upward. Likewise, the side that moved upward before is now moving downward. What happens? The electron flow reverses. Because the direction of flow alternates as the loop turns, the current produced is called *alternating current*.

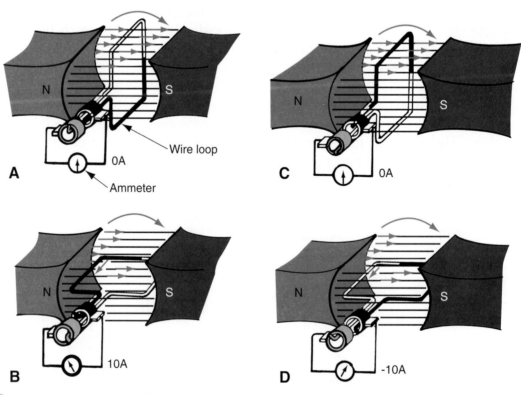

Figure 11-30. A basic AC generator.

Electricity produced by the generator must have a path, or circuit, along which it can flow. Therefore, the terminals (ends) of the loop must always be in contact with an outside wire. This outside wire is stationary. The contact is made with slip rings and brushes.

A separate slip ring is permanently fastened to each terminal of the wire loop. Each slip ring turns with the loop. A brush is placed against each slip ring. As the slip rings turn, the brushes maintain rubbing contact with them. The wire forming the stationary part of the circuit is attached to the brushes. Electrical devices, such as a light bulb, are connected to the external part of the circuit. See **Figure 11-31**.

Current produced in the loop of the generator flows through a slip ring and brush into the external circuit. It travels through the electrical device. Then it returns to the generator through the other brush and slip ring.

Generators at large generating stations are more complex than the loop generator shown in this chapter. However, their basic principle is the same. Generated current can be increased in the following two ways:

- Increasing the rate at which the lines of force are cut.
- Strengthening the magnetic field.

Therefore, many loops of wire are used instead of one. Powerful electromagnets supply the magnetic field.

For practical reasons, the loops are mounted around the inner surface of the generator housing. They remain stationary and are called the *stator*. The electromagnets are mounted around a rotating shaft. This assembly is called a *rotor*. It is placed inside the stator. Current is created by lines of force cutting across conductors instead of by a conductor cutting across lines of force. See **Figure 11-32**.

In alternating current, a *cycle* is a flow or pulse in one direction followed by a pulse in the opposite direction. The number of cycles per second is the *frequency* of the current. Frequency is measured in hertz. One hertz equals one cycle per second. In North America, with few exceptions, alternating current makes 60 complete cycles each second. In many European countries, the alternating frequency is 50 cycles per second.

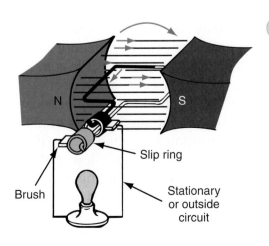

Figure 11-31. An AC generator with an external circuit.

N

S

Slip ring

Brush

Stationary or outside circuit

Figure 11-32. A cutaway view of a large generator. Which part is the rotor, and which is the stator?

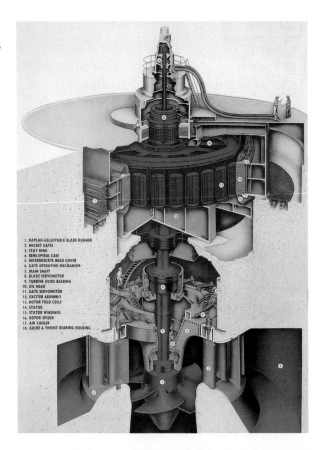

1. KAPLAN-ADJUSTABLE BLADE RUNNER
2. WICKET GATES
3. STAY RING
4. SEMI-SPIRAL CASE
5. INTERMEDIATE HEAD COVER
6. GATE OPERATING MECHANISM
7. MAIN SHAFT
8. BLADE SERVOMOTOR
9. TURBINE GUIDE BEARING
10. OIL HEAD
11. GATE SERVOMOTOR
12. EXCITER ASSEMBLY
13. ROTOR FIELD COILS
14. STATOR
15. STATOR WINDINGS
16. ROTOR SPIDER
17. AIR COOLER
18. GUIDE & THRUST BEARING HOUSING

About 90% of the electricity produced in the world today is alternating current. It is easier to generate in large quantities than direct current. Even more importantly, it is easier to transmit from one place to another.

Direct Current

Direct current (DC) is current that does not change direction in an external circuit. A direct current generator uses a single split ring. It replaces the two slip rings of an alternating current generator. Current in the loop still alternates. However, the split ring, called a *commutator*, sends current only one way through the circuit.

The brushes and commutator of a DC generator are shown in **Figure 11-33**. In part A, terminal 1 is contacting the brush connected to the negative side. Half a turn later, the current changes direction, as shown in part B. Terminal 1 is in contact with the brush connected to the positive side. Current through the external circuit continues in the same direction.

Each half of the commutator is attached to one of the wire loop's terminals. As the current changes direction, the rotating commutator switches the terminals from one brush to the other every half revolution.

Direct current is used in portable and mobile equipment, such as flashlights and car accessories. It is also used in electronic and sound reproduction equipment. **Figure 11-34** shows a DC bicycle generator. A disadvantage of DC current is that it is difficult to transmit over long distances.

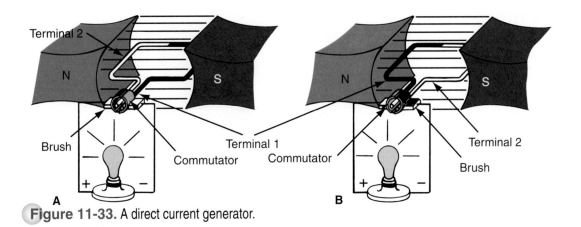

Figure 11-33. A direct current generator.

Figure 11-34. This bicycle light uses direct current supplied by the small generator. What advantage does this generator have over a light operated by a battery?

Electric Motors

In many ways, an electric motor is like a generator. However, while the two have similar parts, their purposes are different. A generator converts kinetic energy to electrical energy. An electric motor changes electrical energy to kinetic energy.

Both a generator and an electric motor apply the laws of magnetism. Both contain magnets and a rotating coil of wire. The coil of wire of an electric motor is placed in a magnetic field, as shown in **Figure 11-35.** The motor spins when a current is applied to the coil of wire.

What makes an electric motor run? Electrons flowing through the coil of wire of an electric motor cause a magnetic field around the coil. Remember the laws of magnetism? Unlike poles attract, and like poles repel. When current is introduced in the coil, the coil's magnetic field reacts with the magnets in the motor. The coil spins as it is attracted and repelled by the motor's permanent magnets.

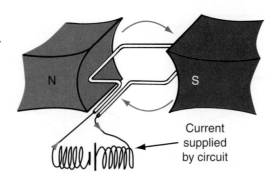

Figure 11-35. An electric motor is much like a generator.

Current supplied by circuit

The coil spins for only part of a turn from the effects of magnetism. For an instant, the current stops and the coil coasts. The rotation would stop except for split copper ring, or commutator, that rotates with the coil. See **Figure 11-36**. When current starts up again, magnetic force keeps the coil turning.

Current passes into and out of the coil through brushes that press against the commutator. Current always passes down on the right side and returns on the left side of the coil. This switches the poles in the coil's magnetic field. The rotation then continues in one direction.

The brushes also serve a second purpose. Since they do not rotate, they prevent the wires from twisting.

Brushes are usually made from carbon. It is a good conductor and produces less friction than metal. The brushes are spring-loaded. Pressure from the spring ensures continuous contact with the commutator.

Cells and Batteries

What most of us call a battery is not a battery at all. It is really a cell. A battery is a set of cells connected together. A *cell*, or *dry cell*, has a single positive *electrode*, a single negative electrode, and an electrolyte. A *battery* is a package containing several cells connected together.

The 9 V batteries used in portable electronic devices and 12 V car batteries are correctly called batteries. Inside a car battery are six 2 V cells. Inside a 9 V battery are six 1.5 V cells connected together. See **Figure 11-37**. When we refer to AA, C and D cells as batteries, we use the wrong term. They consist of only one 1.5 V cell.

Figure 11-36. The brushes of an electric motor are always in contact with the spinning commutator. This allows the electricity to flow into the commutator and into the coil.

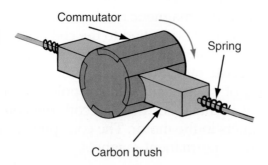

Commutator

Spring

Carbon brush

Figure 11-37. This cutaway of a 9 V battery shows that six 1.5 V cells are connected inside.

Voltaic Cell

The simplest type of cell is the *voltaic cell*. Two rods are immersed in a container filled with a solution of water and sulfuric acid. The mixture is known as an *electrolyte*. One of the rods is copper and the other is zinc. The acid reacts with both of the metals. See **Figure 11-38**.

Some of the atoms from the metals pass into the solution. Each atom leaves behind a pair of electrons. However, the zinc rod tends to lose atoms to the solution faster than the copper rod. Since the zinc rod builds up more electrons than the copper rod, it becomes negative. If the two electrodes are connected by a conductor, excess electrons flow along the conductor from the zinc to the copper. This flow of electrons produces an electric current. Since the flow of electrons is in one direction only, cells and batteries produce DC voltage.

Primary Cells

Cells are classified as either primary or secondary. A *primary cell* is one whose electrode is gradually consumed during normal use. It cannot be recharged. Primary cells are used in flashlights, digital watches, and some cameras.

All of the primary cells used today use the same principles as the voltaic cell. They have three main parts: the electrolyte and two electrodes. See **Figure 11-39**. The electrolyte is a paste made of a very active chemical, such as a strong acid or base.

Inside the battery, two chemical reactions take place. One is between the electrolyte and the *cathode* (negative electrode). The other is between the electrolyte and the *anode* (positive electrode). These reactions change the chemical energy stored in the cell into electrical energy. When the chemical reactions have finished, the cell has no chemical energy left, and it can give no more electricity.

Science Application

Lemon Cell

You won't find a lemon battery powering any of your electronic equipment. The voltage is low compared to other batteries. However, all chemical cells work in the same way. Chemical reactions take place, changing chemical energy to electrical energy.

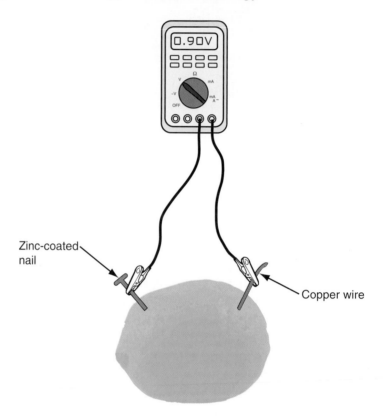

For example, in the lemon battery shown here, two chemical reactions are taking place: one at the copper electrode and one at the zinc electrode. The electrodes are placed in the electrolyte (in this case, lemon juice). In any battery, each pair of metal electrodes, together with the electrolyte, form a single cell. Therefore, the lemon shown is one battery cell.

Science Activity

To make a lemon cell, you will need two terminals:

- Copper, in the form of a copper wire or copper penny
- Zinc-coated nail or screw, or steel wire

Connect the two terminals to a voltmeter. The cell should measure about .5 V. Your lemon cell will not power a motor or even an incandescent light bulb, but what happens when you connect it to an LED? How can you increase the voltage so that the LED lights?

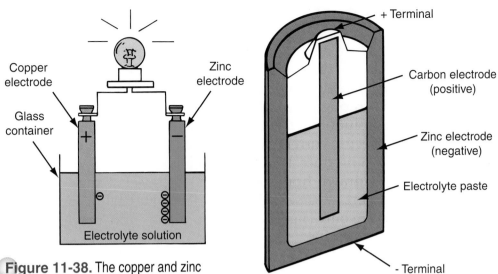

Figure 11-38. The copper and zinc electrodes react with the electrolyte solution at different rates. This results in an uneven buildup of free electrons that can be used to create an electric current.

Figure 11-39. Cross-section of a primary cell.

Carbon-Zinc Cells

The carbon-zinc cell is one of the most common types of primary cell. It is also the least expensive, but it is short-lived. Carbon-zinc cells are produced in a range of standard sizes. These include 1.5 V AAA, AA, C, and D cells. Rectangular 9 V batteries also contain carbon-zinc cells. See **Figure 11-40.**

Alkaline Cells

Alkaline cells are produced in the same sizes as carbon-zinc cells. However, they can supply current longer.

Figure 11-40. These are typical primary cells. Shown from left to right: AA, C, and D cells.

Secondary Cells

A *secondary cell* is one that can store electrical energy to be used as needed, but can also be recharged after the electrical energy is used. See **Figure 11-41**. Lead plate electrodes are placed in a solution of sulfuric acid. A current passing through the lead plates produces chemical changes. The sulfuric acid solution gets stronger, and the cell becomes capable of producing an electric current. This is called "charging a cell." When charged, the cell can produce a current in a circuit.

As electricity is drawn from the cell, the chemical change that took place during charging reverses. However, the materials in the cell are not used up; they are only changed. Therefore, the entire process can be repeated.

Lead-Acid Batteries

Each pair of electrodes in a secondary cell can produce about 2 V. Most motor vehicles require 12 V to operate the starter motor. Therefore, six pairs of electrodes, or cells, must be connected together. **Figure 11-42** shows a number of cells connected together to form a type of automotive battery known as a lead-acid battery.

Even after 100 years of commercial use, lead-acid batteries are still important. These batteries are used to power lights, radios, and all accessories in many vehicles. They also provide backup and emergency power for some electrical installations.

Nickel-Cadmium Batteries

Nickel-cadmium (NiCd) batteries were once the battery of choice for low-power portable products such as calculators and cameras. They can produce an energy density of between 45 and 80 watt-hours per kilogram (Wh/kg). *Energy density* is the amount of energy that can be stored in a battery per unit weight.

Figure 11-41. A secondary cell stores the energy it builds up during charging. It releases the energy as needed to produce an electric current.

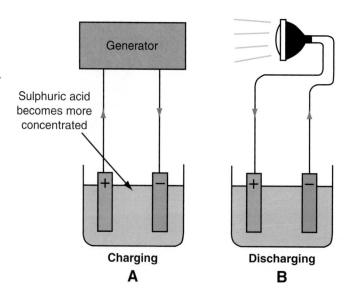

Sulphuric acid becomes more concentrated

Charging
A

Discharging
B

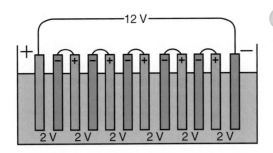

Figure 11-42. Many automobile batteries are made by connecting six cells to produce 12 V.

However, NiCd batteries have some major drawbacks. First, they "remember" the amount of discharge from previous use. This limits the recharge life of the battery. Second, cadmium is considered an environmental hazard and has to be disposed of carefully. For these reasons, NiCd batteries are being replaced by newer, more efficient types of secondary cells.

Nickel-Metal Hydride Batteries

Nickel-metal hydride (NiMH) batteries are similar to NiCd batteries. However, they can produce up to 120 Wh/kg. They are also less harmful to the environment. Therefore, NiMH batteries have replaced NiCd batteries for many applications. The main disadvantage of NiMH technology is that an NiMH battery tends to lose its charge in as little as two weeks, even if it is not used.

Lithium-Ion Batteries

Lithium-ion (Li+) batteries are now widely used to power laptop computers, cell phones, cameras, and many other portable devices. See **Figure 11-43.** They have many advantages over other rechargeable batteries:

- They provide a high energy density (180 Wh/kg), so the products they power can be recharged less often.

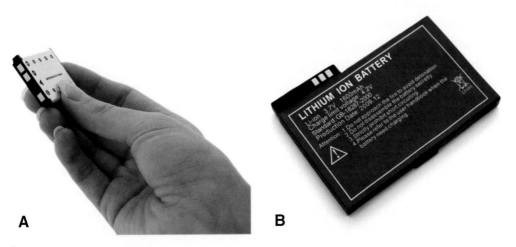

A   **B**

Figure 11-43. Lithium-ion batteries are now used to power most portable electronic devices because they are dependable and have a long battery life. A—Camera battery; B—laptop computer battery.

- They weigh less than other types of secondary cells and batteries.
- They remain rechargeable for a longer period of time.

Earlier Li+ batteries could be damaged by overcharging or overheating. Li+ batteries today incorporate devices that prevent overcharging and overheating, so they are very safe to use. However, because these batteries are more complex than NiCd and NiMH batteries, they are also more expensive.

End Note

People are using electricity for more and more purposes. In the past, electricity was generated mostly by petroleum- or coal-based plants. However, both petroleum and coal are fossil fuels and are nonrenewable resources. What will happen when the fossil fuels run out?

Forward-thinking people have begun to develop other methods of generating electricity. Solar energy and geothermal energy are two alternatives. These systems are already in use in some areas.

In the future, other technologies may also be used. Experiments are underway to produce electricity from such unlikely sources as sea algae and carbon dioxide. Researchers are also working with carbon nanotubes. These carbon molecules have been shown to produce large amounts of electricity under certain circumstances. In the not-so-distant future, perhaps our electricity needs will be met by cleaner, renewable sources.

Batteries, too, are constantly being improved. Researchers are hoping to replace chemical batteries by other advanced forms, such as tiny nuclear batteries. These will be powered by the natural decay of electrons given off by their radioactive source. Known as *betavoltaics*, they would be extremely small and long-lasting. They could be implanted almost anywhere. For example, they could be used in bridges, roadways, and buildings to monitor their condition and safety.

Summary

- Most electricity today is produced by hydroelectric and thermal-electric generating stations.
- Energy conversions are necessary to transport electricity over long distances while minimizing energy loss.
- According to the electron theory, electricity is produced when free electrons flow in the same direction, producing an electric current.
- Like poles of a magnet repel each other, and unlike poles attract each other.
- Generators move wire coils through a magnetic field to produce electricity.
- Electric motors are similar to generators, except that the purpose of an electric motor is to change electrical energy to kinetic energy.
- Batteries are made by connecting cells together to increase the total voltage output.

Preview and Prediction

Copy the following graphic organizer onto a separate sheet of paper. Do not write in this book. In the left column, record at least six predictions about what you will learn in this chapter. After you have read the chapter, fill in the other two columns of the chart.

What I Predict I Will Learn	What I Actually Learned	How Close Was My Prediction?

Test Your Knowledge

Write your answers to these review questions on a separate sheet of paper.

1. How are hydroelectric and thermal-electric generating stations similar? How are they different?

2. Describe where in a transmission and distribution system you might find each of the following voltages.
   A. 230,000 V
   B. 13,800 V
   C. 120 V

3. Briefly explain the electron theory of how electricity is produced.

4. State the two laws of magnetism.

5. What is electromagnetism?

6. What is the difference between alternating current and direct current?

7. How is the purpose of an electric motor different from that of a generator?

8. Explain the difference between a cell and a battery.

9. What is the difference between a primary cell and a secondary cell?

10. How many cells would a lead-acid battery need to produce 24 V?

Critical Thinking

1. Over a period of a year, a portable CD or DVD player uses a large number of cells (often referred to as *batteries*). What things could you do to reduce the amount of money you spend to power this device?

2. Science fiction is a form of literature that often "predicts" technologies of the future. Mary Shelley's book *Frankenstein*, first published in 1818, is one such classic. Obtain a copy of the book and find out how electricity is used in the plot. Then search the Internet to find new ways in which electricity is being used today. How does Shelley's use of electrical principles compare with the latest technologies?

3. Copy the following chart to a separate sheet of paper. Do not write in this book. Think about the advantages and disadvantages of each type of generating station. Then fill in the columns. List as many advantages and disadvantages as possible for each type. Then study your results. Which method of generating electricity do you think is best? Write a persuasive paragraph encouraging the use of this type of generating station.

Type of Station	Advantages	Disadvantages
Hydroelectric Station		
Fossil-Fuel Station		
Nuclear Station		

Apply Your Knowledge

1. Make a model to illustrate one method of generating electricity.

2. Describe the components of a network for the transmission and distribution of electricity. How many of these components can you see in your neighborhood?

3. Make sketches with notes to illustrate the following:
   A. An atom
   B. Electron flow

4. Perform the experiment illustrated in Figure 11-25. You can make iron filings by cutting steel wool into tiny pieces using an old pair of scissors. Expand the experiment by using two bar magnets with like poles together and with unlike poles together. You can fix the pattern of iron filings in place using hair spray.

5. Make a sketch to show how you would make an electromagnet.

Apply Your Knowledge *(Continued)*

6. Battery disposal has become an issue in recent years. Secondary cells, in particular, must be disposed of properly to avoid environmental and health hazards. Find out what the hazards are and how communities are currently dealing with these issues. How can these methods be improved? Develop and propose a new disposal plan that would be safe for humans and the environment.

7. Research one career related to the information you have studied in this chapter. Create a report that states the following:

   • The occupation you selected

   • The education requirements to enter this occupation

   • The possibilities for promotion to a higher level

   • What someone with this career does on a daily basis

   • The earning potential for someone with this career

   You might find this information on the Internet or in your library. If possible, interview a person who already works in this field to answer the five points. Finally, state why you might or might not be interested in pursuing this occupation when you finish school.

STEM Applications

1. **MATH** Ohm's law states that voltage is equal to current times the resistance in an electrical circuit ($V = I \times R$, where V = voltage, I = current, and R = resistance). You will learn more about Ohm's law in the next chapter. However, just by looking at the formula, you may be able to see how Ohm's law applies to transformers. Write a statement explaining how Ohm's law relates to the transmission and distribution of electrical power.

2. **ENGINEERING** Design and make an electrical system to power a head-mounted LED flashlight that could be worn at night on a camping trip. Consider whether to use a generator or battery-operated design. Find out how much voltage is needed and design the flashlight accordingly.

3. **SCIENCE** Although most metals conduct electricity, most electrical wires are made of copper. However, copper is not the metal that best conducts electricity. Find out more about how well common metals conduct electricity. Include copper, tungsten, zinc, aluminum, gold, silver, nickel, and platinum in your research. Make a chart to show your findings. Which metal is the best conductor of electricity? Why is copper used instead for most wiring applications? Which other metals, if any, could be used in electrical wires to conduct electricity efficiently?

12

Using Electricity and Electronics

Better by Design

Leah Buechley designs wearable electronics

Leah Buechley is an American textile researcher who weaves, solders, and sews LEDs and electronics into cloth to build soft, flexible, wearable computers. For example, she constructed the LED bracelet shown here by weaving LEDs and beads together on a traditional beading loom. Computer chips and other electronic components are attached to the bracelet via custom-made fabric circuit boards that Leah developed. Leah also designed a construction kit called the LilyPad Arduino that allows others to build their own computational (electronic) textiles.

An LED bracelet shows up well in dark environments. What e-fashion products would you design and make?

"I believe in learning, inventing, and designing through building and hands-on experimenting."

Reading Target

Connecting to Prior Knowledge

One good way to prepare yourself to read new material is to think about what you already know about the subject. You use electric circuits and electronic devices every day. What do you know about these circuits and devices? Use the Reading Target at the end of this chapter to record your ideas, even if you are not sure of some of the facts.

Key Terms

amperes
AND gate
capacitors
circuit breaker
conductivity
conductors
diodes

electronics
farads
fuse
insulators
integrated circuit
laser
load
logic gates
NAND gate
NOR gate
NOT gate (inverter)
ohms
Ohm's law
OR gate
overload
parallel circuit

potentiometers
rectifiers
resistance
rheostats
schematics
semiconductors
series circuit
series-parallel circuit
short circuit
superconductor
switch
transistor
voltage
volts
watt
Watt's law

Objectives

After reading this chapter, you will be able to:

- Design, draw, and build different types of electric circuits.
- Identify the characteristics of conductors, insulators, and semiconductors.
- Use Ohm's law and Watt's law to calculate current, voltage, resistance, and power.
- Name and state the function of common electronic components.
- Discuss the impact of electronics technology on personal privacy and security.

Useful Web sites:
www.media.mit.edu/~leah/
www.lumalive.com/
www.numetrex.com
www.instructables.com/id/turn-signal-biking-jacket/

Think about the things that electricity does. It operates motors found in many large and small appliances. The motors run electric mixers, blowers, pumps, dishwashers, washing machines, and many other appliances. They power subway trains and golf carts. Electricity also operates the lights in movie theaters, sports stadiums, and your home. It operates alarm systems, your MP3 player, and kidney dialysis machines.

Prior to the late 1800s, electricity was a curiosity. Electrical appliances were almost unknown. This situation changed in 1879 when Thomas Edison invented a reliable light bulb that lasted a long time. People started switching from gas lights to electric lighting.

Today, factories use electricity and electrical circuits to start and stop machines automatically. Electricity controls assembly lines and robots. Whole factories can be run electrically. A few people working at computers can control machines, lights, assembly lines, packaging, and loading of products. Indeed, it is hard to imagine what we would do without electricity.

SAFETY: Remember that every time you work with electricity, you must think about the hazards and risks involved.

- Water and electricity are a dangerous combination. Keep all electrical devices away from sinks, swimming pools, and bodies of water.
- Do not fly kites near power lines.
- If a ball or other object gets stuck in a tree near a power line, do not try to get it down.
- Use only 1.5 V cells or 9 V batteries to power your experiments in this course.

Electric Circuits

An electric circuit is a closed path for electric current. The path starts from a source, such as a cell or battery. It continues through a resistance (load), such as a lamp, before it returns to its source. All power systems, including electrical circuits, must have a source, a process, and at least one load. To make a simple circuit, connect a lamp, two pieces of wire, and a 1.5 V cell. In the circuit shown in **Figure 12-1**, a wire is connected to the negative terminal of the cell. Electrons flow from the negative terminal and continue through the wire and the lamp to the positive terminal.

Figure 12-1 is a pictorial view of this circuit. However, it takes too long to make a picture of each component. This is especially true when the circuit is complicated. Therefore, designers use symbols rather than pictures. See **Figure 12-2**. Symbols can be drawn quickly and are understood everywhere. Diagrams of circuits using symbols are called *schematics*.

The circuit in **Figure 12-2** has voltage, resistance, and current. Voltage is supplied by the cell. The resistance is the glowing element in the lamp. The current is the flow of electrons in the circuit.

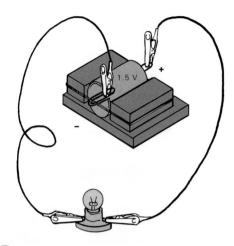

Figure 12-1. A simple circuit showing a lamp lit by a 1.5 V cell.

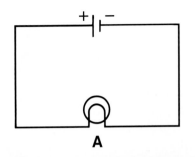

A

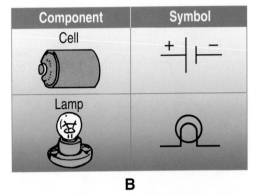

B

Figure 12-2. A—The circuit pictured in Figure 12-1 shown as a schematic. B—Circuit components and their symbols.

To turn the lamp on and off, you could connect and disconnect a wire at any one of the four places shown in **Figure 12-3**. However, an easier method would be to add a switch to the circuit.

A *switch* is a device that allows the circuit to be turned on and off. With the switch closed, current flows and the lamp lights up. When the switch is open, the flow stops, and the lamp turns off. Mechanical switches are also used to direct current to various points. **Figure 12-4** shows the simplest type of switch: a single-pole, single-throw (SPST) switch. **Figure 12-5** shows six of the many types of switches.

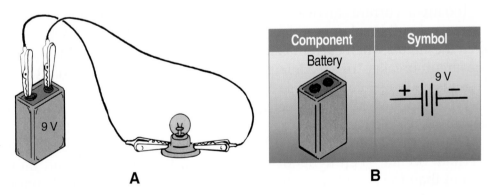

A   **B**

Figure 12-3. A—To turn off the lamp, release any of the four clips. B—A 9 V battery and its symbol.

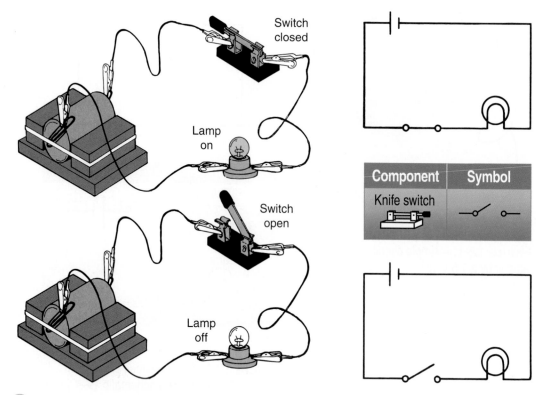

Figure 12-4. Study the open and closed switches and their diagrams. What is the difference between the two circuit diagrams?

Not all switches are purely mechanical. Fiber optics can also be used to create a switch. For example, in automatic faucets found in public washrooms, a dual fiber optics line is used. When hands are placed under the faucet, a beam of light from one line is reflected off the hands to the other line. The second line is connected to a sensor. The sensor sends a signal to a solenoid valve to release the water.

Protecting Circuits

Too much current can overheat and damage circuits. To prevent this, a fuse or a circuit breaker is added. See **Figure 12-6**. These are devices that open the circuit when current is too high. The fuse "blows" or burns out. The circuit breaker trips a contact. In either case, it opens the circuit. The circuit is no longer complete, so current stops.

When applied to using electricity, the term *load* means a source of resistance in the circuit. The bigger the load, the more current it needs.

High current can be caused by an overload or a short circuit. An *overload* occurs when lights or appliances in the circuit demand more current than the circuit can safely carry. A good example of this is when too many appliances are plugged into one outlet. A *short circuit* is a path of low resistance. It results in high current because the current avoids a section of a circuit, such as a lamp or a motor (load). See **Figure 12-7**.

Toggle   Rocker   Rotary

Knife   Pushbutton   Slide

Figure 12-5. Six types of switches. Where have you seen these switches used?

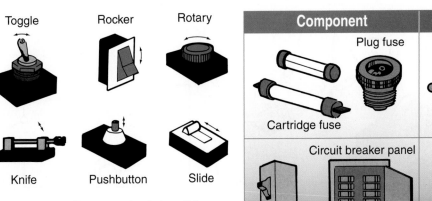
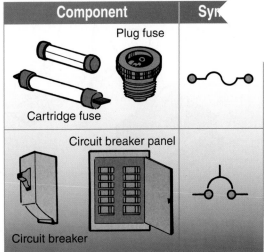

Component	Syn
Plug fuse	
Cartridge fuse	
Circuit breaker panel	
Circuit breaker	

Figure 12-6. Overload protection devices such as fuses and circuit breakers protect the wiring and devices in a circuit.

How Fuses and Circuit Breakers Work

A *fuse* is a current-limiting device. When too much current reaches a fuse, a thin wire called a *filament* melts inside the fuse. See **Figure 12-8.** This stops the current flow before the circuit wires can be damaged. Once a fuse blows, it must be replaced.

When high current enters a *circuit breaker*, it heats a bimetal (two metals) strip. The strip bends because one metal expands more than the other. This opens contacts so no current can pass. Unlike a fuse, a circuit breaker can be reset and reused.

Figure 12-7. Overloading a circuit causes the wires to heat up, creating a fire hazard.

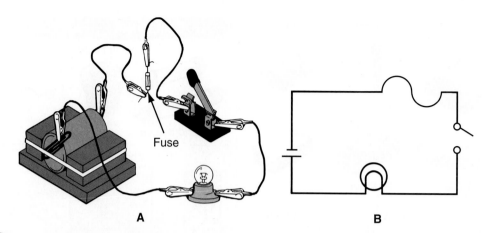

A   B

Figure 12-8. A fuse interrupts the flow of electricity when an overload occurs. A—Circuit with a fuse. B—Schematic of the same circuit.

Direction of Current

The direction of electricity in a circuit can be shown either as "electron flow" or "conventional current." See **Figure 12-9**.

Electron flow is based on the electron theory. This theory states that current moves from negative to positive. Conventional current is based on an older theory of electricity. Early scientists assumed a current moved from positive to negative. Both theories are acceptable. In this book, however, all explanations will be based on electron flow.

Types of Circuits

A circuit is a pathway along which electricity travels. The circuit shown in **Figure 12-9** contains only one lamp. If two or more lamps are put into this circuit, they can be connected in one of two ways, depending on the desired result.

Series Circuits

A *series circuit* is a circuit in which all loads are connected one after another so the same current enters each of them in turn. There is only one path for electron flow, as shown in **Figure 12-10**.

When lamps are connected in series, each gets an equal part of the voltage. For example, three lamps connected to a 1.5 V cell each receive 0.5 V. Therefore, each bulb is dimmer than if only one lamp is in the circuit. However, current remains the same across each lamp.

Figure 12-9. These circuit diagrams show the two theories of electric current. Electron flow moves from negative to positive. Conventional current moves from positive to negative.

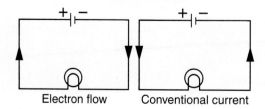

Electron flow   Conventional current

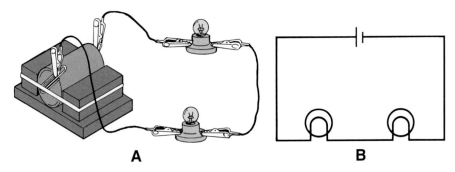

Figure 12-10. A—These lamps are connected in series. Electricity flows through each lamp in turn. B—The schematic for the series circuit shown in A.

In a series circuit, if one lamp burns out, all of the lamps go off. This is because the burned-out lamp opens the circuit, so current stops flowing. For this reason, very few series circuits are used in homes.

Parallel Circuits

A *parallel circuit* is a circuit that has more than one path for electron flow. See **Figure 12-11.** You can see that when lamps are in parallel, the current splits. It goes through each of the lamps without passing through any others first. In parallel circuits, the current varies in each path. However, voltage is always the same. If one bulb burns out, the circuit is not broken, and the other bulbs continue to burn. That is why most circuits in the home are parallel circuits.

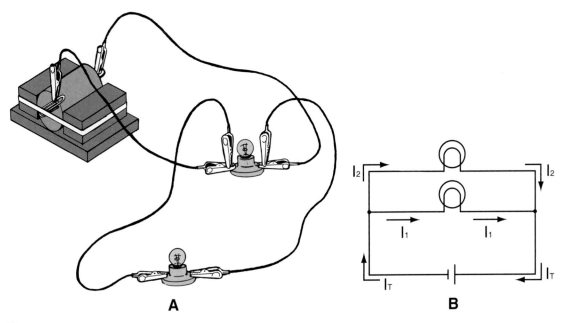

Figure 12-11. A—These lamps are connected in parallel, so there are two paths for current to follow. At the left intersection, the current splits into the two branches. At the right intersection, the separate currents add back together and equal the original current. B—The schematic for the parallel circuit shown in A.

Series-Parallel Circuits

The third kind of circuit is the *series-parallel circuit*. In these circuits, some loads are wired in series, and others are wired in parallel. See **Figure 12-12**. They therefore exhibit characteristics of both series and parallel circuits. The different branches react to current and voltage in different ways. If a branch is a series branch, it behaves like a series circuit. Current is the same throughout that branch of the circuit, with the voltage dividing between the components. In a parallel branch, voltage remains the same throughout the branch. Current is divided between the parallel components.

In the circuits just described, lamps were connected in series, parallel, or series-parallel. Switches may also be connected in these ways. Look at **Figure 12-13**. For the bulb to light, both switch A and switch B have to be closed. In **Figure 12-14**, how many switches must be closed to close the circuit?

Figure 12-12. A series-parallel circuit has separate branches that are either series or parallel. Which of the loads in this circuit are wired in parallel? Which are wired in series?

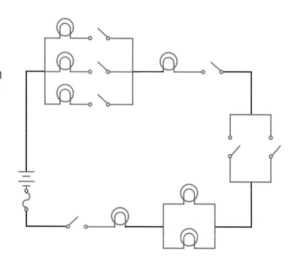

Figure 12-13. Because the switches are connected in series, this circuit conducts current only when both switches are closed.

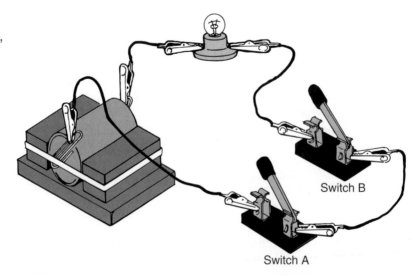

Switch B

Switch A

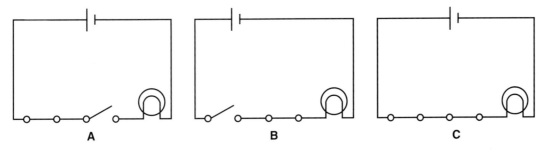

Figure 12-14. A, B, and C show the same circuit with the switches in different states. In which circuit or circuits is the bulb lit?

What happens when switches are connected in parallel, as shown in **Figure 12-15**? Switch A or switch B completes the circuit and turns on the light. Look at the diagrams in **Figure 12-16**.

Conductors, Insulators, and Semiconductors

As described in Chapter 10, materials that allow electric current to flow easily are called *conductors*. Copper, aluminum, silver, and most other metals are examples of good conductors. Copper is used most often for house wires. Strength, low cost, and low resistance to current make it a good choice. See **Figure 12-17**.

Figure 12-15. This parallel circuit conducts current when either switch A or switch B is closed.

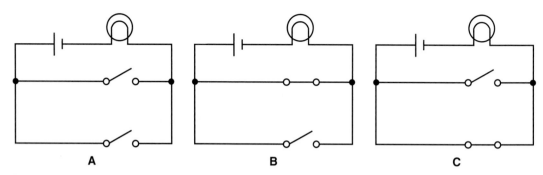

Figure 12-16. In which circuit or circuits is the bulb lit?

Figure 12-17. Copper is the most widely used conductor for electric wiring.

Materials that do not allow current to pass are called *insulators*. Glass, rubber, plastic, porcelain, and paper are good insulators. Insulators play an important part in controlling electricity. They are wrapped around a conductor to prevent it from passing current to another conductor. This keeps the current in the correct path.

One of the most important uses of insulators is to protect us from electric current. See **Figure 12-18**. Our bodies conduct electricity, especially when wet. If you touch a live wire by accident, you get a dangerous shock. The shock may kill you.

Some materials are better conductors than others, and almost all materials have some resistance to the flow of electricity. Electricity produces heat as it forces its way through this resistance. The greater the resistance, the more heat produced. This can be used to our advantage. A resistance wire can be used to produce heat. The most common type of resistance wire is an alloy of nickel and chromium. It is called *nichrome wire*. Sometimes the resistance wire becomes red-hot. For example, resistance

Figure 12-18. Insulators protect against shock. Where is insulation being used in this photo?

Figure 12-19. An electric heater works because an electric current makes a high-resistance wire red hot.

wires provide the heating in electric stoves, toasters, and other heating appliances. See **Figure 12-19.** Sometimes the conductor becomes white-hot. This is the case with an incandescent light bulb.

Semiconductors are materials that do not conduct as well as copper or silver. Nor do they insulate as well as rubber or glass. They have some characteristics of each. **Figure 12-20** provides examples of conductors, semiconductors, and insulators.

For many years, technologists have dreamed of producing a material that will conduct electricity without resistance. Such a material is called a *superconductor.* The advantage is that current flowing in a superconductor can flow forever.

Until recently, superconductivity was possible only at low temperatures close to absolute zero, –459° F (–273° C). When mercury is cooled to a few degrees above absolute zero, it conducts electricity with no resistance.

Recently, materials have been discovered that conduct at higher temperatures. When these superconductors become widely available, they will revolutionize the electronics industry. The absence of electrical resistance reduces the amount of heat produced. This allows components to be packed more closely together, reducing the size of components. Superconductors will also enable computers to operate at much greater speeds.

Science Application

Testing Conductivity

A material's *conductivity* is its ability to conduct electric current. The difference between a material that conducts well and one that conducts poorly or not at all is related to the material's structure. You already know that electrons move around the nucleus of an atom. In some materials, the electrons are held more tightly than in others. Electrons that are not held tightly can jump from one atom to another, producing electric current.

Science Activity

To test the conductivity of materials, you will need a battery, battery holder, two metal paper clips, and test leads, as shown in the illustration. Also collect small samples of a variety of materials. Include wood, glass, and various kinds of metal and plastics. Some of the materials should be examples of small products, such as coins and hardware items. Set up your test equipment as shown. Then connect each material to the circuit. Test each material and record your findings. Write a report to document your experiment, and include the report in your portfolio.

Class	Materials	
Conductors	• Copper • Silver	• Tungsten • Nichrome
Semiconductors	• Silicon • Germanium	
Insulators	• Rubber • Glass • Porcelain	• Nylon • PVC • Mica

Figure 12-20. Classification of materials commonly used in electrical and electronics applications.

Measuring Electrical Energy

When measuring electrical energy, three terms are important: volts, ohms, and amperes. One way to understand their meaning is to compare electricity to water. Water flows through a hose under pressure. If pressure is increased, more water flows than before. Electricity inside a wire is also under pressure. With electricity, the electrical pressure (E) is called *voltage* and is measured in *volts* (V). See **Figure 12-21**.

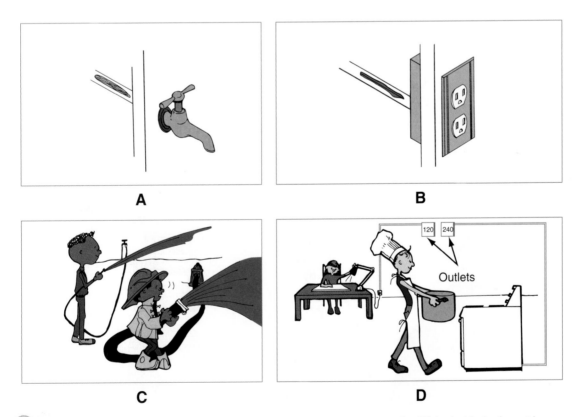

Figure 12-21. Comparison of electricity pressure and water pressure. A—Water behind a faucet is always under pressure. B—Electricity behind an outlet is also under pressure. C—To use water, you attach a hose and turn on a faucet. D—To use electricity, you connect a wire to an outlet and usually toggle a switch.

As the water flows through the pipe, it meets resistance. The smaller and longer the pipe, the greater the resistance. So it is with electricity. Flow is affected by diameter and length of the wire. Also, electricity flows more easily through some materials than through others. Electrical resistance (R) is measured in *ohms* (Ω).

The amount of water that flows out of the end of the pipe in a given period depends on both pressure and resistance. See **Figure 12-22**. The higher the pressure and the weaker the resistance is, the greater the amount of water leaving the hose.

Ohm's Law

The amount of electricity that passes a point in a conductor in a given period also depends on pressure and resistance. The movement of electrical current through circuits can be described in the same way that water flows through pipes. Current is similar to water flow speed. Voltage is similar to the water pressure. Resistance can be compared to the diameter of a water pipe, which can act as a restriction to flow. Electrical current (I) is measured in *amperes* (A): the higher the pressure and the weaker the resistance in ohms, the higher the current.

This relationship between voltage, resistance, and current is described by a formula known as *Ohm's law*. **See Figure 12-23.** Ohm's law is written as:

$$Voltage = Current \times Resistance$$

$$(or)$$

$$Volts = Amps \times Ohms$$

$$(or)$$

$$E = I \times R$$

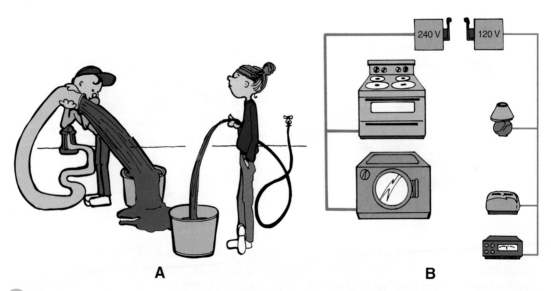

Figure 12-22. Comparison of electricity and water pressure and resistance. A—High pressure and a large hose equal heavy water flow. B—High voltage and a large conductor equal high electron flow.

Figure 12-23. To remember Ohm's law, just think of listening to your instructor with your EIR (pronounced like "ear"). Voltage is often represented using the letter E for electromotive force.

Watt's Law

Another unit of electrical measurement is the watt. A *watt* is the unit used to measure the work performed by an electric current. To calculate the power (P) in watts, use *Watt's law* to multiply the voltage by the current. See **Figure 12-24.**

$$\text{Power} = \text{Current} \times \text{Voltage}$$

$$\text{(or)}$$

$$\text{Watts} = \text{Amperes} \times \text{Volts}$$

$$\text{(or)}$$

$$P = I \times E$$

A monthly electricity bill is based on the number of watts used. See **Figure 12-25.** The utility company provides a meter for each home. The meter measures how many watts are used. The watt is a small unit, so the basic unit used by power companies is a kilowatt. One kilowatt is equal to 1000 watts.

Figure 12-24. Remembering Watt's law is as easy as PIE.

Figure 12-25. A utility meter measures the amount of electricity used in kilowatt-hours (kWh).

Appliances are frequently switched on and off in most homes, so the electrical usage varies. The electricity used is measured over periods of one hour. The unit is therefore one kilowatt-hour. A kilowatt-hour means 1000 watts used for a period of one hour.

Figure 12-26 shows the dials of a typical electrical meter. Be careful when reading the dials. Some of them revolve clockwise, and some revolve counterclockwise. To read a dial, write down the number the pointer has passed. In **Figure 12-26**, the correct reading is 23,642.

Electronics

In this chapter, you have learned about the flow of electrons in a circuit. *Electronics* is the use of electrically controlled parts to control or change current in a circuit. It is the technology of controlling electron flow. Electrons can be used to control, detect, indicate, measure, and provide power. A variety of electronic materials and components carry out these functions.

Resistors

Resistors are to electronics what friction is to mechanics. The friction of brakes limits your speed on a bike. Resistors limit the amount of current flowing through a circuit. Resistors make it more difficult for current to flow. Many electric circuits have only one power source, for example a 9 V battery. But a different current may be needed by different components. Resistors change the current in parts of the circuit to match the needs of one or more components. In **Figure 12-27**, bulb A will be brighter than bulb B. The current in bulb B is smaller because there is a resistor in that loop of the circuit.

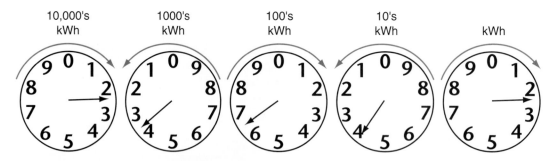

Figure 12-26. You can tell which way each pointer revolves by looking at the numbers. If the 1 is to the right of the 0, the pointer revolves clockwise. If the 1 is to the left of the 0, the pointer revolves counterclockwise. What is the reading of the meter in this figure?

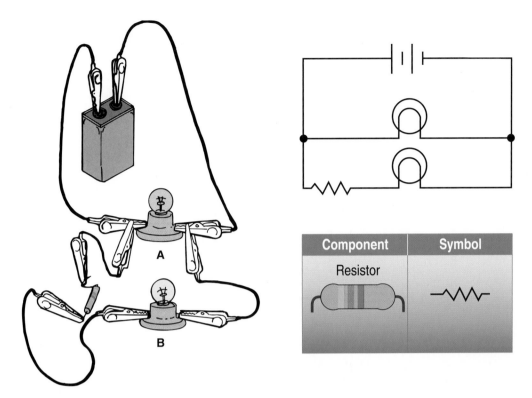

Figure 12-27. This parallel circuit has a resistor in one loop.

Resistors are made in many sizes and shapes. See **Figure 12-28**. All resistors do the same thing: they limit current. In a typical carbon composition resistor, powdered carbon is mixed with a glue-like binder. Changing the ratio of carbon particles to binder changes the resistance. The greater the amount of carbon used, the less resistance the resistor has. See **Figure 12-29**.

Resistors are made in a wide range of values. The value corresponds to the degree to which they limit the flow of electrons, their *resistance*. Resistors are often quite small, which makes it difficult to show their values. To overcome this problem, resistors are usually marked with four colored bands. See **Figure 12-30**.

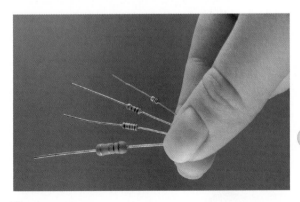

Figure 12-28. Resistors come in different shapes and sizes. They are often quite small.

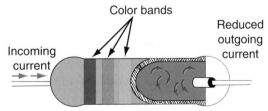

Figure 12-29. Resistors limit current flow.

Color		1st Band	2nd Band	3rd Band	4th Band
Black		0	0	Ω (no zeros)	
Brown		1	1	1 zero	
Red		2	2	2 zeros	
Orange		3	3	3 zeros	
Yellow		4	4	4 zeros	
Green		5	5	5 zeros	
Blue		6	6	6 zeros	
Violet		7	7	7 zeros	
Gray		8	8	8 zeros	
White		9	9	9 zeros	
Gold					5
Silver					10
None					20
		BAND 1	BAND 2	BAND 3	BAND 4

Figure 12-30. Calculating the value of a resistor.

You can calculate the value of a resistor from the first three bands. To read the value, hold the resistor with the colored bands to the left. Then use the table in **Figure 12-30** to calculate the value of the resistor.

In **Figure 12-31**, the first band of the resistor is violet. The table shows us that a violet first band number stands for 7. The second band is green, which stands for 5. The third band is orange. An orange third band means that three zeros follow the first two numbers. The value of this resistor is 75,000 Ω, or 75 kΩ. (The k stands for "kilo" or thousand.) A resistor whose value is 75,000,000 Ω is written as 75 MΩ. (The M means "mega" or million.)

Resistors have a fourth band that is usually silver or gold. It indicates the accuracy or tolerance of the resistor. The fourth band of the resistor in **Figure 12-31** is gold, indicating that the resistor has a tolerance of ±5%. Five percent of 75,000 is 3,750. Therefore, the actual resistor value is 75,000 ± 3,750, or between 71,250 Ω and 78,750 Ω.

Figure 12-31. Use the chart in Figure 12-30 to calculate the value of this resistor.

Two types of variable resistors are rheostats and potentiometers. *Rheostats* lower or raise the current in a circuit. See **Figure 12-32**. The dimmer used for car dashboard lights is one example. *Potentiometers* lower or raise the voltage. See **Figure 12-33**. They are used as volume controls in sound systems.

Diodes

Diodes are devices that allow current to flow in one direction only. Diodes have two ends: positive (anode) and negative (cathode). A dark band indicates the negative end, as shown in **Figure 12-34**.

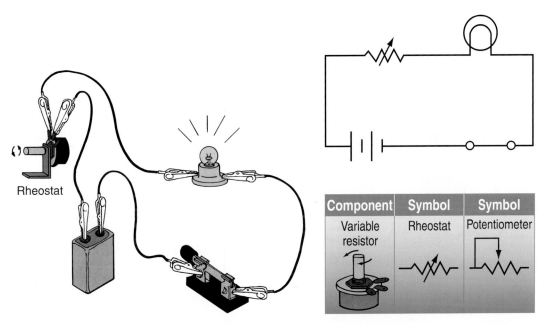

Figure 12-32. A rheostat is connected in series.

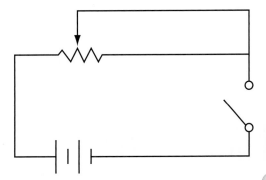

Figure 12-33. A potentiometer is connected in parallel.

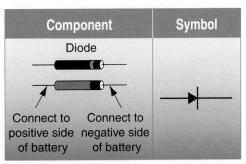

Figure 12-34. A dark band indicates the negative end of a diode.

Figure 12-35 shows two bulbs in parallel. Each branch of the circuit has a diode. Only bulb A will light, because the diode next to it is positioned correctly to allow electrons to flow. Diodes are most commonly used in *rectifiers*, which change alternating current to direct current. Some different types of diodes are shown in **Figure 12-36**.

Another type of diode is a light emitting diode (LED). LEDs also conduct in only one direction. They need less current to make them glow than most bulbs, but they are usually not as bright. Therefore, LEDs are used where brightness is not important. For example, they are used to show that electrical equipment is turned on and working, as shown in **Figure 12-37**. LEDs are also used in remote controls for TVs and other audio visual equipment. When a button is pressed on the remote, a pattern of low frequency infrared light is detected by the receiver. The receiver interprets the pattern to perform the desired task.

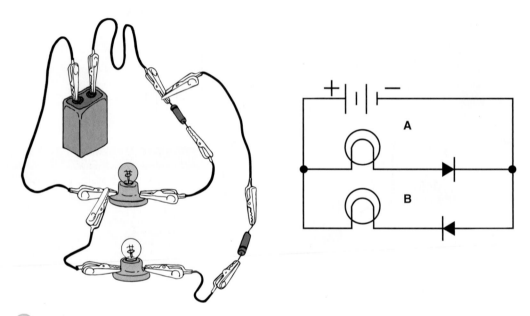

Figure 12-35. A circuit set up with two lamps in parallel. Each lamp has a diode connected in series, but only one lamp will light. Which lamp will light, and why?

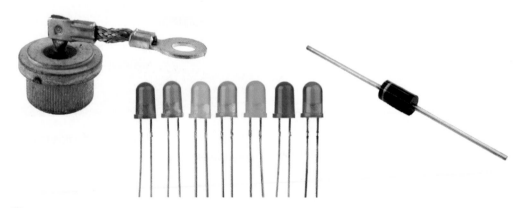

Figure 12-36. Different types of diodes.

Figure 12-37. LEDs are often used to show the status of electronic devices. How many LEDs do you see in this sound system?

Capacitors

Capacitors are designed to store an electrical charge and release it all in one quick burst. This extra storage of voltage is sometimes needed in circuits that need sudden increases in voltage. One example is a camera flash. A battery cannot supply a sudden high voltage surge for the flash mechanism. However, a capacitor can deliver energy fast enough. See **Figure 12-38.** A capacitor's storage potential is measured in units called *farads*.

SAFETY: A capacitor connected in a direct current circuit can store a charge for a considerable time after the voltage to the circuit has been switched off. **NEVER** touch the leads of a capacitor before it has been discharged.

Figure 12-38. A capacitor is a "sandwich" made up of conductors and insulators.

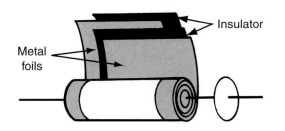

Insulator

Metal foils

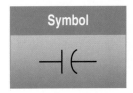

Symbol

Transistors

A *transistor*, **Figure 12-39**, contains a semiconductor material that acts like a switch. The most common semiconductor materials are silicon and germanium. The semiconductor allows electron flow only under certain conditions. By turning on and off, transistors signal the ones and zeros that combine to signify information stored in a computer.

All transistors have three terminals: a collector (c), a base (b), and an emitter (e). An electric current flows through the transistor only when an electrical voltage is applied to the base. For the base of an NPN transistor to have a voltage, it must be connected to the positive side of the battery. See **Figure 12-40**. If it is connected to the negative side of the battery, the base has a low voltage and the lamp will not light. See **Figure 12-41**. Therefore, the lamp can be switched on and off by changing the voltage on the base of the transistor.

Transistors are used as fast switches in timing, counting, and computer circuits. In these circuits, the signal is either on or off. Amplifying transistors are used in radios and stereos where a weak signal must be amplified in order to be heard over a speaker. Hearing aids also use transistors to amplify sound.

Figure 12-39. Negative voltage controls the output of a PNP transistor. Positive voltage controls the NPN transistor.

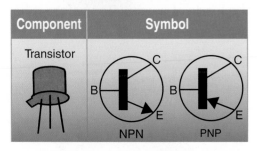

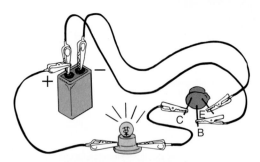

Figure 12-40. The base terminal of an NPN transistor is connected to the positive terminal of the battery. The lamp will light.

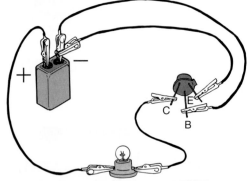

Figure 12-41. With the base terminal of the NPN transistor connected to the negative terminal of the battery, the lamp will not light.

Logic Gates

Logic gates are digital circuits formed by transistors, resistors, and diodes. They process one or more input signals in a logical way. Logic gates are based on the binary number system and have two states: off and on. These states correspond to the binary digits 0 and 1. **Figure 12-42** shows how the five basic logic gates work. The first column shows the symbol for each logic gate. The second column contains truth tables, which show the output states for every possible combination of input states. The third column explains how each logic gate functions.

Symbol	Truth			Description
A ⊃ **C** **B** AND Gate	Input		Output	Both the input value AND the output value must be 1 (on) in order for the output value to be 1. This is like having two switches in series.
	A	B	C	
	0	0	0	
	0	1	0	
	1	0	0	
	1	1	1	
A ⊃ **C** **B** OR Gate	Input		Output	One OR more of the input values must be 1 (on) to create an output value of 1. This is like having two switches in parallel.
	A	B	C	
	0	0	0	
	0	1	1	
	1	0	1	
	1	1	1	
A ⊳○ **C** NOT Gate (Inverter)	Input	Output		If the input value is 1 (on), the output value is NOT 1; it is 0 (off). If the input value is 0 (off), the output value is NOT 0; it is 1 (on). Because NOT gates reverse an input value, they are often called inverters.
	A	C		
	0	1		
	1	0		
A ⊃○ **C** **B** NAND Gate	Input		Output	Produces a 1 (on) only if all inputs are 0 (off). Produces a 0 if any or all of the inputs are 1. Notice the little circle at the end of the symbol, similar to the circle on the NOT gate. The NAND gate combines the function of the AND gate and the NOT gate.
	A	B	C	
	0	0	1	
	0	1	1	
	1	0	1	
	1	1	0	
A ⊃○ **C** **B** NOR Gate	Input		Output	Produces a 0 (off) only if all inputs are 1 (on). Produces a 1 (on) if any or all of the inputs are 0 (off). The NOR gate combines the OR gate and the NOT gate.
	A	B	C	
	0	0	1	
	0	1	0	
	1	0	0	
	1	1	0	

Figure 12-42. Characteristics and symbols for logic gates.

Integrated Circuits

The electronic circuits described so far in this chapter have been built using separate components. An *integrated circuit* is a more advanced circuit in which all the components, including transistors, resistors, capacitors and diodes, are placed together on silicon wafers. Many integrated circuits are about the size of your smallest fingernail, or smaller. They are also called *chips* or *ICs*. See **Figure 12-43**. The earliest ICs, built in the 1960s, only contained a few components. By the 1980s, hundreds of thousands of components were integrated onto a chip. Today a chip may contain several billion.

The world is full of integrated circuits. If you have ever received a greeting card that plays a song, you will know that the chips can be so small they are hard to find. Hidden between the layers of paper is a miniscule battery, a speaker, and a chip on which the sound is recorded. Communications, manufacturing, and transportation systems all rely on ICs.

The three types of integrated circuits are analog, digital, and mixed. Digital ICs work by computing binary mathematics electronically. Analog ICs work by processing continuous signals and perform functions such as sound amplification and mixing. A mixed integrated circuit combines both analog and digital circuits on a single chip. They are often used to convert analog signals to digital so that digital devices can process the signals.

The two main advantages of ICs compared to circuits with separate components are cost and performance. Cost is lower because the components are not added one at a time. Instead, they are printed onto a silicon wafer. Performance is higher because the components are small and close together. They consume little power.

A more recent development has been the introduction of programmable ICs. These devices contain circuits that can be programmed by the user rather than being fixed by the manufacturer. One type, PICAXE chips, are used for school microcontroller projects. They can be programmed using a computer language that is a variant of BASIC.

Figure 12-43. Integrated circuits, or ICs, are produced in various shapes and sizes and contain different circuitry for specific purposes.

Lasers

The term *laser* is an acronym for "Light Amplification by Stimulated Emission of Radiation." Laser light is a form of radiation that has been boosted to a high level of energy. It produces a strong, narrow beam of light. Lasers have made possible such conveniences as automatic supermarket checkouts, fiber optic communication, and a new generation of printing devices. Lasers are also key to the development of compact discs used in entertainment. Laser light is used to read or burn digital information onto the CD.

When a recording is made on a CD, the laser "burns" a pattern of pits on the center layer of the disc. During play, when the laser strikes a pit, it detects rapid flashes of high and low intensity signals that it reads as zeros and ones. This binary information is translated by a microprocessor (chip) back into music.

Blu-ray and HD DVD discs can contain more data than CDs or DVDs. The data pits can be smaller because they are read by a blue-violet laser. The blue-violet laser has a shorter wavelength than the red laser used to read CDs.

Lasers are also used for many other applications, including surgery, reading bar codes, and cutting metal and other materials. **Figure 12-44** shows an industrial laser cutter being used to make accurate cuts in metal.

Think **Green**

Recycling Electronic Devices

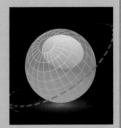

With electronics technology improving every day, our electronic devices become "old" quickly. New, improved products are constantly being introduced. When we replace our existing electronics with new versions, what happens to the old ones? How responsibly do we act?

The first question to ask yourself is whether you really need the latest and greatest products. Will they really work better than your existing electronics, or are they just a status symbol?

When you do decide to upgrade your electronic devices, the next question is what to do with the old ones. If the devices are in working condition, one good solution is to give them to charity. Many organizations collect electronics for use in schools and community programs. National organizations such as Goodwill and CollectiveGood have electronics recycling and reuse programs. Your community may also have donation and recycling opportunities. The next time you replace an electronic device, research opportunities for donating or recycling your old device. You may be able to help people in need, as well as doing your part to help the environment.

Figure 12-44. Lasers are used in industrial cutting machines because of their high level of precision.

Privacy and Security

All information transmitted electronically can potentially be accessed by spy agencies or hackers. Furthermore, data may be altered or deleted by those who spread a computer virus, whether that data belongs to you, a bank, or the government.

Electronic devices can be used to protect private homes. A video camera might be hidden behind the face of a clock or the eye of a teddy bear. However, outsiders may access other cameras installed in your home. For example, using the right software, hackers could turn on the small camera that is attached to a computer. If your computer has an "always-on" high-speed connection, someone else could use it as an Internet server.

It is likely that you are being watched when you use public electronic devices. At an ATM, the bank machine might record time-and-date-stamped video images of you as you withdraw cash. See **Figure 12-45.** When you use a cell phone, your location (in which wireless cell) can be calculated to within a few yards. At home on your computer, your Internet Service Provider (ISP) can record information related to your e-mails, the contents and source of every Web site you visit, and the length of time you spend there. Any member of your family can also check which Internet sites you access using special software downloaded from the Internet.

You can take some basic steps to help ensure your privacy. For example, for computer applications, create and use passwords that are difficult to break. Install security and firewall software to help keep unauthorized people from accessing your computer or Internet service. When you need to discuss confidential information on the telephone, use a wired (landline) telephone instead of a cell phone.

Figure 12-45. In many cities, traffic signals are enforced by security cameras. The cameras are programmed to take pictures of vehicles that fail to obey stoplights.

The era of electronics really started in 1947 when transistors were invented at the Bell Laboratories. Transistors provided a small, inexpensive, and efficient way to amplify electrical currents. Early transistors were used in transistor radios that were so much smaller that for the first time a person could carry a radio anywhere.

Today, one silicon chip can have millions of microscopic transistors, resistors, and conductors on it. Many products we use every day contain microprocessors that automate many tasks to make our lives easier.

Electronics can be used both to protect and to invade your privacy. How much access to personal data do we want government and business to have? This is an important question. The answer has to respect the right of an individual to privacy versus the legitimate needs of others to gather information for security purposes. Governments are concerned about national security issues, so the trend toward collecting even greater amounts of information will likely continue. Also, advertising and marketing companies are searching to find people's spending habits. As technology improves, so does the capability for electronic surveillance. How much is enough? How much is too much?

○ An electric circuit is a continuous path from a source, through a load, and back to the source. The three basic types of circuits are series, parallel, and series-parallel.

○ Materials that allow current to flow are called *conductors*. Those that do not allow current flow are called *insulators*. *Semiconductors* allow current under certain conditions.

○ Electrical energy is measured in terms of volts, ohms, and amperes. Work performed by an electrical current is measured in watts.

○ Electronics is the use of electrically controlled parts to control or change current in a circuit.

○ Advances in electronics technology provide ways to collect information about people without their knowledge.

Connecting to Prior Knowledge

Copy the following graphic organizer onto a separate sheet of paper. Allow space in each row for one or more sentences. Before you read the chapter, write sentences in the first column to record your current knowledge about electric circuits and electronic components and devices. As you read through the chapter, record new concepts you learned by reading the chapter. When you are finished, review the chart to see how much you have learned. If you have questions about topics in the chapter, record your questions at the bottom of the chart. Ask your questions in class or do research on your own to find the answers.

What I Know (or Think I Know) about Electricity and Electronics	What I Learned in This Chapter
Electric Circuits	
Conductors, Insulators, and Semiconductors	
Measuring Electrical Energy	
Electronics	
Privacy and Security	
Further Questions:	

Test Your Knowledge

Write your answers to these review questions on a separate sheet of paper.

1. Define the term *electric circuit*.

2. What might happen if too many appliances are plugged into the same electrical outlet?

3. Explain how a fuse protects a circuit from overload.

4. Briefly explain the differences between a series circuit, a parallel circuit, and a series-parallel circuit.

5. Name four materials that are conductors of electric current and four insulators.

6. A portable electric heater with a resistance of 15 ohms is connected to a 120 V AC outlet. What is the current flow in the circuit?

7. What is an integrated circuit?

8. Name the five basic types of logic gates.

9. Calculate the value of each of the following resistors.

	1st Band	2nd Band	3rd Band	4th Band
A.	red	green	yellow	silver
B.	orange	blue	brown	gold
C.	white	brown	red	none
D.	violet	green	orange	silver

10. Name two things you can do to help prevent unauthorized access to your private information.

Critical Thinking

1. Suppose you are fixing a snack after school one day, and suddenly all the kitchen lights go out. You check the refrigerator, and it has no power. Then you check the rest of the house, and you find that the TV still works in the family room. Where would you begin looking for the source of the outage in the kitchen? Why?

2. Some of the needs and wants people express are created by the use of inventions and innovations. Explain how this is true of the development of integrated circuits and microprocessors. What needs and wants have they caused? How would our society be different without them?

3. Advances in electronics technology has led to widespread use of robots in manufacturing. Probe the pros and cons of using robots to manufacture products. What are the advantages? What are the disadvantages?

4. Technologists work constantly to design new products and improve existing products in many different fields. They do not work alone, however. Write a paragraph explaining how technologists in various fields (such as electronics and manufacturing) work together with technologists, scientists, and other workers in other fields.

Apply Your Knowledge

1. Draw a circuit diagram for a circuit that contains a lamp, dry cell, fuse, and switch connected in series.

2. Draw a circuit diagram for a circuit that contains two lamps and a dry cell connected in parallel.

3. Airport security systems use a variety of security measures to help increase passenger safety. These measures increase security at the expense of privacy. Research the various security measures being used at airports. Explain how these measures have been influenced by past events, and predict what measures may be added in the future.

4. Describe how you could build a circuit that would turn on a light whenever someone enters a room and turn the light off 5 minutes after the person leaves the room. What components would you need? Sketch the circuit.

5. Search the Internet for information about electronic privacy. Can organizations collect, use, or disclose personal information about you without first telling you their intentions and obtaining your consent? How clear must their explanations be as to what they intend to do with the information? Write a report explaining your findings.

6. Research one career related to the information you have studied in this chapter. Create a report that states the following:

   - The occupation you selected
   - The education requirements to enter this occupation
   - The possibilities for promotion to a higher level
   - What someone with this career does on a daily basis
   - The earning potential for someone with this career

   You might find this information on the Internet or in your library. If possible, interview a person who already works in this field to answer the five points. Finally, state why you might or might not be interested in pursuing this occupation when you finish school.

STEM Applications

1. **MATH** Ohm's law states that voltage equals current times resistance, or $E = I \times R$. Rewrite this equation to find the current in a circuit if you know the voltage and resistance. Then use your equation to find the current in a 12 V circuit if the total resistance is 8 Ω.

2. **ENGINEERING** Design a lighting circuit for a public restroom in which the exterior light above the door stays lit all the time, but the interior light comes on only when the door is closed. Sketch a diagram for your design.

13

Information and Communication Technology

Better by Design

Nicholas Negroponte leads the One Laptop per Child design team

Nicholas Negroponte is a computer scientist who wants to provide every child worldwide with a laptop with software designed for joyful, self-empowered learning. To achieve this objective, the *One Laptop per Child* association has developed the XO laptop. The XO uses open-source software and has a video/still camera. Importantly, it contains no hazardous materials. The laptop can be recharged by human power using a crank, a pedal, or a pull-cord. The *Give One, Get One* program asks donors to pay $400 for two XOs: one XO laptop is sent to the purchaser and a second is sent to a student in a poor country.

Students in Khairat, India, learn by interacting with the computer.

The *Give One, Get One* program helped make a computer available for this student in a remote area of Nigeria.

"Computing is not about computers any more. It is about living."

Reading Target

Finding the Main Idea

As you read this chapter, look for the key points, or main ideas, in each part of the chapter. Then look for important details that support each main idea. After you have read the entire chapter, use the Reading Target graphic organizer at the end of the chapter to organize your thoughts about what you have read.

Key Terms

analog signals
binary digital code
bit
Bluetooth® technology
byte
central processing unit (CPU)
cloud computing
communication

communication technology
data
digital signals
distributed computing
feedback
haptic device
hardware
hypertext transfer protocol
  (HTTP)
information
information technology
media
microelectronics
modulate
motherboard
radio frequency
  identification (RFID)

retrieval
robotics
router
server
social media
software
storage
telecommunication
  technology
triangulation
virtual reality
Voice over Internet Protocol
  (VoIP)
Wi-Fi™ technology
World Wide Web
  (Web or WWW)

Objectives

After reading this chapter, you will be able to:

○ Describe the components of a typical communication system.

○ Explain difference between data and information.

○ Give examples of information and communication technology systems.

○ Compare conventional and modern examples of telecommunication technology.

○ Describe the role of computer technology in information and communication systems.

○ Explain why the Internet is considered the world's largest telecommunication system.

Useful Web sites:
wiki.laptop.org/go/Home
www.ted.com/index.php/talks/Nicholas_negroponte_on_one_laptop_per_child.html

Communication is any exchange of information. People, animals, and even machines communicate regularly. Talking is one form of communication. Writing is another. In fact, our society depends heavily on *communication technology*—the technology that allows us to communicate.

Communication Systems

Some communication systems consist of the people, machines, and methods that allow us to communicate. As you may recall from Chapter 3, some communication is verbal, and some is nonverbal. However, all communication systems have certain things in common. As shown in **Figure 13-1**, a complete communication system includes:

- **Source**—The person or machine that has a message to be delivered.
- **Encoder**—Device that changes the message into another form for transmission.
- **Transmitter**—Device that sends the encoded message toward its destination.
- **Medium**—The wired or wireless means used to send the information.
- **Receiver**—The device that accepts the encoded information and relays it to the decoder.
- **Decoder**—Device that translates the encoded message into an understandable form.
- **Destination**—The person or machine that receives the decoded message.

These seven steps are used to transmit and receive a message. In addition, messages may be placed in *storage* so that they can be retrieved at a later time. Voice mail is an example of a storage device.

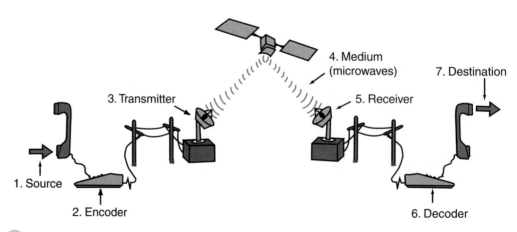

Figure 13-1. Talking to a friend on the phone is an example of using a communication system.

Let's look at an example of a communication system. Suppose that you have an idea to meet in a specified chat room online with a friend from another country. You call your friend on the phone and say, "Let's meet online tonight in this chat room." Do you see the system? You are the source. The telephone encodes the message. Satellites transmit your message. A satellite dish acts as the receiver of your information and relays the message to your friend's phone, which decodes the message. Finally, your friend is the destination of the message.

Suppose your friend cannot meet online tonight. "Sorry, I can't. I must study tonight." Your friend provided you with *feedback*, which is a response to the receiver's question or statement. This proves that your message was heard and understood. Your friend then says, "But let's remember to do it Friday night." Then both of you will store the information in your mind. When Friday comes, you will remember to meet online. This is an example of *retrieval*.

Types of Messages

The type of communication system we use depends on the type of message to be sent. Some communication systems allow humans to communicate with other humans. Others transmit information from humans to machines, and still others allow machines to communicate information to humans or to other machines.

Messages intended for machines are very different from those meant for people. They may be computer programs to run computerized machine tools, or access codes or passwords for a secure Internet site. Machines can also send messages to humans. For example, if you leave a car door open while the engine is on, a light appears in the dashboard to inform you that the door is open. In some cars, a computerized voice tells you that the door is open.

The purpose of the message also affects its design. Politicians often state certain views on issues in order to gain the favor of the majority of the voters. Movie directors and writers make scary movies to draw people who love horror films to the movie theaters. Stand-up comedians draw people that love to laugh to comedy clubs. Musicians write and play certain styles of music to get more people to buy their songs.

Types of Media

You and millions of other people around the world use the Internet every day. You may use it to communicate with others, keep up with the news, check the weather, make travel plans, conduct business, shop, buy a ticket to a rock concert, entertain yourself, or learn new things. Staying connected has become so important that it is hard to get away from your computer and your Internet connection. You feel that you might miss an important e-mail message or some news you need to know.

It is likely that you have several other forms of *media*, or communication and information sources, in your home. These may include radios, televisions, newspapers, magazines, photographs, store catalogs, and a telephone directory. In these sources, information flows in one direction. It starts at a source, such as a newspaper, the television or radio, and flows to an audience. The audience cannot immediately interact with or comment on what is provided. See **Figure 13-2.**

Social media is any digital tool or service that uses the Internet. Examples include YouTube, Flickr®, Facebook®, MySpace™, message forums, message boards, blogs, wikis, and podcasts. These media lead to the formation of communities around shared interests. Using online tools, members of the audience can become involved, communicating and exchanging content. Each member in the conversation is both a member of the audience and an author. He or she can create, comment on, share, or remix content.

Today, social media is replacing older forms of media. For example, a critique of a concert appears not only in a newspaper, but also in blogs on a multitude of Web sites. In various ways, these sites become interactive. See **Figure 13-3.**

Information Technology

Imagine that your teacher has asked you to measure the height of all the students in your class and to record these heights in a list. You must also record, for each height, whether the person you measured is male or female. When you have completed this task, you have a set of facts, or *data*, about height and gender.

Now your teacher asks you to do three things with this data. First, divide the list into two separate lists: one for females and one for males. Second, calculate the average height of the females and the average height of the males. Third, write a statement that describes your results.

Figure 13-2. Reading a newspaper is an example of traditional media. What are the advantages and disadvantages of this form of communication?

Figure 13-3. Getting news and information from the Internet is an example of social media.

Suppose your results are as follows: The average height of the females is 5'6", and the average height of the males is 5'8". You now have some information. Your statement could read as follows: The average height of the sixteen males in my class is two inches greater than the average height of the fourteen females. See **Figure 13-4**.

Information is data that has been organized in a meaningful way. For example, your calculations in the above scenario helped you draw conclusions about the difference in height between the males and females in your class.

We are bombarded daily with huge amounts of data and information. We can remember some of this information, but much of it must be stored, processed, and communicated by machines. Today, computers are used for these purposes. *Information technology* is the computer-based technology used for storing and processing information.

Females (14)	Height (inches/cm)	Males (16)	Height (inches/cm)
Female 1	65/165	Male 1	70/178
Female 2	69/175	Male 2	69/175
Female 3	63/160	Male 3	72/183
Female 4	66/168	Male 4	67/170
Female 5	68/173	Male 5	68/173
Female 6	62/157	Male 6	69/175
Female 7	61/155	Male 7	69/175
Female 8	68/173	Male 8	66/168
Female 9	66/168	Male 9	71/180
Female 10	68/173	Male 10	66/168
Female 11	70/178	Male 11	62/157
Female 12	69/175	Male 12	68/173
Female 13	67/170	Male 13	65/165
Female 14	62/157	Male 14	67/170
		Male 15	70/178
		Male 16	69/175
Total (Σ)	Σ = 924/2346		Σ = 1088/2763
Average = Total ÷ # of females or males	924 ÷ 14 = 66" 2346 ÷ 14 = 168 cm		1088 ÷ 16 = 68" 2763 ÷ 16 = 173 cm

Each height is one piece of data

Information

The average height of the sixteen males in my class is two inches (5 cm) greater than the average height of the fourteen females.

Figure 13-4. When you arrange data in a meaningful way, it becomes information.

Information and Communication Technology Systems

Information technology is closely related to communication technology, and the two are often used together. Computers are one of the most common forms of information and communication technology. Other forms include televisions, cameras, cell phones, and similar items. These technologies help us communicate both sounds and images.

The growth of information and communication technology is largely due to the development of microelectronics. *Microelectronics* is the use of very small versions of traditional electronics. Switches and circuits are made incredibly small. This has resulted from the invention of new manufacturing processes and the use of new materials.

Information and communication technology systems are found in many places. One common example is a modern office. In the office, documents are created and corrected on a computer using a word processing program. They are stored electronically. A document can be added to an e-mail message as an attachment and sent instantly to anywhere in the world. It can be sent electronically to a photocopy machine, many identical copies can be printed. Meanwhile, the employee may be speaking directly to someone on the other side of the world using a telephone. If the number is busy, the phone automatically redials until the line is free.

Department stores are another example of how machines can be linked in a system. Purchases are scanned by a sales assistant using a bar code reader. The scanner sends information from the bar code to a computer connected to the cash register. The system totals the amount owed. At the same time, it sends information on the sale to the store's headquarters. There, another computer uses the information to monitor inventory. Payment for the purchases is also processed electronically.

Figure 13-5. ATM machines allow us to access cash 24 hours a day. What impact does this have on people's shopping habits?

Even the banking industry uses electronic communication systems. See **Figure 13-5**. In addition to ATMs, where you can deposit checks and withdraw money, most banks now allow customers to do most banking tasks online. Customers can check their balances, transfer money between accounts, and pay bills using online banking. Some banks even allow customers to deposit checks using iPhone technology.

Information and communication today rely heavily on microelectronics, computer, and telecommunication technologies. These are combined into systems that:

- Create, collect, select, and transform information
- Send, receive, and store information
- Retrieve and display information
- Perform routine tasks

For example, a night watchman responsible for the security of a large building can monitor a closed-circuit television system. It receives, displays, and stores information about the condition of different parts of the building.

Hospitals are another example. They use electronic instruments to measure a patient's pulse, heartbeat, and other vital signs. See **Figure 13-6**. Most hospitals now track patient medications using electronic methods also.

The most important result of microelectronics is the microprocessor, or computer chip. See **Figure 13-7**. As you learned in Chapter 12, a computer chip is a tiny flake of silicon covered with microscopic electronic circuits. Chips can be mass-produced in the tens of thousands, which reduces the price per chip. A single chip costing less than a video game contains most of the switches and circuits needed by a computer.

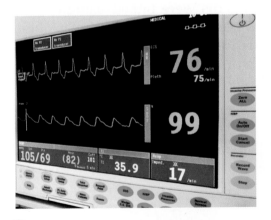

Figure 13-6. A cardiac monitor in a hospital tracks a patient's pulse, heart rhythm, and blood pressure, among other things. The monitor sounds an alarm if the patient's vital signs become abnormal.

Figure 13-7. A single microprocessor may contain thousands of tiny components.

It may surprise you to learn that more than 90% of all microchips are *not* in desktop or laptop computers. They are in cars, homes, and industrial machines. They are around us to such an extent that we rarely notice them. For example, a microchip controls the antilock brakes on cars. In pacemakers, the miniature components on a chip time heartbeats. Microprocessors also set thermostats, switch DVDs on and off, pump gas, and control car engines. Robots and on-board computers in satellites rely on them. See **Figure 13-8.**

Telecommunication Technology

The other technology that has led to the growth of information and communication technologies is telecommunication. *Telecommunication technology* is a general term for technologies that can send information over distances. Radio, television, cell phones, landlines, and the Internet are all aspects of telecommunication.

Conventional Technologies

Conventional telecommunication technologies include landline telephone systems, local radio stations, and local and network television stations. These technologies are still in use today, along with newer technologies that may one day replace them.

Telephone

Originally, "landline" telephone systems were linked using copper telephone lines. Many of these have now been replaced by fiber optic cables. Fiber optic cables can carry light-coded messages over long distances. A fiber optic strand four-thousandths of an inch (0.1 mm) in diameter is capable of carrying 2000 two-way telephone conversations at once.

Figure 13-8. A GPS system is an example of a communication device that uses microchips to process information in real time.

See **Figure 13-9**. At the transmitter end of a fiber optic-cable network, telephone signals are converted into pulses of light. These pulses travel through the glass fibers to the receiver end. There, the pulses are converted back into electrical signals that are carried to a telephone or computer.

Radio

Local radio programs are transmitted using electromagnetic waves. The radio transmitter codes the sounds by changing the electromagnetic waves in some way. For example, some types of radio *modulate* (change) the amplitude, or height, of the waves. Others change the frequency of the waves, or number of waves per second. See **Figure 13-10**. The modified waves move through the air to a receiver that contains an antenna. The antenna converts the changes, or modulations, in the sound waves back into sound. The radio then amplifies the sound and sends it through a speaker so the listener can hear it.

Radio signals tend to weaken over distance. They can also be distorted by noise, or other electromagnetic fields in the area. This is why radio technology was originally a local medium only.

Television

The first United States standard for sending television audio and video signals was created by the National TV Standards Committee (NTSC) in 1941. The NTSC standard signals were sent as a continuous stream of information. These signals are called *analog signals*. Today, all television signals are transmitted in digital form. The audio and video are converted to binary code to form *digital signals*.

The two types of digital television are standard digital television (SDTV) and high-definition television (HDTV). SDTV is an improvement over the old analog television systems because the signal does not degrade, or weaken, over distance. Colors are truer and images are crisper, too. HDTV takes digital technology one step further. It uses more bits per second than SDTV. This results in better sound and picture resolution.

Figure 13-9. Fiber optic cables carry information coded as pulses of light. They can transmit data faster than copper wire, and they have a higher capacity.

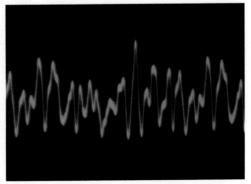

Figure 13-10. Electromagnetic waves can be modulated, or changed, to carry sounds.

Cellular Technology

Cell phones are really radios. In a typical analog cell phone system, the cell phone carrier divides the city into "cells" or areas. Each cell uses radio frequencies to transmit signals to and from your phone. The cells are small, usually only about 10 square miles (25.9 square kilometers). This size allows a city to reuse the same frequencies in different cells. Each cell has a transmission tower that receives and sends the radio signals. See **Figure 13-11**. As you move around in the city, the cell phone carrier's wireless network tracks the signal. When the signal starts getting weak, the network switches your phone automatically to another cell where the signal is stronger.

In a digital cell phone, voices are converted to binary code. The code is transmitted wirelessly to the phone on the receiving end. There, the binary code is converted back into voice so the person can hear what the caller is saying.

Satellite Technology

Satellite telephones use a technology that is similar to cellular technology. However, they send signals to satellites instead of cellular transmission towers. Satellite technology has the advantage that it can cover the entire Earth. See **Figure 13-12**. Signals may be transmitted from satellite to satellite before reaching their final destination. When they reach their destination, a satellite dish collects the signals and processes them, as shown in **Figure 13-13**.

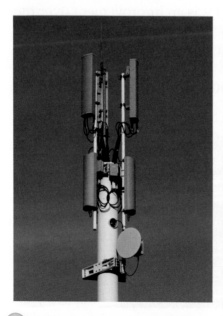

Figure 13-11. A transmission tower with cellular antennas receives and transmits radio signals on the cellular frequencies assigned to its cell.

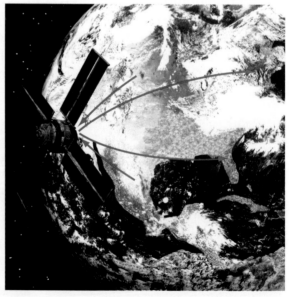

Figure 13-12. Satellite technology allows digital phone signals to be received by satellites orbiting the Earth. The satellite may send the signals to another location on Earth or to another satellite that is in a better position to transmit the signal back to Earth.

Communication satellites are now used for many different telecommunication needs. They can handle more information than cables. They can also transmit it faster. For this reason, many television and radio stations now use satellite signals.

Satellites are also used by global positioning systems (GPS). The GPS system in a vehicle calculates its position by measuring the amount of time satellite signals take to reach it. The system uses three satellites. Comparing the signals from three satellites narrows the possible location to a point where the signals from the three intersect. This method is called *triangulation*.

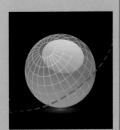

Think **Green**

Space Trash

In 1957, the former Soviet Union sent the first satellite into orbit around the Earth. Since then, hundreds of satellites have been launched. Many of them broken or obsolete. When they are no longer useful, satellites become "space trash." What happens to them?

If they are in a low orbit (close to Earth), they eventually fall back to Earth. Although some of the smaller pieces burn up before they reach the ground, other pieces hit the Earth's surface. This poses a danger to humans if the satellite hits a populated area. Satellites in a higher orbit continue to circle the Earth. This may not sound like a big issue at first. However, it does cause problems. The biggest danger is that it may hit satellites that are still working or collide with spacecraft traveling to or from the International Space Station. Space trash, like working satellites, travels at speeds exceeding 20,000 miles per hour (32,200 kilometers per hour). Even a small particle traveling at such speeds can damage sensitive equipment.

NASA keeps track of space trash to help guide our current satellites and the space station around it. This is a good short-term solution. As we abandon more and more equipment in space, however, the problem will only get worse.

What can we do to prevent the buildup of trash in space? Governments all over the world are studying this problem. They have not yet found a solution that is practical, but the problem will have to be dealt with soon. Do some research and find out what has been proposed so far. Evaluate the different proposals. Can you think of any other ideas that might work? As a student, you may not be able to do much to solve this problem right now. You *can* remain informed, however. Someone in your generation may find the answer that will help keep Earth and the space surrounding it clean and safe.

Computer Technology

A computer is really just a machine that can accept information, process the information, and then output the result. It can perform these actions at very high speeds. Computers can also store and retrieve both raw and processed data.

Figure 13-13. When satellites direct signals to Earth, the signals are collected by a dish-shaped satellite receiver.

Computer Processes

Information processed by a computer must be input in a form that the computer can handle. Computers use a simple code of electrical signals. There are only two signals in this code, on and off. These are written as 1s and 0s. This is called a *binary digital code*. Binary means two, and digital means number. Inside the computer, an on condition is used to represent a 1. An off condition represents a 0. This is similar to the logic gates described in Chapter 12.

To provide for more than two possible states, signals are combined to produce patterns. Imagine a set of four lightbulbs, as shown in **Figure 13-14**. Each one can be turned on or off. Each of these bulbs can be assigned a value.

In the binary system, each digit has a value twice as large as the one on its right. The bulbs are therefore assigned values of 8, 4, 2, and 1. When a bulb is off, it represents 0. When it is switched on, a bulb represents the value assigned to that position. The values are added together. In **Figure 13-14**, the first bulb (1) and the third bulb (4) are on. This represents the number 5, because $1 + 4 = 5$. If the first bulb is off and the second, third and fourth are on, the number represented is 14, because $2 + 4 + 8 = 14$. This group of four bulbs can represent any number from 0 (all off) to 15 (all on).

Figure 13-14. You can think of binary values as a series of light bulbs that are either on (1) or off (2).

Each on or off signal is known as a *bit*, which is short for "binary digit." When bits are used in groups, the groups are called *words*. The example in **Figure 13-14** uses a four-bit word. Since computers need to work with numbers larger than 15, a system of eight-bit words is used, as shown in **Figure 13-15**. Each eight-bit word is a *byte*.

Types of Computers

Until the 1940s, computers were largely mechanical devices. World War II presented a new challenge: how to crack secret enemy codes that changed three times a day. Using the ideas of mathematician Alan Turing, a computer called Colossus was made in 1943. It used 2000 vacuum tubes. Intercepted messages were fed into Colossus as symbols on paper tape. The computer processed them at the rate of 25,000 characters per second. However, the Colossus machine was limited to breaking codes. The first electronic computer capable of tackling many different jobs was ENIAC in 1945. See **Figure 13-16**.

Place Value	128	64	32	16	8	4	2	1
Binary No.	1	1	0	1	1	0	1	0

```
128  ─────────┐
64   ────────┐│
16   ──────┐ ││
8    ─────┐│ ││
  2  ────┐││ ││
─────
 218
```

Figure 13-15. The byte 11011010 represents the number 218.

Math Application

Converting Binary Numbers

To understand how to read binary numbers, first look at our normal system of numbers. We use the base 10 number system. It is called that because it consists of ten numbers: 0, 1, 2, 3, 4, 5, 6, 7, 8, and 9. The columns determine whether the numbers are multiplied by 1, 10, 100, 1000, and so on. For example, consider the number 265. The 5 is in the 1s column (5 × 1). The 6 is in the 10s column (6 × 10), and the number 2 is in the 100s column (2 × 100).

The binary system is similar in that each column represents a value. However, it also has a major difference. In our base 10 system, we can count from 0 to 9 before we move to the next column. The binary system has only two numbers, so there are only two possibilities for any column: 0 or 1.

To convert a base 10 number to binary, make a table similar to the one below. It consists of the top row (Place Value) of the table in **Figure 13-15**. Starting from the left, compare each number to the number to be converted. Select the first column that shows a number that is equal to or smaller than the base 10 number to be converted.

Take the number 5, for example. Four is the largest number that is smaller than 5, so write a 1 in the 4s column and subtract 4 from 5. This leaves you with 1. Still reading from the left, what is the next column that is equal to or smaller than 1? It is not the 2s column, so write a 0 in the 2s column. The 1s column is equal to 1, so write a 1 in the 1s column. Subtracting 1 from 1 leaves 0, so the conversion is now complete. The number 5 in base 10 equals 101 in binary: 1 + 0 + 4 = 5. Check your work by adding up the column values, as shown.

Place Value	128	64	32	16	8	4	2	1
						1	0	1

```
  4 ←────────────────┐
  0 ←──────────────┐
+ 1 ←────────────┐
─────
  5
```

Converting Binary Numbers (continued)

To convert a binary number to a base 10 number, first create the same kind of table you made for the conversion to binary. This time, however, write the binary number in the columns, as shown in the example below. In this example, we are using the binary number 11001. Write the 1s and 0s in the columns, starting at the right (1s) column and working to the left. Then add the values of the columns to arrive at the base 10 number.

Place Value	128	64	32	16	8	4	2	1
				1	1	0	0	1

```
16  ←
 8  ←
 0  ←
 0  ←
+1  ←
───
25
```

Math Activity

Practice converting numbers between base 10 and binary.

1. Convert the following binary numbers into base 10 numbers.

   A. 111   C. 01010

   B. 1011   D. 00110011

2. Convert the following base 10 numbers into binary numbers.

   A. 4   C. 35   E. 101

   B. 20   D. 70

Figure 13-16. ENIAC stands for "Electronic Numerical Integrator and Computer." It was designed to calculate ballistic firing tables. However, it was fully programmable, so it could perform many other tasks as well.

Today, computers are available in a range of sizes and abilities. Most homes have at least one laptop or desktop computer. Businesses often have networks of powerful desktop computers that are sometimes called *workstations*. Larger businesses that need more computing power may use a more powerful minicomputer. Research companies, college research facilities, and other organizations that need to work with a large amount of data may use a mainframe or supercomputer. For example, weather forecasting requires a supercomputer.

In the past, supercomputers were very large, expensive computers. Today, the Virginia Tech Terascale Cluster is one of the fastest computers in the world. It is not really a single computer, however. It is actually a group, or cluster, of 1,100 Apple Macintosh computers. Problems are broken into smaller chunks and distributed among the computers. This is known as *distributed computing*. This approach is used by companies such as Akamai, which provides distributed file storage and retrieval around the world. When you download a movie clip or a big file from a Web site, you are probably receiving it from this type of distributed system.

Computer Components

Computing involves three stages: input, processing, and output. **Figure 13-17** shows a variety of devices used at each stage. Increasingly, the input and output devices are wireless. Computer components include both hardware and software.

Hardware includes all the physical parts of the computer. This includes the *central processing unit (CPU)*, input devices, and output devices. It also includes temporary and permanent memory and the physical parts necessary to support sound and graphics. These parts are often located on circuit boards that plug into the *motherboard*, which contains the main circuitry of the CPU. You may hear these circuit boards referred to as *boards* or *cards*.

Input	Processing			Output
Mouse				Monitor
Keyboard			Data stored in memory	Printer
Modem	Input	Software programs		Modem
Game controller				Other computers
Touch-sensitive screen				CD/DVD
Graphics tablet				Flash drive
Voice		Calculations		Speakers

Figure 13-17. Computers can have many different input and output devices. The devices that are used depend on the user's needs and preferences.

Computers that are not networked must have some type of storage device, such as hard disk drive. Networked computers may also have storage devices. However, some businesses use a common *server* on which all work is stored. The individual computer consoles, or workstations, access the server using the network.

Software includes all of the programs used to direct the operation of a computer. It includes the operating system and application software. The operating system, such as Windows 7 or Mac OS 10, provides an interface between the hardware and the application software. Application software, such as Adobe® Photoshop® or Microsoft® Office, allows you to use the computer to perform specific tasks.

Bar Code and RFID Technologies

Bar codes are also being used to access information. See **Figure 13-18.** Bar codes are a series of thin and thick lines that can be read by a photo diode in a laser scanner. The barcode reader reads the black spaces, which absorb the light, as 1s. It reads the white spaces, which reflect the light, as 0s.

A more recent method of identification is called *radio frequency identification (RFID)*. RFID tags are tiny microchips equipped with radio transmitters. They can be attached to product packaging. They can also be implanted in dogs and cats, or even in humans. The chips store information ranging from product color and expiration date to medical records and ownership details.

Wireless Technology

Computers, entertainment systems, and telephones can communicate with each other using a variety of connections. These include physical elements such as electrical wires and component cables. They also include wireless technologies such as Bluetooth and Wi-Fi.

Figure 13-18. A bar code contains encoded information about a specific product. Warehouse workers often use bar code systems to help track orders and inventory.

Bluetooth technology is a short-range technology that connects electronic devices without the clutter of cables. It has a maximum range of only about 100 feet. It is therefore used to connect devices over short distances, such as within a person's home. One advantage of Bluetooth is that it keeps transmission power extremely low. This helps save battery power in portable devices such as cell phones and GPS units.

Many people also use *Wi-Fi technology* to connect computers in different rooms. Wi-Fi has a larger range than Bluetooth—as much as 300 feet. It is used for high-speed Internet access when cables are not an option. For example, many restaurants, hotels, and even airports now offer Wi-Fi connections to their customers. See **Figure 13-19**. It is also used to connect computers to a local area network (LAN). A LAN may cover a small physical area, like a home or office. It can also cover a small group of buildings, such as a university or an airport.

Wireless technology carries signals on radio waves. A computer with wireless capability sends and receives radio signals from a router. A *router* is a device that allows two or more different networks to communicate. It manages communication between the computer and the Internet.

More and more devices are using wireless connections. Some digital cameras can upload photographs to a computer wirelessly. The keyboard and mouse used with many desktop computers are wireless. Smart meters send readings of how much electricity a house consumes. The Kindle™ is an electronic book that can download books wirelessly. Using services such as iTunes, you can download music and rent or buy movies whenever and wherever you have a wireless device and a wireless connection.

Another recent development in wireless technology is virtual input using lasers. The Magic Cube and evoMouse shown in **Figure 13-20** are laser devices that act as a virtual keyboard and mouse. The Magic Cube laser projects an image of a keyboard onto a flat surface such as a desk or table. When you touch a key on the keyboard, an optical sensor interprets the position of your finger. It sends the information about which key you pressed to the computer. With the evoMouse, your finger acts as the "mouse." The optical sensor tracks the movement of your hand and finger.

Figure 13-19. Offering a Wi-Fi Internet connection helps restaurants attract businesspeople who might otherwise not have time to eat at a restaurant.

Figure 13-20. A laser projection keyboard combines laser technology with an optical sensor and infrared technology.

It provides all the same functions as a typical mouse, including scrolling and zooming as well as the standard clicking and double-clicking. Because it tracks your finger or hand movement, you can even use this device as a handwriting recognition device.

Cloud Technology

Imagine that you have recently purchased a new laptop computer. Instead of installing software programs for word processing, spreadsheets, and databases, you need only load one application: a Web browser. Welcome to cloud computing! *Cloud computing* is a form of distributed computing. See **Figure 13-21.** It uses the Internet to provide the software users need for all their applications, including text files, spreadsheets, presentations, e-mail, and Web page development. The "cloud" network stores all of the files and content centrally. To make content available to many users at the same time, it distributes the tasks among many different source computers. Because hundreds of computers work in parallel to provide the service, cloud computing speeds up processing and access time.

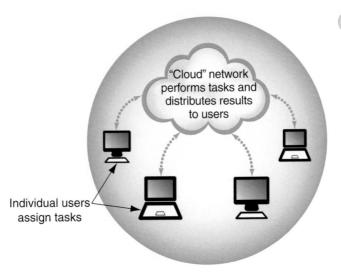

Individual users assign tasks

Figure 13-21. In cloud computing, the "cloud" network performs tasks by distributing them to networked computers. The results are sent to the controlling computer in the cloud. From there, the results are sent to the individual who requested the information.

Robotics

A robot is a computer that can interact with its environment. A true robot is one that can perceive its environment and respond automatically, without human control. The field of *robotics* is the development of technology and applications for robots.

Today, robots can look very different from one another. Some are mobile, and others are stationary. See **Figure 13-22**. Industrial machines often look like arms. Robots used to explore the ocean floor may look like miniature submarines. Drones look like small airplanes.

All robots have three main parts:

- Mechanical parts to do the work
- Computer electronics to control movement
- Software to provide instruction

Advances in robotics allow us to use robots instead of people for jobs that are dangerous or boring. For example, robots can be used in deep underwater explorations. They were used to help clean up the Gulf oil spill in 2010. They are also used in nuclear power plants to clean up radioactive spills and perform other high-risk tasks.

In industry, robots are used in many different ways. Robots with machine vision check bottles and jars to make sure they are filled to the right level. They can package products and stack them onto pallets ready for shipment. They can also perform high-precision work, such as placing tiny parts on circuit boards. In some cases, different types of robots are used together to perform tasks. For example, in some car assembly plants, a large robot can lift a small one inside a vehicle to assemble parts.

Figure 13-22. This industrial robot forms the soles of athletic shoes. The end effector, or working end, of the robotic arm is shaped to accept the shoe form.

Other robotic technologies provide robots for use in the home. For example, robotic vacuum cleaners are now commonly available. Robots can be used to wash floors, clean swimming pools, clear out gutters, and cut the lawn. Home entertainment robots, such as robotic pets, contain advanced vision, sound, and movement systems. Robotic "assistants" can be used for housekeeping and home care. See **Figure 13-23**.

Virtual Reality

Virtual reality is an environment that is created by a computer. The computer delivers sights, sounds, and sometimes tactile (touch) inputs that make an artificial environment seem real. The inputs can be real or imaginary.

In some systems, the user enters a virtual world by putting on special glasses or a head-mounted display. See **Figure 13-24**. The display contains two screens. By showing images on both screens, the display produces an image that appears to be three-dimensional (3D). The display also contains a stereo speaker system. Pilots take training in flight simulators while wearing helmet systems that immerse them in computer-generated airspace. Athletes can be immersed in an environment that simulates the course they will be steering or jumping.

Figure 13-23. Some retirement homes now have robotic assistants that can move around, interact with people, and perform simple tasks for them.

Figure 13-24. Many virtual reality systems use a headset or glasses to place the user in an artificial reality.

Virtual reality systems may also include a *haptic device* such as a wired glove. See **Figure 13-25.** The device provides a physical feedback to user motions. This technology allows the user to work with objects in a virtual environment. The user can "touch" and even move virtual objects. The computerized haptic device provides mechanical pressure that makes the experience seem real.

Another type of virtual system, called CAVE, consists of an entire room. CAVE stands for "Cave Automatic Virtual Environment," but the room itself is also called a CAVE. The CAVE is similar to a theater, except the input surrounds you. Computer-controlled images appear on the walls of the room, and sometimes on the floor and ceiling as well. These images use 3D graphics. Users wearing 3D glasses can see objects not only on the walls, but also freestanding in the middle of the room. See **Figure 13-26.** Multiple speakers provide sound input. Users can walk around the room and experience the "reality" from any point of view. Computers track the location of the user and update the displays and sounds accordingly.

The Internet

The Internet connects millions of computers around the world. It is currently the world's largest and most diverse telecommunication system. In fact, the Internet is a network that is made up of many other networks. Information is transmitted over the Internet using computer languages called *protocols.*

World Wide Web

One such protocol is the *hypertext transfer protocol (HTTP)*. This protocol allows us to send and receive all kinds of files over the Internet. It can transmit text, graphics, photographs, sound, and video files, among others.

Figure 13-25. The sensors on a haptic glove allow you to touch objects in a virtual world.

Figure 13-26. This man is playing a virtual game of football using a CAVE virtual reality system.

The *World Wide Web (Web* or *WWW)* consists of all of the processes that use HTTP to transmit information over the Internet. We use Web browsers such as Internet Explorer or Safari to search for and display Web sites. See **Figure 13-27.**

The Web contains millions of Web sites. The fastest way to search for specific Web sites is to use a search engine, such as Alta Vista™, Google™, or Yahoo!™. By entering key words or phrases, you can display a list of matching Web pages almost instantaneously. Not all search engines use the same method to search the Web. Therefore, they may display different results. For the most possible matches, you can use a meta-search engine. This is a search engine that enters your key words into other search engines to find matches. Examples include Mamma® and Dogpile®. To access a search engine, type its address at the top of your browser. The address is usually the name of the site plus ".com." For example, to access the Dogpile meta-search engine, type "dogpile.com."

Other Internet Uses

We also use other kinds of Internet protocols. For example, we use SMPT, or simple mail transfer protocol, to send and receive e-mail. File transfer protocol, or FTP, allows the exchange of large files between computers.

Figure 13-27. The World Wide Web allows us to find, display, and even create Web pages. This man is accessing templates developed to help people create their own Web pages.

Another protocol that is gaining popularity is *Voice over Internet Protocol (VoIP)*. This protocol converts voice signals into a digital format that can be sent over the Internet. VoIP allows people to use their Internet connection to place telephone calls. The audio quality is similar to that of a cell phone. However, VoIP is much less expensive for making long-distance calls.

The Internet is also helping people who are hearing impaired. Instant messaging (IM) is a convenient way to carry out online conversations. Chat rooms are becoming one of the main ways that members of the deaf community communicate among themselves or with the hearing population. IM is available for the laptop, pager, or cell phone.

Copyright Issues

The availability of so much information on-line has led to questions about what is free and what is copyrighted. This includes text, graphics, and even computer programs available on the Internet. While you must pay for software produced by Microsoft, other software, including the Linux operating system, can be used by anyone without charge. This free software (also known as open-source software) is becoming increasingly available. Applications including the Firefox web-browser and Wikipedia, the on-line encyclopedia that anyone can edit, are examples of applications that are free.

However, most of what you find on the Internet is copyrighted material. Using this material without permission is called *plagiarism*. Plagiarism is against the law and carries heavy fines. Therefore, if teachers find out that students have plagiarized material, they may give a failing grade for that work. If you do not know whether something is copyrighted, assume that it is. You can use the material for reference, but do not copy it word-for-word.

In the last century, communication technology has changed dramatically. At the beginning of the 20th century, telephones were relatively new. Computers and even televisions had not been invented yet. People depended on radio for news. They processed most types of information without the help of technology.

Today, almost all of our communication and information sources involve complex technologies such as cellular and satellite technology. We use the Internet to download movies, ringtones for our phones, and music for our MP3 devices. We use e-mail to stay in touch with family and friends. A cell phone is no longer simply a telephone. It is a multipurpose computer. For example, a smart phone can take photographs. It includes games and text services that allow people to send and receive e-mail and text messages.

Experimental optical microchips are now on the market. Particles of light, rather than electrons, are used to control and power circuits. Switches conduct light instead of electricity. Optical technology is much faster than the technology used in computers today. The use of optical technology will make computers faster. This will pave the way for more realistic virtual worlds, among other exciting inventions and innovations.

Computer electronics are currently limited by the size of their microelectronics. In the future, microprocessors embedded in computers and other machines will likely be made of materials other than silicon. Future computers will depend on nanotechnology. Today's computer chips contain approximately 40 million transistors. Through nanotechnology, that same chip would hold about one billion transistors. As these chips become more powerful, the size of components will shrink. It will then be possible to embed them in any device, even the clothes that we wear.

- All communication systems contain some elements in common.
- Information technology helps us store and process the large amount of data and information we receive every day.
- Information and communication technology systems are made possible by microelectronics.
- Telecommunication includes conventional technologies such as radio and television, as well as newer technologies such as satellite technology.
- Computer technology enables all communication and information systems.
- The Internet is the world's largest and most diverse telecommunication system.

Finding the Main Idea

Create a bubble graph for each main idea in this chapter. Place the main idea in a central circle or "bubble." Then place the supporting details in smaller bubbles surrounding the main idea. A bubble graph for the first part of the chapter is shown here as an example.

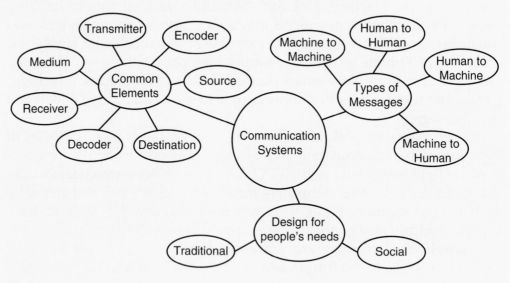

Test Your Knowledge

Write your answers to these review questions on a separate sheet of paper.

1. What seven elements do all communication systems have in common?

2. What four types of communication systems do we use to send messages?

3. What is the difference between data and information?

4. Explain the importance of microelectronics to communication and information technology.

5. What is telecommunication technology?

6. Name an advantage of satellite communication.

7. What is the difference between a bit and a byte?

8. What is the difference between Bluetooth and Wi-Fi technology?

9. What is the purpose of the hypertext transfer protocol used on the Internet?

10. What is plagiarism, and how can you avoid it?

Critical Thinking

1. Cloud technology is so-named because the actual distribution of tasks is hidden from the user. What are the advantages of this? What are the disadvantages?

2. As virtual reality systems become more affordable, it is possible that people will spend more and more time in virtual "worlds." For example, you could take a virtual vacation without leaving your home. What implications might this have for society? What effect might it have on communication?

3. Information available on the Internet may be true. However, it may *not* be true. How can you tell the difference? Describe a method of checking information you find on the Internet.

4. Many people like to play the type of computer game that simulates the real world. Why do they like to simulate the real world, when they can experience what is real?

Apply Your Knowledge

1. From your own community, give three examples of old information technology and three examples of new information technology.

2. Look through one issue of a newspaper. Cut out all the references to micro-electronics. Try to find references from each section of the newspaper, including the advertisements.

3. Use a computer equipped with drawing software to draw a block diagram to show the major components of a telecommunication system. Explain how the computer can be used to manage information about system components.

4. Selecting the right television for your home is not an easy task. Assume that a friend or relative has asked for your advice on the purchase of a new television. Make a summary of the advantages and disadvantages of as many different types of TVs as are available in your local stores.

5. Humanlike robots that can communicate with their owners are sometimes shown in newspapers and magazines. Identify robots designed for use in homes and describe their capabilities.

6. Video games have been around for decades. Find out who made the first video game and when. Then make a timeline of the most important developments in video games since then.

Apply Your Knowledge (Continued)

7. Research one career related to the information you have studied in this chapter. Create a report that states the following:

- The occupation you selected
- The education requirements to enter this occupation
- The possibilities for promotion to a higher level
- What someone with this career does on a daily basis
- The earning potential for someone with this career

You might find this information on the Internet or in your library. If possible, interview a person who already works in this field to answer the five points. Finally, state why you might or might not be interested in pursuing this occupation when you finish school.

STEM Applications

1. **ENGINEERING** Home theaters achieve great sound and visual effects when they are well designed. Research the conditions that make an ideal room for a home theater, including the floor plan, furnishings, natural and artificial light, and wiring and ventilation requirements. Then design a room for a home theater. Specify the equipment that would be used in the room.

2. **SCIENCE** Research the basic components of a communication satellite. Use common materials such as toothpicks and printer paper to create a model of a communication satellite. Be ready to explain what each part of the satellite does.

3. **MATH** With a friend or classmate, design a secret code based on the binary number system. Be careful not to tell anyone else what your code is. Then design a message written in your code and display it for the class. Can anyone "break" your code?

4. **ENGINEERING** Design a robotic device that can skim leaves from the surface of a body of water, such as a swimming pool. Think carefully about all of the design considerations. Will it be free-moving or stationary? How will it be powered? How will it work? What materials will be needed? Write a report describing your device. Include sketches to show what it will look like.

How might information and communication technology change our idea of what an office should be? What effects might this have on businesses, the environment, and society in general?

Agricultural, Medical, and Biotechnology

Better by Design

Dickson Despommier and Blake Kurasek design vertical farms

What if you could eat food grown only a few blocks from where you live? Dr. Dickson Despommier has proposed building vertical farms in cities. Crops would be grown using mineral nutrient solutions without soil (hydroponics). Some crops could also be grown in air by spraying the plant's roots with a nutrient-rich solution (aeroponics). Vertical farms have many advantages. For example, crops could be grown year-round. Crops would not fail because of bad weather. Vertical farms would also require less land. Building on Dickson's scientific work, Blake Kurasek has designed "living skyscrapers." In these buildings, produce grown in the building can supply its occupants and a surrounding population.

Shipping food over long distances can be avoided by growing food near the consumer.

"Over the next 50 years, population growth will require an additional area the size of Brazil to grow sufficient food. This amount of high-quality land is simply not available."

Creating an Outline

An *outline* is an orderly statement of the main ideas and major details of a text passage. Each main idea or detail is written on a separate line. Creating an outline can help you understand and remember what you read. Read each section of this chapter carefully. Then use the Reading Target graphic organizer at the end of the chapter to create an outline of the chapter.

antibodies
aquaculture
biotechnology
cloning
deoxyribonucleic acid (DNA)
gene
genetic code
genetic engineering
greenhouse
hydroponics
implants

irradiation
laparoscopic surgery
minimally invasive surgery
monoculture
precision farming
prosthetic devices
reproductive cloning
telemedicine
transplants
vaccines

After reading this chapter, you will be able to:

- Describe how food is grown, harvested, and processed.
- List medical technology and devices that can improve quality of life.
- Discuss the use of biotechnology in foods, medicines, and other products.
- Debate the implications of biotechnology.

Useful Web sites:
www.verticalfarm.com
www.blakekurasek.com

Several types of technology affect our health and wellness either directly or indirectly. Agricultural technology, for example, is responsible for producing the foods we eat to remain healthy. Medical technology includes surgical procedures, implants, and prosthetics (artificial limbs) that keep us healthy and make our lives better. A related field, biotechnology, works hand-in-hand with agricultural and medical technologies to improve our health.

Agricultural Technology

Farmers own and operate farms that grow plants that provide us with food and many other products. They prepare the land, plant seeds, and care for the plants. When the crop is mature, they harvest the food. In some cases, they also package and sell the products they grow. Other farmers and ranchers raise animals such as cows, sheep, and poultry. They care for the animals and shelter them in barns or other farm buildings.

Many other agricultural specialties also exist. For example, some agricultural specialists grow plants in greenhouses. A *greenhouse* is a building made mostly of glass or transparent plastic. It helps control conditions such as humidity and temperature. Greenhouses provide an artificial ecosystem that mimics the climate necessary for growing specific plants.

Hydroponics is a type of agriculture in which plants are grown in a water-based system, without soil. See **Figure 14-1**. *Aquaculture* farmers raise fish and shellfish. They raise the animals in ponds, floating net pens, or other systems in which they can feed and protect them. See **Figure 14-2**.

Farming is hard work, and crop growth is unpredictable due to weather conditions. Having too much rain can cause as much damage as having too little. An early frost in autumn or a late frost in spring can destroy an entire crop. Weeds sometimes grow faster than the crops. The farmer also has to fight insects and plant diseases.

Figure 14-1. In hydroponic farming, the plant roots are bathed in a water solution that contains the nutrients they need.

Figure 14-2. Aquaculture farmers often protect their stock by enclosing them with nets or fences in their native habitat.

The prices farmers can get for their crops may also vary from year to year. Many farmers plant more than one type of crop. If prices drop for one crop, the farmer may be able to make up the difference with another type of crop. This practice also helps farmers avoid weather-related crop failures. While too much rain might destroy one crop, another crop may survive or even thrive in the additional rainfall.

Much of modern farming is automated. Specialized machinery helps reduce some of the work. See **Figure 14-3**. It is used to till the soil, spray fertilizers and insecticides, irrigate fields, and harvest crops. These techniques allow farmers to produce a larger crop using less time and fewer resources. As world population increases, these improvements become more important on a global scale.

Farming and Computer Technology

Today, farmers use computers for many tasks. For example, by using a database on the Internet, they can learn about weather forecasts. They can connect with the board of trade to see the current price of corn or other crops. This information helps them decide the best time to plant or to sell their crops.

Some farmers now use *precision farming* methods. They gather data from satellites to locate problems in their fields. For example, they can find an area that is not draining correctly or that is being destroyed by insects. They can then target that specific area for repairs or pest control. This helps farmers use less pesticide, which is better for the environment. It also decreases the amount of time they spend working in the field.

Food Processing

The foods we eat include both natural or processed foods. Natural foods include the corn, tomatoes, and peppers you see on the shelves of supermarkets. Processed foods are produced synthetically. Margarine, candies, pies, and many drinks are processed using both chemicals and natural ingredients. See **Figure 14-4**.

Figure 14-3. Automatic irrigation systems water crops on a large scale.

Figure 14-4. A—Natural foods include many items you can find in the produce section of a grocery store. B—Cheese is an example of a processed food. Other examples include breakfast cereals, canned soups, and many types of meat products.

To process food, wastes such as skins, bones, cobs, and shells, are removed first. The food might be cut, crushed, or ground. It is then sorted, and any substandard products are removed. Next, it is mixed and blended with additives, such as colorings, spices, and preservatives. For example, sausage can be made from a combination of pork, fat, spices, applesauce, corn syrup, and water. Heating, pickling, dehydrating, refrigerating, or freezing may be used to help preserve the food.

These processes, along with irradiation, help provide food with safe long-term storage capability and minimize the risks associated with eating spoiled food. *Irradiation* is a method of exposing foods to a low level of radiation to kill bacteria and other parasites. After many studies, scientists have concluded that this is a safe method of keeping food safe from bacteria for an extended time.

Finally, foods might be formed into a certain shape, such as forming chocolate into bars or pasta into uniform shapes. Foods are then packed to both protect them and keep them clean. See **Figure 14-5**.

Figure 14-5. Pasta is formed into many different shapes before it is packaged.

Math Application

Food and Calories

You have probably heard the old saying, "You are what you eat." This is true in many ways. If you eat foods that are good for you and maintain a balanced diet, you will feel better than if you eat junk food routinely. If you eat too much or fail to get enough exercise, you will gain weight. If you do not eat enough to supply your daily needs, you will lose weight. Balancing the calories you eat with your level of physical activity can help you maintain a desirable weight.

The United States Department of Agriculture has created a tool you can use to find out what foods are best for your health and how much of each food you should eat. You can find this tool at www.mypyramid.gov. You can find the number of calories in almost any food using several sites on the Internet.

Math Activity

In this activity, you can find out if your diet is appropriate for your needs. Follow these steps:

1. Determine the number of calories you need every day. The number depends on your current weight and your level of activity. Use the following chart to find the ideal number of calories for you.

Activities	80 lb.	100 lb.	120 lb.	140 lb.	160 lb.	180 lb.
None	960	1200	1440	1680	1920	2160
A few activities	1280	1600	1920	2240	2560	2880
Many activities	1600	2000	2400	2800	3200	3600

2. Keep a record of the foods you eat in one day.

3. Find the number of calories for each portion of food you eat. One way to do this is to refer to the Nutritional Facts listing on the packages of the food you eat. Another method is to search the Internet. For example, the Web sites for most fast food restaurants now list the calories in their foods. You can also go to one of the many "calorie counter" sites. Add up the number of calories you consumed in one day to find your total calorie intake.

4. Compare the number of calories you ate with the chart above. If there is a big difference between your ideal calorie needs and the amount you consume, make a plan to change your eating or exercise habits.

5. Go to www.mypyramid.gov and review the recommended amount of each food group. Compare this with the list of foods you ate in one day. How healthy is your diet? How can you improve it?

6. Use MyPyramid to plan an entire week of healthy meals that meet your calorie requirements. Show the number of calories in each food you choose, and add the calories for each day to make sure you are planning the correct number of calories.

Medical Technology

What are the requirements to remain healthy? It helps to have comfortable surroundings, quiet moments for ourselves, and people who love and support us. In addition, fresh air, water, sunshine, rest and sleep, are vital for a healthy body. We know that smoking is very harmful to our health and that too much exposure to the sun can be dangerous. We understand the importance of proper nutrition and exercise.

Sometimes, however, we have to depend on technology to keep us healthy. We need medicines to cure diseases or ease symptoms. Surgical techniques are sometimes necessary to repair injuries or remove cancers. Other devices, such as hearing aids, eyeglasses, and artificial body parts, are also part of medical technology. See **Figure 14-6.**

Medicine

Medicines, or drugs, act in many different ways to fight disease. *Vaccines* stimulate the body's immune system to develop special proteins called *antibodies*. Other medicines, such as penicillin, stop the growth of bacteria. Still others work with our bodies to lower blood pressure or cholesterol levels.

Some of the most widely used medicines were originally derived from plants. For example, aspirin was first extracted from the bark of white willow trees. Today, like many other medicines, aspirin is made synthetically. Still, natural plant material is important in medical research. The search for new medicines is often concentrated in areas, such as rainforests, that have many plants whose properties have not yet been explored. See **Figure 14-7.**

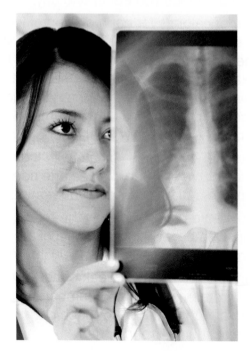

Figure 14-6. X-rays and other diagnostic tools are other examples of medical technology we use on a regular basis.

Figure 14-7. Our search for new medicines often takes place in rainforests because of their many as yet unknown plant and animal species. What impacts might this have on the environment? How can we search these areas responsibly?

Repairing the Damage

Medical technology is not just about treating diseases. It also helps when we are injured, or when our natural body parts begin to wear out.

Surgical Techniques

For many procedures, patients used to spend several hours in surgery, a week in the hospital, and a month recuperating at home. With the latest technologies, however, these times are being reduced. Surgeons can do procedures in a fraction of the time, and recovery time is minimal.

One reason is that new procedures are less invasive. Many types of abdominal surgery can be done through small incisions instead of the large incisions used by surgeons in the past. This type of surgery is known as *laparoscopic surgery*, or *minimally invasive surgery*.

Robotic surgery is another example of technological improvements in surgical techniques. For example, heart surgery has traditionally required the surgeon to saw through the patient's sternum and crack open ribs to create a space wide enough for hands to reach inside.

Today, a thin, remotely controlled robotic arm can be inserted between a patient's ribs through tiny incisions. The surgeon sits at a computer to control the instruments at the ends of the robotic arm. The instruments include tiny surgical tools, a light, and a camera. The camera sends views of the internal organs to a video monitor. The surgeon controls the instruments while watching the images shown on the monitor. See **Figure 14-8**.

Robotic surgery has become routine because it can be more precise than traditional surgery. The technique is now used for many major and minor operations, including gall bladder removals and pacemaker operations. Patients are exposed to less risk because the incisions are very small. Less tissue has to heal, so the patient recovers more quickly. The surgery is also less painful.

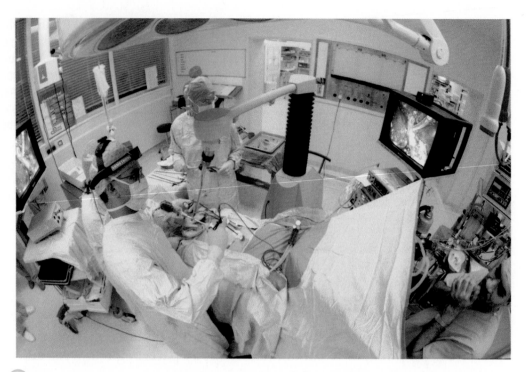

Figure 14-8. This surgeon is using a foot switch and special headband to control a robot, which holds and directs the tiny camera inside the patient's body.

Prosthetic Devices

Doctors use replacement surgery when parts of the body wear out. They replace natural parts with artificial parts. *Prosthetic devices* are implants or replacement body parts that mimic the original human body part. These devices can reduce the pain from arthritis, injuries, or simply years of use. Hips and knees are examples of human joints that wear out over time. These joints also frequently suffer from arthritis as our bodies age.

Prosthetic devices must be made from certain metals and plastics that do not react with the body's chemistry. See **Figure 14-9.** For example, an artificial hip uses a ball-and-socket joint. The artificial ball is often made of stainless steel or ceramic. The ball fits into a cup made of a high grade polyethylene. The stem that holds the ball in place is inserted into a space that has been drilled into the femur (thigh bone).

Most artificial replacement parts are successful, if only for a period of time. For example, the amount of wear of the cup against the ball in a knee or hip joint is a major factor in how long the parts will last. Titanium is generally used for artificial hip joints. Carbon fiber is often used for artificial legs. Artificial eyes are made of acrylic plastic. Various types of silicone can be molded into shapes for reconstructing a face or replacing a finger. See **Figure 14-10.**

Through medical technology, artificial parts can be made to replace natural ones, but are they better than natural ones? This has been debated at length. Newer prosthetic devices come very close to allowing people to function as well or better than they did using their natural body parts.

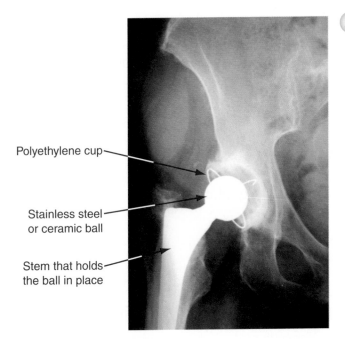

Figure 14-9. A hip replacement can improve the mobility of an older person whose hip joint has begun to fail.

Polyethylene cup

Stainless steel or ceramic ball

Stem that holds the ball in place

This is especially important in sporting events. For example, does a runner with carbon-fiber prosthetic legs have a better chance of winning a race? Experts have studied the question, but so far, they have not been able to give a clear answer. Some people think the prosthetics give the athlete an unfair advantage. Others think the athlete is at a disadvantage. As medical technology continues to improve, this question may become very important in sports and other areas.

Transplants and Implants

Replacement body parts can be mechanical, electronic, or organic. Mechanical transplants such as metal hip joints, plastic valves, or electronic pacemakers are called *implants*. Living organs that replace faulty ones are called *transplants* and are usually from another human donor. Some

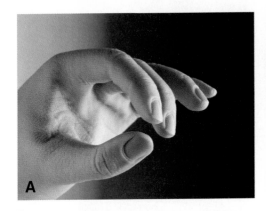

Figure 14-10. Prosthetics may or may not look like the original human body part. Some, such as the hand in A, look natural. Others, like the artificial leg shown in B, are built more for function than style.

body parts, such as the heart and kidneys, are far too complex for artificial parts to be made at a reasonable cost. These organs are transplanted. See **Figure 14-11**. Other body parts are now routinely transplanted or implanted, including:

- Heart pacemakers
- Heart valves
- Hearing aids
- Speech devices
- Lenses to correct eye cataracts
- Knee and hip joints
- Dentures
- Artificial limbs

Other Advances in Medical Technology

Would you swallow a pill that contains a camera? A new camera-in-a-pill has been invented that, when swallowed, can move or stop, according to what a doctor needs to see. The pill is radio-controlled and can be used to replace a colonoscopy, a procedure that can be very uncomfortable.

Technology is helping millions of people stay in touch with their doctors and nurses using *telemedicine*. Patients can use the Internet and other communication devices to transmit data from their homes to a database containing their medical files. Satellite-Internet connections from remote areas to 24-hour emergency medical centers allow patients to consult specialists using video links. Using telemedicine, a nurse can ask a patient health questions, receive information, and give advice.

Figure 14-11. This heart is being prepared for transplant. The human heart is too complex to be created artificially with our current knowledge and tools.

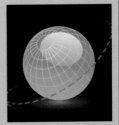

Think **Green**

Medical Technology and the Environment

Have you ever thought about what happens to used medical supplies? We know that blood and other body fluids can spread disease. Every day, clinics and hospitals use a huge amount of disposable supplies, such as gauze. What happens to the gauze that is used to clean blood from an injured arm, for example?

Blood and other body fluids are considered hazardous waste. Gauze, needles, and other disposable supplies that come in contact with these fluids are collected separately from other trash. They are disposed of according to strict government guidelines. This helps prevent other people from coming in contact with contaminated material and possibly catching the disease. For more information about hazardous waste, refer to Chapter 16 and **Appendix B** at the end of this textbook.

Biotechnology

Biotechnology is the use of living organisms to produce new foods, medicines, and other products. It is closely related to both agricultural and medical technology.

The modern age of biotechnology started in 1953 when James Watson and Francis Crick announced the discovery of *deoxyribonucleic acid (DNA)*. See **Figure 14-12**. Strands of DNA are considered the basic building blocks of life. This discovery opened the way for a more precise understanding of how living things work. Scientists and technologists started to take the basic blocks of living creatures apart and put them back together in new ways.

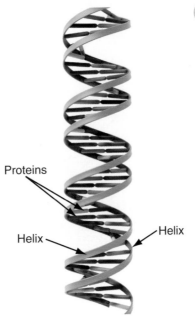

Proteins

Helix

Helix

Figure 14-12. DNA is made up of two helixes, or spirals, bound together by proteins. Notice the different colors used for the proteins in this model. Each color represents one of four proteins. The arrangement, or sequence, of the proteins defines the traits of all living things.

Genetic Engineering

In the past, improvements in animals and plants occurred over hundreds of years. For example, to increase the yield of soybeans, a farmer would crossbreed plants. The farmer might cross plants with the highest yield with another variety of the same species that had a resistance to drought. This method of improving crops took generations to produce results. Hardier plants and animals are now possible in a single generation by recombining genes in a process known as *genetic engineering*. A *gene* is a sequence of DNA that defines a specific trait, such as hair color or eye color.

Agricultural Applications

Every living thing, from the smallest beetle to the tallest oak tree, has a set of genes, which we call a *genetic code*. This code determines precisely what traits it will have. In agriculture, biotechnologists look for the most beneficial traits in plants and animals. For example, they may try to increase nutrition, flavor, and disease resistance by moving traits from one organism to another.

Genetically Engineered Animals

The Enviropig™ is one example of genetic engineering. See **Figure 14-13**. Manure from normal pigs has a high phosphorous content that pollutes rivers and lakes. Phosphorous increases the amount of algae in water. Algae rob fish and other organisms of oxygen. That is why reducing pig pollution is very important. Genes from other organisms have been introduced in the Enviropig to allow it to process its food more efficiently. This reduces the amount of phosphorous in its manure, which helps protect water supplies.

Figure 14-13. The saliva of genetically engineered pigs contains phytase, an enzyme that allows the pigs to digest phosphorus more completely.

Cloning is another example of genetic engineering. Technically, *cloning* is the process of creating an exact copy of any kind of biological material. Scientists in research laboratories have been cloning DNA segments into bacterial host cells for more than 30 years.

The type of cloning that you may have heard most about is *reproductive cloning*. This is the process of creating a genetic "twin" of an existing animal or microorganism. Reproductive cloning first came into the spotlight in 1997 when Scottish scientists created "Dolly," a cloned sheep. The controversy over this type of cloning continues. Many people are concerned about the ethical implications of cloning.

Genetically Modified Crops

Genetically modified crops are another example of the use of genetic engineering. The crops are modified to be more resistant to insects and other pests. These crops are being used on many farms in the United States and around the world. We can produce fortified vegetables, or crops that are less affected by frost. See **Figure 14-14.**

Many genetically engineered products are highly controversial, however. In the past, the diversity of genes and species has provided a variety from which the best combinations of seeds have been drawn. Some people question whether *monoculture*, the spreading of a single species over a large area, will result from genetically engineered seeds. Monocultures are extremely vulnerable to new parasites and diseases.

Figure 14-14. Genetically engineered corn may produce larger, disease-resistant crops. What are the disadvantages of using genetically engineered plants as food sources?

Many questions remain about genetically modified products. Some questions include:

- What effect will genetically modified organisms have on wild animals and plants? Are the risks greater than the benefits?
- Are we sure that it is safe to mix genetic material from different life forms? Have the tests been carried out over an adequate period of time?
- Could transferring genes across natural boundaries cause epidemics of diseases such as mad cow disease?
- Could there be pressure to change less desirable genetic traits in humans?
- What effect will the higher cost of genetically modified seeds have on farmers?
- Will the smaller number of varieties be a source of increased risk for farmers because monocultural fields are more vulnerable to disease and pest attacks?

Biotechnology and Humans

Biotechnology and genetic engineering also affect humans directly. The Human Genome Project (HGP) was finished in 2003 and shows the complete genetic blueprint for building a human being. It was found that humans possess only about 20,000 genes. Identifying this genetic information became possible only recently, when computers became powerful enough to process the data. Billions of pieces of data are needed to describe a single human.

Scientists are working to discover which human genes are responsible for certain hereditary disorders. They hope to be able to test for specific genetic diseases and perhaps even cure them using genetic engineering.

Like many technologies, genetic medicine has both good and bad sides. If doctors can screen people for the likelihood that they will develop heart disease or cancer, they can take steps to prevent the diseases. This information would also allow people to make good lifestyle choices. They could select the right diets or exercise levels to increase their chance of remaining healthy.

This knowledge, however, could also result in genetic discrimination. For example, will a person, family, or group be treated differently because tests show they have a reduced life span or a tendency toward certain behavior?

Some questions that might have to be answered in the future include:

- Should people be forced to undergo genetic testing?
- Should third parties have access to the genetic information of others?

While certain defects could be removed and attractive features added, as far as we know, no gene gives us the important characteristics that make us human. No particular genes make a human being kind-hearted, law-abiding, or capable of loving. Also, keep in mind that no one can guess exactly what effect genetically engineered changes might have on humans in the future.

The Future of Biotechnology

You have probably noticed that much of the material in the agricultural and medical sections of this chapter could also have been placed in the biotechnology section. As scientists and technologists continue to improve medical and agricultural technologies, these technologies will become even more linked to biotechnology.

Improvements in biotechnology can help people live fuller, healthier lives. See **Figure 14-15.** Improvements to prosthetic devices will allow more and more people with missing limbs to do things they would normally be unable to do. A retina implant is being developed to give sight to people who are blind. Cochlear (ear) implants may help deaf people hear. Many medical procedures will be possible, but they may also be controversial. Many advances will have ethical and moral implications.

As discussed in Chapter 1, technology itself is neither good nor bad. Remember that all technology can be used for good purposes or bad purposes. In biotechnology as in other areas of technology, we must use technological advances wisely. We must think of both intended and unintended consequences. Information is the key to making wise technology decisions.

Figure 14-15. With the help of various medical and biotechnologies, seniors can participate in activities that were once reserved for younger people.

For thousands of years, farmers have been improving their crops. They have saved seeds from the best plants in this year's harvest to use next year. They used trial and error to find the best crops and the best methods for growing them. Medical technology, too, has advanced largely by trial and error. Someone accidentally discovered that aloe soothes burns and that the bark of the white willow tree can ease pain.

Today, technology has dramatically speeded up the development of new crops, drugs, and processes. We grow larger crops and have the opportunity to live longer and healthier lives. These advances come at a price, though. Many chemicals that are used to increase food production they can cause danger to other creatures. For example, some agricultural runoff can spawn a deadly oxygen-eating bacterium that kills fish and other species living in or around lakes and streams. Can we develop technology to fix these effects?

As new technologies such as genetic engineering are developed and improved, they will raise even more pressing questions. Will humans experience increased genetic uniformity, a narrowing of the gene pool, and a loss of genetic diversity? Will human efforts cause other species to become extinct? Who will make the decisions as to what is a "good" gene and what is a "bad" gene that should be eliminated? These and similar questions may need to be answered not once, but on a continuing basis now and in the future.

- Agricultural technology is responsible not only for growing foods and other products, but also for processing and packaging them.
- Medical technology provides medicines, surgical techniques, and even prosthetics to help us stay healthy and to treat injuries and illnesses.
- Biotechnology, which is based on the use of living organisms, is becoming increasingly connected to both agricultural technology and medical technology.
- As these technologies continue to advance, we must keep informed about new ideas and consider both the potential good and the potential harm they may do.

Creating an Outline

Use a graphic organizer similar to the one below to create an outline. Write your outline on a separate sheet of paper. Remember to use Roman numerals for the main ideas and letters for supporting ideas. Try to supply at least two supporting details for each main section. Add more detail lines if necessary to describe all of a section's important details.

I.   Agricultural technology includes growing, harvesting, and processing food and other items we use every day.
A.
B.
C.
D.
II.
A.
B.
C.
D.
III.
A.
B.
C.
D.

Test Your Knowledge

Write your answers to these review questions on a separate sheet of paper.

1. Name at least two types of farming specialties and explain what these farmers do.
2. What are the advantages of precision farming?
3. What is irradiation?
4. How do vaccines help people fight disease?
5. What are the two basic sources of new medicines?
6. Name at least two advantages of robotic surgery over manual surgery.
7. What is a prosthetic device?
8. What is the difference between an implant and a transplant?
9. What is biotechnology, and how is it related to agricultural and medical technologies?
10. Briefly describe the process of genetic engineering.

Critical Thinking

1. In your own words, explain why is genetic engineering such a hotly debated topic.

2. If biotechnology could produce a robot that truly thinks, would it be ethical to "pull the plug" on such a robot? Explain.

3. One possible use of genetic engineering is in the field of law enforcement. How can genetic engineering help police and crime labs? Do you think this is a good idea? Explain.

Apply Your Knowledge

1. Choose a crop that is commonly grown in the United States, such as soybeans, corn, or wheat. Research to find out the average price farmers have received for this crop each year for the last 10 years. Make a graph to show the results. Note any trends that you see and suggest possible reasons for them.

2. Investigate the process of making cheese. Prepare a multimedia presentation to share your findings with the class.

3. Use the Internet to investigate new techniques and methods that are being used to prevent or cure a currently incurable illness of your choice. Prepare a three-minute speech on the subject and deliver the speech to your class.

4. Find out more about what happens to hazardous medical waste after it is collected. How is it processed? How does this keep it from harming other humans and animals?

5. Research the latest advances in genetic engineering and choose one topic that seems controversial. As a class, debate the pros and cons of the new product or technique.

6. Research one career related to the information you have studied in this chapter. Create a report that states the following:

   - The occupation you selected
   - The education requirements to enter this occupation
   - The possibilities for promotion to a higher level
   - What someone with this career does on a daily basis
   - The earning potential for someone with this career

   You might find this information on the Internet or in your library. If possible, interview a person who already works in this field to answer the five points. Finally, state why you might or might not be interested in pursuing this occupation when you finish school.

STEM Applications

1. **ENGINEERING** Design and make a hydroponics unit for a location that needs attractive plants without continuous maintenance. Document your design and include it in your portfolio.

2. **TECHNOLOGY** Find a recipe for making applesauce from apples. With your parent or guardian's permission, follow the recipe. Record the technology you used in the course of preparing the applesauce. Brainstorm technological improvements that might make the process easier, faster, or more fun.

3. **SCIENCE** Research the construction of a greenhouse. Individually or as a class, design and make a greenhouse that will allow you to grow a type of plant that does not ordinarily grow in your location. Test the greenhouse by using it to grow the specified plants from seeds. Document your design and record the results of your growing efforts. Did the plants sprout? Did they flourish? What kind of care did they require? At the end of the growing season, publish your results in a scientific reporting format.

Better by Design

Metaphase Design Group develops user-centered designs

Work gloves provide important protection, but often they make handling objects more difficult. At Metaphase Design Group, a team of designers explored how work gloves are used. They then designed a range of gloves for specific tasks. Constructors' gloves have a small magnet in one finger. The magnet is strong enough to pick up one nail, but weak enough to prevent picking up more than one. People who work in shipping, receiving, and delivery need bare fingertips to use their electronic tracking devices. Their gloves have fingertip "hats" that can be flipped back and held with Velcro®. The wedge-like thumb design slides easily under boxes for handling.

A small magnet in the index fingers of these gloves allows carpenters to pick up one nail at a time.

The tips of these gloves can be folded back and secured by Velcro when employees are working with electronic equipment.

"The best way to find out what a product should do is by watching people use it in the real world."

Reading Target

Summarizing Information

A *summary* is a short paragraph that describes the main idea of a selection of text. Making a summary can help you remember what you read. As you read each section of this chapter, think about the main ideas presented. Then use the Reading Target graphic organizer at the end of the chapter to summarize the chapter content.

Key Terms

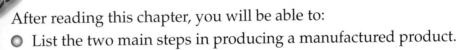

artisan
assembly line
automation
computer numerical control (CNC) machine tool
continuous improvement
degree of freedom
division of labor
durable
efficiency
factory
flexible manufacturing system

interchangeable parts
just-in-time delivery
lean manufacturing
manufacturing
mass production
nondurable
production technology
rectangle
subassemblies
sustainable manufacturing
tolerance
work cell
zero defects

Objectives

After reading this chapter, you will be able to:

○ List the two main steps in producing a manufactured product.
○ Describe the development of manufacturing processes in the last 200 years.
○ State the purpose of automation and identify its advantages and disadvantages.
○ Explain the differences between traditional and modern management systems.
○ Describe the elements of a production system.
○ Explain how sustainable manufacturing is different from traditional manufacturing.

Useful Web sites:
www.metaphase.com/
www.chi-athenaeum.org/gdesign/index.html

Manufacturing is the process of using raw materials to create products. Some of the products we produce are *durable*. They are meant to last for a long time. A bicycle is an example of a durable product. Other products, such as paper plates, are meant to be used once and then discarded. These items are *nondurable*. Durable and nondurable products are manufactured using similar manufacturing and production processes.

The wood used to build the chairs in your home came from a tree growing in a forest. What is involved in creating a chair from a tree? From **Figure 15-1**, you can see that the steps in producing any product are:

- Obtaining and processing raw materials
- Changing raw and processed materials into a product we can use

1
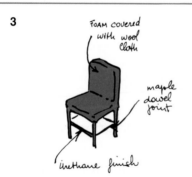

Trees are felled in the forest, transported to a sawmill, and converted into boards.

2

Designers submit alternative chair designs. The best is developed in detail. Working drawings are produced.

3

Models and prototypes are made. Style, materials, and construction techniques are reviewed.

4
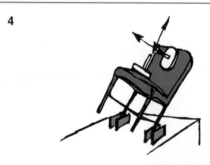

Prototypes are tested to determine the chair's strength and durability. Weaknesses in the design are corrected.

5
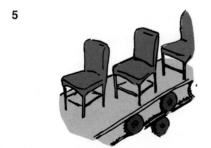

A mass production system is planned. Skilled workers operate and maintain machines to mass-produce the chair.

6

Office staff keeps track of materials and supplies. Salespeople receive orders and notify the warehouse of addresses for delivery.

Figure 15-1. Making a chair from a tree takes much planning and work.

Processing Raw Materials

All materials for products come from nature in one form or another. Some materials are renewable and can be reproduced continually. Others are nonrenewable; once used, they cannot be replaced.

Renewable Materials

Renewable raw materials come from plants or animals. Some are found in a wild state, and others are produced on farms. For instance, forests and tree farms provide wood for the lumber industry. See **Figure 15-2**. The fishing industry harvests fish and other marine life from oceans, lakes, and waterways, as shown in **Figure 15-3**. Wild animals are hunted and trapped for their furs, hides, and meat.

Nonrenewable Raw Materials

Nonrenewable raw materials include fossil fuels, nonmetallic minerals, and metallic minerals. Fossil fuels include coal, peat, petroleum, and natural gas. Refer to Chapter 10 for more information about fossil fuels.

Nonmetallic minerals include construction materials such as sand, gravel, and building stone. They also include abrasive materials, such as corundum. Among the metallic minerals are those from which iron, copper, and aluminum are extracted.

Figure 15-2. Lumberjacks harvest trees, a renewable raw material. Environmentally sound harvesting practices help preserve the forests and ensure that trees remain a renewable raw material in the future.

Figure 15-3. Commercial fisheries harvest fish and other seafood.

Manufacturing Products

The second step in production is to change raw and processed materials into useful products. Today's manufactured products include computers, jet planes, glues, lasers, plastics, medications, photocopying machines, and bubble gum. Construction products include structures that people use for living, working, traveling, and playing. Among these structures are houses, office towers, and sports stadiums. Also included is the construction of road tunnels, bridges, towers, and dams.

Manufacturing has changed in the last 200 years. It has evolved through three stages:

- The individual artisan
- Mechanization and mass production
- Automation

The Individual Artisan

Before the eighteenth century, the production of articles was entirely in the hands of individuals. One person, an *artisan*, made products. Artisans used hand methods. Each product evolved by trial and error. One generation passed on acquired skills to the next through an apprenticeship system. The artisan was responsible for every step in the process. He or she did everything from obtaining the raw materials to completing the finished product. For instance, **Figure 15-4** illustrates the work of a chair maker. This artisan, working alone, produced Windsor chairs.

Mechanization and Mass Production

In the second stage in the evolution of manufacturing, the production process was divided into specialized steps. Machinery replaced handwork. This was done to reduce the unit cost of the product. For the first time, products could be made in large quantities. This change, which occurred first in England, was known as the Industrial Revolution.

Early steam-powered machines replaced the muscle power of workers and animals. See **Figure 15-5**. Burning coal produced the steam. Watt's steam engine was the first machine to convert the chemical energy of coal into steam and then into mechanical energy. The mechanical energy powered the machinery.

As a result of the increased use of steam-powered machinery, production had to be located in larger buildings called *factories*. Towns and cities developed rapidly as people moved to live near these factories.

At first, factories had to be near the coalfields. Moving the coal long distances was too costly. After the mid-nineteenth century, however, transportation became cheaper. Coal could be more readily transported. Factories could be built almost anywhere.

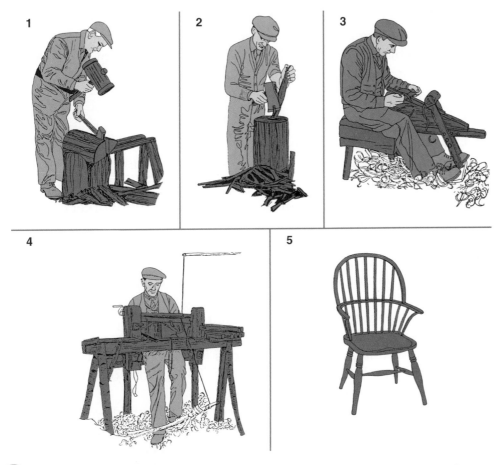

Figure 15-4. Steps in making the legs and spindles for a Windsor chair. A single artisan working alone made the entire chair.

Figure 15-5. Machinery in early factories was powered by steam.

The factory system expanded because it was efficient. *Efficiency* means that good use is made of energy, time, and materials. Efficiency was further achieved through the use of new methods of production. These included:

- Division of labor
- Use of specialized machinery to build parts
- Use of interchangeable parts
- Assembly lines and mass production

Division of labor means each person is assigned one specific task in the making of a product. Through constant repetition, the worker becomes skilled in that task. As the worker becomes skilled, he or she can perform the task at a more rapid rate.

With specialized machinery, the same part could be made again and again with little variation in size or shape. Because more products could be produced in the same amount of time, the cost for producing each item dropped. This meant the item could be sold for less, too.

Interchangeable parts are parts that are the same shape and size. They are made within *tolerances* (specifications for size and shape) to ensure that a new part can be substituted for one that is worn or broken. Each new part fits just like the old one.

An *assembly line* allows assembly of parts in a planned sequence. This is possible only when parts are manufactured to uniform standards. Use of assembly lines made factories more efficient. A continuous assembly line could quickly produce a large number of identical items. Assembly begins with one major part. Other parts are added as the product moves to other stations along the line.

Making a large number of products on an assembly line is called *mass production*. Henry Ford first used it in 1914. Ford's assemblers worked side by side in long lines. See **Figure 15-6**. The parts were brought to them. Each worker had only one assigned task as the Model T cars moved slowly by on the assembly line. As a result, the time needed to produce a car dropped from 12 hours 30 minutes to 1 hour 33 minutes.

Figure 15-6. Henry Ford perfected the first assembly line to produce his Model T car.

Modern automobile plants are huge assembly points. See **Figure 15-7**. Each plant contains a number of small assembly lines that produce *subassemblies*. These are components having a number of parts. Engines, gearboxes, and suspensions are examples. Subassemblies are later assembled into complete automobiles.

Automation

While the assembly-line techniques pioneered by Ford increased the rate at which products could be produced, they were still limited. Humans could work only so fast. To produce products faster, automated assembly lines were developed.

Today, instead of making products, workers build, monitor, and maintain the machines that make products. *Automation* is the use of computers or other machines control machine operations. Machines have been developed that perform many different operations. They can receive a number of parts and assemble them at high speed in the correct sequence.

For example, the machinery shown in **Figure 15-8** is used to package macaroni. The display packages have to be opened, filled, sealed, weighed, and boxed. Almost all of the operations are automated. Even on final inspection, machines check each package for the correct amount of macaroni.

Figure 15-7. Modern automobile assembly lines are automated to improve speed and precision.

Math Application

Calculating Material Needed

The steel used to make food cans is rolled into huge coils and then shipped to a tinplate mill. There it is given a fine coating of either tin or chromium oxide. This coating protects the steel from rust. The rolls are then delivered to can-making factories.

To manufacture cans efficiently, the manufacturer needs to know how much coated steel to order for each batch. When unrolled, a can has the shape of a long rectangle. A *rectangle* is a four-sided shape in which the opposite sides are parallel and the adjacent sides are at right angles to each other.

The length of the rectangle is equal to the circumference of the can. If you know the diameter of the can, you can calculate the circumference using the formula $\pi \times D$. D equals the diameter of the can, and π, or pi, equals approximately 3.1416. For example, a small can of soup measures 4″ high and has a diameter of 2.5″. The circumference of the can equals $3.1416 \times 2.5″ = 7.8535″$.

You can now calculate the area of the rectangle that makes up the sides of the soup can. Multiply length (which equals the circumference of the can) × the height of the can: $7.8535″ \times 4″ = 31.414$ square inches.

Math Activity

1. Find the surface area of the sides of a can of soup that measures 4.5″ high and has a diameter of 4″.

2. How many square feet of sheet metal would be required to produce the rectangles for 1,000 soup cans? (*Hint:* To convert square inches to square feet, divide by 144.)

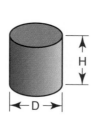

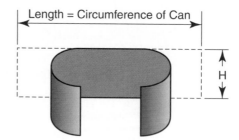

D = Diameter
H = Height

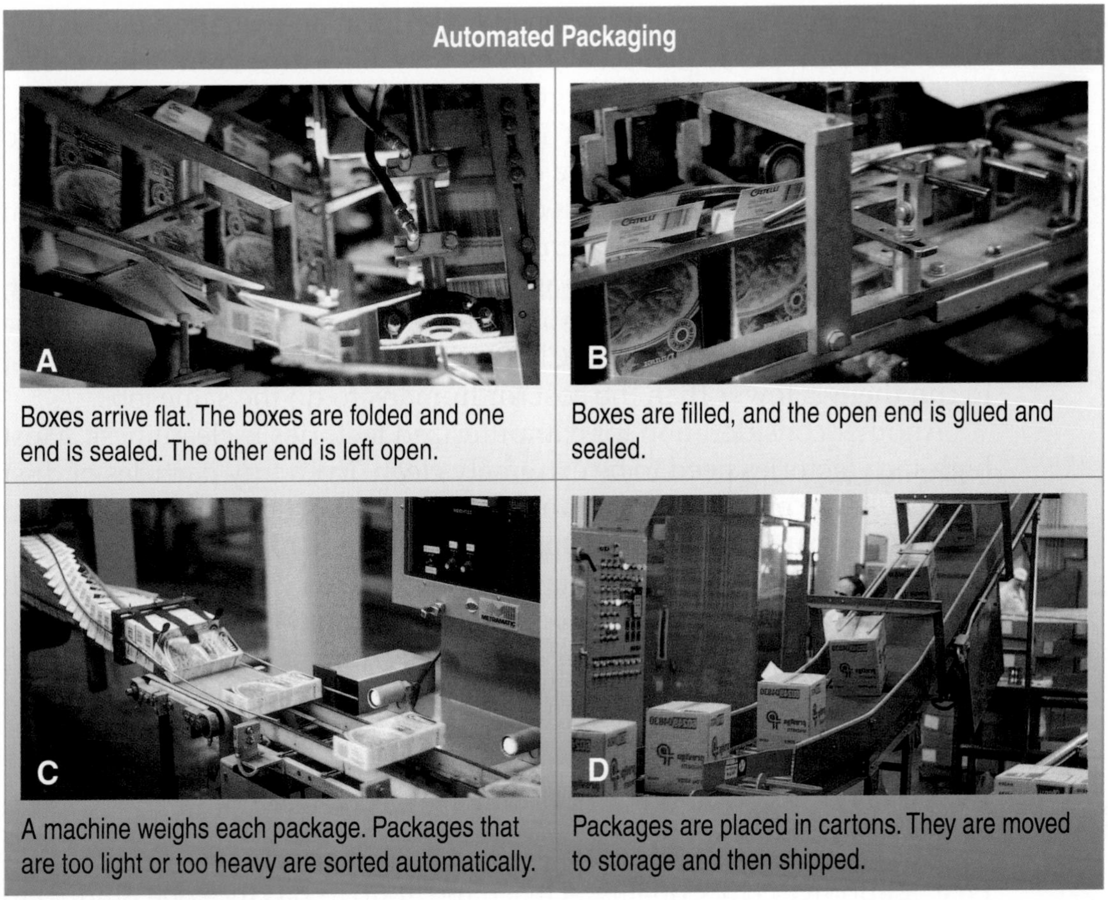

Automated Packaging

A Boxes arrive flat. The boxes are folded and one end is sealed. The other end is left open.

B Boxes are filled, and the open end is glued and sealed.

C A machine weighs each package. Packages that are too light or too heavy are sorted automatically.

D Packages are placed in cartons. They are moved to storage and then shipped.

Figure 15-8. An automated packaging line.

Robots

The 1980s brought further development in automated manufacturing. Industrial robots revolutionized production lines. Computer-controlled robots can be programmed to perform various functions, including:

- Handling—loading and unloading parts and assemblies on machines
- Processing—machining, drilling, painting, and coating
- Assembling parts and subassemblies
- Breaking down an object into its component parts
- Fixing objects permanently in place by welding or soldering
- Performing dangerous tasks and operations
- Transporting materials and parts or delivering mail

To understand the operation of an industrial robot, imagine that you have been blindfolded and tied to a chair. You are able only to move one arm and to rotate at the waist. Your one arm has joints at the shoulder, elbow, and wrist. Robots also have joints much like a waist, shoulder, elbow, and wrist that can move in two or three directions. Each joint, or direction of movement, in a robot arm is called a *degree of freedom*. Most robots have five or six degrees of freedom. See **Figure 15-9.**

Some robots can be programmed to perform an operation by leading them through the sequence of moves they have to follow. This is like taking someone by the hand to guide him or her through a strange place. As the robot arm is moved, it is possible to record the positions into computer memory by pushing a button or trigger on the robot arm.

In many factories, computers control the machinery. Once its program of instructions has been written, a *computer numerical control (CNC) machine tool* can do the same job again and again. It can work day and night, seven days a week. However, such machines are expensive to purchase and use. The use of robots and CNC machines is effective only if the cost to run them is lower than the cost for humans to do the same jobs.

Another consideration for computerized factories is cleanliness. Most high-tech factories need to be extremely clean. Even tiny particles of dust may cause the machines to malfunction. This is particularly important in the manufacture of computer chips. See **Figure 15-10**.

Robots have several advantages over humans. They work better in hot, noisy, or dangerous situations. Robots do not take coffee breaks, go home ill, or sleep. They continue working 24 hours a day and operate for thousands of hours before they require maintenance. See **Figure 15-11**.

Advantages of Automation

Automation, including robotics, has certain advantages for industry. First, it improves work quality. A machine set up to produce one product to a high standard will continue to produce parts to the same standard. Also, automation results in increased production. Each worker can produce more products in the same amount of time. Thus, the cost to produce each item is reduced.

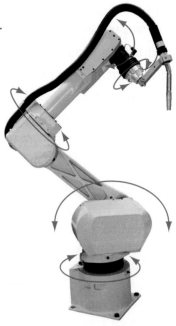

Figure 15-9. This industrial robot has five degrees of freedom.

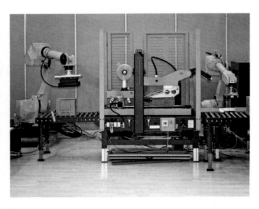

Figure 15-10. The enclosed workspaces contain special filters to keep dust and other particles out. Even outside the filtered areas, workers wear head coverings and "booties" to prevent contamination in this "clean room."

Figure 15-11. These robots automate the final packing and shipping of manufactured items.

Disadvantages of Automation

In spite of its advantages, automation also has some disadvantages. Machines sometimes malfunction. If a machine has been set up incorrectly, all of the products it produces are inaccurate or substandard. Such mistakes, if not detected early, can be very costly. Thousands of defective items may have been produced.

Another concern is the loss of jobs as machines and robots take over tasks in the workplace. The prospect of being replaced by machines frightens or angers many people. They are concerned about being unemployed or having to retrain for another type of job.

This is not a new issue. Until 1880, more than half of all workers in advanced nations worked on farms. At that time, tractors and other machines were developed. Farm mechanization displaced more than two thirds of the farmhands. These machines enabled one person to produce what ten or more had done previously. In the last 100 years about 90% of the farm jobs have disappeared.

Many of the workers leaving the farms went to work in factories, but there, too, machines were replacing people. Ford's mass production caused fears that the assembly line would result in loss of jobs. However, mass production created mass markets. Most people could now buy goods once affordable only to the rich. Products began to sell in the millions.

Math Application

Mathematics and Machining

A hundred years ago, machine operators used lathes to produce machined parts like the one shown here. Today, most production lathes are computer-controlled. They work by reading thousands of bits of information stored in the computer's memory. To place this information in the machine's memory, the programmer creates a series of instructions that the machine can understand.

All of the information needed to make a product must be present in the machine programming. Otherwise, the numerically controlled machine cannot complete its task.

Math Activity

The drawing shown below is missing some horizontal measurements. Calculate the measurements for A, B, C, D, and E. Start by noting the given dimensions. Then add or subtract as necessary to obtain each value. (For this exercise, you may ignore the vertical measurements.)

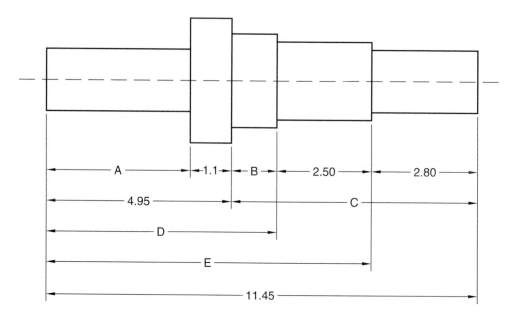

Production and Management Systems

Production technology consists of the systems and processes that are used to manufacture products. Factory management and production systems have evolved slowly over the years. New ideas and better technology have improved efficiency and working conditions. In some cases, they have also lowered the costs to make products.

Traditional Management Systems

The traditional management system for a mass production factory is shown in **Figure 15-12**. Each person is given responsibility for a specific job. However, this system has some drawbacks.

For example, in this management system, any problems that occur must be sent up the chain of command for a solution. The problem is passed along from a machine operator to a supervisor to a department head to a manager, and even higher in some cases. Meanwhile, the machine operator waits for a solution.

Another limitation of the traditional system of management is that workers look after only one machine. They are not skilled in the operation of other machines. If the machine a worker uses is not needed, the worker cannot go to work on a different machine, resulting in "down-time."

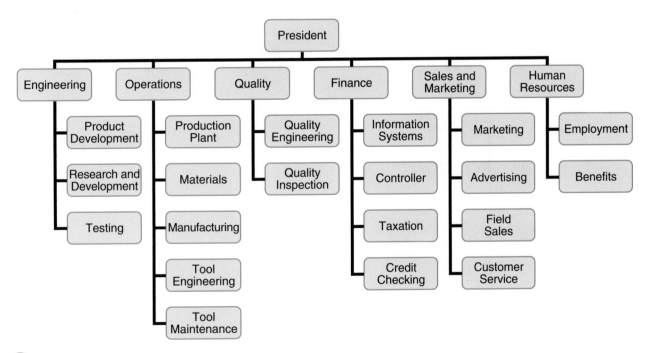

Figure 15-12. Typical organization of a traditional factory.

Lean Manufacturing

Many factories have changed to a new organization called *lean manufacturing*. The purpose of lean manufacturing is to make products more efficiently. It is more cost-effective because fewer resources are needed. It focuses on eliminating waste, ensuring quality, and maximizing employee involvement in the manufacturing process.

In this system, employees work in small, product-oriented teams. See **Figure 15-13**. They work in *work cells* that consist of all the equipment and processes used to make a specific part. All of the equipment is located in the same small area. Team members are cross-trained to use all of the equipment in their work cells. They become skilled in the operation of many different machines. Because anyone on the team can use any piece of equipment, down-time is rare.

Leadership comes from the workers. Individual teams and team members are encouraged to look for better solutions to manufacturing issues. Solutions to problems are sought by the workers, rather than by top-level management, as was the case with the old organization. The entire team is involved in problem solving, making changes, and even the design of new products.

Continuous improvement is the theory that all products and processes can and should be improved on a continual basis. This is an important part of lean manufacturing, for both the product and the manufacturing process. To encourage continuous improvement, lean manufacturing systems reward employees for suggesting improvements to the production process.

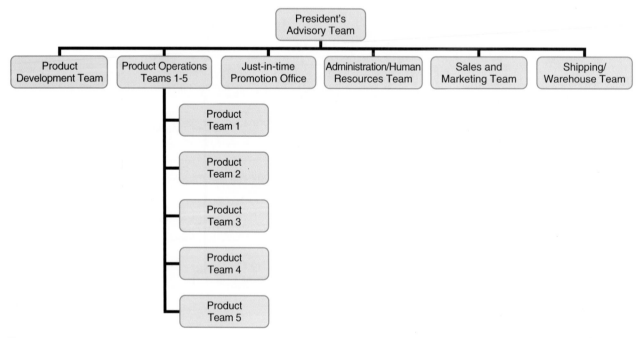

Figure 15-13. In lean manufacturing, workers are grouped into small, product-oriented teams. Each team is encouraged to make improvements to manufacturing processes.

Lean manufacturing works best when it incorporates three features:

- Just-in-time delivery
- Zero defects
- Flexible manufacturing system

Together, these three changes result in a reduction in product development time, a reduction in production time, and elimination of waste.

Just-in-Time Delivery

In the traditional method, known as "batch and queue," parts were created in advance. Then they were stockpiled, or stored near the equipment, to await future use at an undetermined time. This method resulted in higher inventory costs. Also, more factory space was needed to store the inventory.

In *just-in-time delivery*, parts are not stockpiled. Only the parts that are actually needed for the current day's production are made. Those that must be ordered are ordered daily. Each order consists of just enough parts to fill the daily need. This reduces the floor space needed and lowers the cost of producing the final product.

Zero Defects

Defective parts are expensive. They cannot be sold, so the time and materials used to make them is wasted. Automatic machines or an assembly line may have to be stopped if a defective part is noticed, resulting in a further waste of time.

The concept of *zero defects* was first introduced by Philip Crosby in 1979. He defined zero defects as "doing things right the first time." If parts are made correctly the first time, no defective parts are passed forward. Assembly lines do not have to be stopped, and there is less need for troubleshooting. In addition, customer satisfaction is improved because the product works. All of these things result in better efficiency and increased profit.

Flexible Manufacturing Systems

Manufacturing systems that are designed to handle changes easily are known as *flexible manufacturing systems*. See **Figure 15-14**. A work cell in a flexible manufacturing system may consist of a number of automated machines that make parts or assemble products. The equipment can be changed quickly to produce a different type of product.

Reduction in setup time is important because it helps to reduce the total time from the beginning to the end of making a product. This is called the total cycle time. When the total cycle time is reduced, a company can meet changing customer needs more quickly.

Figure 15-14. Programmable tools like this CNC milling machine can be reprogrammed and retooled quickly. This helps manufacturers meet changing needs without a loss of productivity.

Production Systems

All production systems involve five basic operations.

- Designing
- Planning—organizing a system in which personnel, materials, and equipment work together
- Tooling up—acquiring and setting up tools and machines for production
- Controlling production—using machines to make the product
- Packaging and distribution—packaging, storing, and transporting the products to wholesalers and retailers

Designing

Designing involves making the original plans and drawings of products that satisfy consumer demands. Before starting to design a product, a manufacturer determines what buyers want. This involves market research. The product will not be made if the potential sales are not large enough to make a profit after costs are paid.

Designing is the responsibility of industrial designers. The designers work with engineers, market researchers, and sales personnel. First, they learn the needs of potential customers. Relying on research, they generate ideas and make preliminary sketches of the product. Next, they rework these sketches and make detailed drawings. See **Figure 15-15.**

Designers consult with the engineering staff to determine what production methods and processes to use. When a basic design has been approved, drafters produce scaled or full-size drawings. A three-dimensional model may also be made.

The information gained from the sketches, drawings, and models is then used to make working drawings and specifications. These drawings show the exact size and shape of each part. Specifications include information about:

- The materials to be used
- The number of parts to be made

Figure 15-15. Many designers create sketches and refine drawings using computer-aided design (CAD) software.

- The operations needed to produce the part
- The level of accuracy required

Full-scale working models and prototypes are sometimes made. Physical prototypes are often used, but many products can also be tested using a digital 3D prototype and computer testing software. See **Figure 15-16.** These prototypes help identify weaknesses or errors in the design.

Planning

Production must be planned. Personnel, materials, and equipment must be specified carefully to ensure a smooth operation. This planning includes:

- Selecting and ordering equipment, machines, and processes
- Finding the best way for people to work together
- Determining how long each manufacturing operation will take
- Gathering information on production costs

An engineer or production planner considers a variety of ways to complete each operation. See **Figure 15-17.** He or she selects the most efficient one. The decision is based on the time involved, cost, and the quality and quantity of the product.

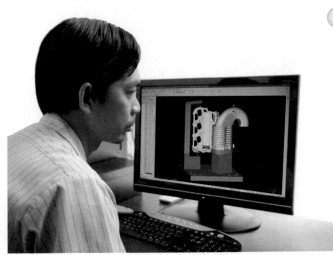

Figure 15-16. Testing 3D digital prototypes saves the cost of building a physical prototype. In most cases, building and testing a digital prototype is also faster, decreasing total development time.

Figure 15-17. Production planners calculate the most efficient and cost-effective methods for making a new product.

Tooling Up

Tooling up is the process of gathering or making the tools and equipment needed to manufacture a new product. Some of the tools and machines may be purchased. Tool-and-die makers design and create others especially for the product. The machines are installed according to the plan created by the engineer or production planner. See **Figure 15-18**. Then a trial production run is conducted to test the setup.

Controlling Production

The production process is organized so that materials and parts move efficiently from one operation to the next with as little waste as possible. This sequence begins in the receiving area, where raw materials are kept. The inventory manager's job is to control the flow of all materials and products. He or she prevents the purchase of unnecessary raw materials, as well as the production and storage of unnecessary products. Materials or parts arrive only a day, or even hours, before they are needed.

Figure 15-18. During the tooling-up stage, engineers check machine setups to make certain parts will be machined accurately.

From receiving, materials move through various stages of processing and assembling. The focus is on avoiding waste of materials, resources and personnel. All unproductive actions and motions are minimized. For example, time is saved when parts are stored closer to the place where they are used. Processes are fine-tuned to minimize the wasted time that occurs when workers or machines have nothing to do.

Throughout the production process, a variety of inspection tools are used. See **Figure 15-19**. Gauges are used to check the sizes of parts. X-rays check the internal structure of metal parts. The amount of inspection varies. In the manufacture of aircraft, it is important to check every part before it is installed in the aircraft. However, for most consumer products, it is enough to check a small number of items from a large batch.

Packaging and Distribution

Various forms of packaging are used to protect products during shipping. Bubble packaging, boxes, cartons, and crates are used to protect the product. They provide insulation and protection against moisture, weather, and rough handling. Packaged products must also be labeled so that the consumer can recognize the contents. Labels and other kinds of markings show the product, the name of the manufacturer, quantity, and directions for use and care. They also provide other special information.

Packaged and labeled products are usually stored in a warehouse to await shipment. See **Figure 15-20**. They are organized in quantities that are convenient for handling, sorting, and counting. The machinery for handling bulk shipments includes conveyor belts, forklift trucks, and pallets. Pallets are wooden platforms on which the packaged products

Figure 15-19. Quality is checked throughout the manufacturing process. In addition, a quality control inspector checks a certain number of parts from each batch produced.

Figure 15-20. In many warehouses, inventory is tracked using a bar code attached to each storage box or container. The boxes are stacked on pallets and placed on racks until they are needed.

are placed. A forklift then picks up the loaded pallet and moves it to and from its storage location. When they are sold, products are loaded onto trucks, railroad cars, and ships. These vehicles transport the products to wholesalers and retailers.

Sustainable Manufacturing

Economic growth all over the world has meant an increasing demand for products. This, in turn, has led to an increased demand for raw materials. The result has been an almost intolerable burden on our planet's resources. Can we continue to take from the Earth without eventually running out of raw materials? The answer is no. We must change the way we make our products.

Sustainable manufacturing is the use of processes designed not only to reduce waste, but also to conserve resources such as electrical energy and water. The processes result in little or no pollution. In addition, considerable thought is given to reusing the product after it is no longer useful. **Figure 15-21** compares sustainable manufacturing with traditional manufacturing practices.

The cradle-to-cradle (C2C) design philosophy challenges designers to create products with a positive footprint. This means that not only should every component of the items be biodegradable, nontoxic, and fully recyclable, but their waste should be reusable, too. In nature, a tree grows leaves and later sheds them. The leaves decompose, fertilizing the ground on which they fall. In cradle-to-cradle design, manufactured products are designed to be used in this kind of perpetual, closed loop.

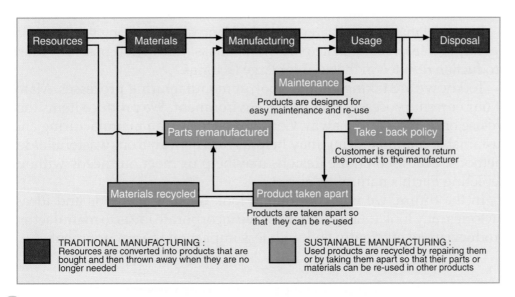

Figure 15-21. Comparison of traditional and sustainable manufacturing.

Ideally, in order to preserve resources for future generations, products should be designed to be:

- Energy-efficient
- Easily and affordably maintained so they last longer
- Easily disassembled into component parts
- Recycled for other uses or reused after repair or updating
- Packaged and returned to the manufacturer when they are no longer needed

Think Green

Aluminum: A Sustainable Material

Recycling aluminum cans is nothing new. People have been doing it for many years now. What you may not know is that aluminum recycling is a perfect example of sustainable manufacturing. Aluminum is 100% recyclable, and it can be recycled again and again.

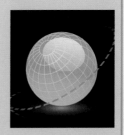

The aluminum recycling process is not complex. It consists of removing contaminants, such as paint, and then melting the aluminum. It can then be molded into storage shapes such as ingots (bars) or rolls. These products are supplied to the manufacturers for use in new products.

Recycling aluminum also saves energy. The recycling process requires only 5% of the energy needed to mine aluminum from bauxite ore. In addition to helping the environment, this lowers the cost of producing aluminum products.

Many communities also accept other types of aluminum for recycling. Accepted items may range from aluminum foil to used lawn furniture. This is important, because although aluminum is easily recycled, it decomposes very slowly. If it ends up in a landfill, it may take 400 years or more to break down. Check with the recyclers in your community. What aluminum items are accepted?

Industrial manufacturing arose from advances made during the Industrial Revolution. Interchangeable parts made mass production possible. Mass production resulted in the need for large factories.

Today, we are taking another look at manufacturing processes. Many of our current processes pollute the environment. We produce items that we use once, then throw away. Concepts such as lean manufacturing and sustainable manufacturing may help us move away from wasteful older methods. Cradle-to-cradle designs may help us meet our needs without depleting Earth's natural resources.

In the future, we will continue to look for new methods and ideas. Factories may look much different as we incorporate ideas to manufacture products more efficiently and with less impact on the environment.

- The first step in producing a product is to process renewable or nonrenewable raw materials.
- The second step in producing a product is to change processed raw materials into useful products.
- Automation of assembly lines and other manufacturing processes has resulted in higher productivity and better accuracy.
- Traditional production and manufacturing systems are being replaced by more efficient lean manufacturing methods.
- Sustainable manufacturing incorporates many of the concepts of lean manufacturing, but emphasizes lowering pollution levels and designing products to be reused or recycled rather than thrown away.

Summarizing Information

Copy the following graphic organizer onto a separate sheet of paper. For each chapter section (topic) listed in the left column, write a short, one-paragraph summary of the topic in the right column. Do not write in this book.

Chapter Section (Topic)	Summary
Processing Raw Materials	
Manufacturing Products	
Automation	
Production and Management Systems	
Sustainable Manufacturing	

Test Your Knowledge

Write your answers to these review questions on a separate sheet of paper.

1. What are the two main steps in producing a product from raw materials?
2. List the three stages through which the manufacturing process has evolved.
3. What is the advantage of products made with interchangeable parts?
4. Describe the difference between a traditional mass production assembly line and an automated assembly line.
5. List five jobs that robots can perform in the manufacture of a product.
6. List the advantages and disadvantages of automation.
7. How is lean manufacturing different from traditional manufacturing?
8. What three features are often incorporated into lean manufacturing systems?
9. List the five basic steps in a production system.
10. How is sustainable manufacturing different from traditional manufacturing?

Critical Thinking

1. This chapter briefly discusses the potential loss of jobs caused by automation. However, other factors also cause job loss in manufacturing. Research and explain the role of the global economy on jobs in the manufacturing sector.
2. Analyze the following statement: "Sustainable manufacturing will further reduce jobs in manufacturing because products will be reused indefinitely." Is the statement true? Explain your answer.

Apply Your Knowledge

1. Select a raw material to research. Find out:
   A. Where is it found?
   B. In what form is it found in its natural state?
   C. How is it extracted, harvested, or farmed?
   D. How is it transported?
   E. How is it processed or refined?

2. Create a chart comparing robots to human workers. Design the chart to show advantages and disadvantages of using humans and robots in each stage of the production process.

3. Imagine that in your working life, you started as an artisan. You then moved to a factory and worked on a mass production line. Your last job was in a fully automated factory in which robots did the work. Describe the advantages, disadvantages, and working conditions in each of your three jobs.

4. Research how new technologies have replaced, outdated, or created new jobs in your community.

5. Give one example of leading-edge products in each of the following areas. Find examples that are *not* described in the textbook. Where and by what company is each product manufactured?
   A. Computers and electronics
   B. Telecommunications and networking
   C. Media and entertainment
   D. Nanotechnology and materials
   E. Energy and the environment
   F. Biotechnology and health care
   G. Transportation and cities
   H. Privacy, security, and defense

6. From a science-fiction book, comic, TV show, or movie in which events occur in the future, identify three technical objects or systems that do not exist today. Describe how each one operates.

7. The following list shows how change took place from the beginning to the end of the twentieth century. Find or draw a picture that illustrates each of the changes. For example, for part A, you could have a picture of a steam train and another of an electric train.
   A. Steam to electric
   B. Rural to urban
   C. Cottage industries to factories
   D. Cart tracks to railroads
   E. Personal contact to telephone
   F. Horse to car and airplane
   G. Brick and stone to steel and aluminum
   H. General store to department store
   I. Natural materials to synthetic materials

8. Research one career related to the information you have studied in this chapter. Create a report that states the following:
   - The occupation you selected
   - The education requirements to enter this occupation
   - The possibilities for promotion to a higher level
   - What someone with this career does on a daily basis
   - The earning potential for someone with this career

   You might find this information on the Internet or in your library. If possible, interview a person who already works in this field to answer the five points. Finally, state why you might or might not be interested in pursuing this occupation when you finish school.

STEM Applications

1. **TECHNOLOGY** Design a plant stand that has no more than four or five pieces. The plant stand should be made of commonly available, inexpensive materials. Plan a production system to mass-produce the plant stand. Create a graph or chart to show all of the steps in the production process. If the materials are available, ask classmates to help you test your production plan by producing a batch of plant stands.

2. **ENGINEERING** Identify a product you use every day that is manufactured using traditional methods. Plan a sustainable manufacturing system in which the product could be manufactured. Your plan should minimize waste and account for what happens to the product after its useful life is over. Present your plan to the class.

3. **ENGINEERING** Using the cradle-to-cradle design philosophy, design and make a cabinet that will hold up to 40 CDs or DVDs. Create a sustainable manufacturing plan to produce the cabinet. Document your plan in a report, using charts and graphs to show the various steps of production. Make a poster showing the planned life cycle of the cabinet.

4. **TECHNOLOGY** Research to find a manufacturing company that has switched from traditional to lean manufacturing. Assess the products output before and after the switch. Write a report explaining what is different in the new system. State whether the company's goal of reducing waste was achieved. List the overall positive and negative effects of the change.

16

Technology and the Environment

The LifeStraw Personal contains filters that make polluted water safe to drink.

Better by Design

Mikkel Vestergaard Frandsen and the LifeStraw® Personal

Safe drinking water is hard to find in many Third World countries. Many people in undeveloped areas suffer from diseases they catch by drinking polluted water, and many of them even die. Mikkel Vestergaard Frandsen developed a straw called LifeStraw Personal that makes polluted water safe to drink. Contaminated water is drawn in through the lower end. The straw kills 99.999% of water-borne bacteria and viruses. By the time the water reaches the person's mouth, it is safe to drink. The filters in the straw can process about 185 gallons (700 liters) of water. This is enough to provide drinking water for one person for about a year. It is inexpensive, light, and portable and can be carried on a string around the neck. This makes it perfect for people in Third World countries who need a reliable source of drinking water.

LifeStraw Personal does not require batteries and can be used anywhere.

"We must use our innovative skills to save the lives of millions of people who are dying needlessly."

Preview and Prediction

Before you read this chapter, glance through it and read only the heads of each section. Based on this information, try to guess, or predict, what the chapter is about. Use the Reading Target graphic organizer at the end of the chapter to record your predictions.

Key Terms

acid rain
carbon monoxide (CO)
clear-cutting
composting
hazardous materials
hazardous waste
household hazardous waste
landfill
nitrogen oxides (NO$_X$)

ozone
radioactive waste
recycling
smog
sulfur dioxide (SO$_2$)
sustainable lifestyle
watersheds
wetlands

Objectives

After reading this chapter, you will be able to:

- Describe the effects of technology on air quality.
- Identify sources of land and soil pollution.
- Explain how the use of technology can affect water quality.
- Identify different types of hazardous waste and explain their effects on the environment.
- Explain how people can reduce harmful effects of technology on the environment.

Useful Web sites:
www.lifestraw.com
www.controllingpollution.com/water-pollution-facts/

People have been using technology to improve their lives for hundreds of years. For many years, people did not realize that the technology they used might have unintended, and often harmful, effects. Today, we know that technology can have both intended and unintended effects. Some of the unintended effects can be harmful.

For example, the cars and airplanes that provide us with fast, reliable transportation also produce a large amount of air and noise pollution. Batteries leak chemicals into the soil, contributing to land and water pollution. In fact, concern for the effects of technology on the environment is growing rapidly. We are starting to see just how much our technological activity is affecting Earth.

Technology has an impact on every part of Earth: land and soil, water, and air. It also has other harmful effects, such as increased noise. This chapter looks at the effects of technology on the environment. It also explores how we can use technology to help reverse some of the harmful effects.

Air Quality

Until the Industrial Revolution, people never thought much about air quality. There was no need, because the air was relatively clean. The major pollutants were smoke and ash from wood and coal fires used for heating and cooking.

Advances in technology that made the Industrial Revolution possible also greatly increased the amount of smoke, ash, and chemicals in the air. Some of the larger cities started to experience smog. *Smog* is a combination of smoke and fog. See **Figure 16-1**. When the smoke comes

Figure 16-1. The smog that often blankets major cities can be harmful to humans, animals, and even plants.

from factories or automobile exhaust, smog can contain chemicals that are harmful to our respiratory systems. It may cause symptoms ranging from coughing and sneezing to nausea. It can also irritate our eyes and throats. For people who are at higher risk, such as children, seniors, and people with chronic diseases, the effects can be even more harmful.

Air pollution damages more than human health, however. It also has harmful effects on the environment. It affects forests and national parks, as well as agricultural crops. The Environmental Protection Agency (EPA) has identified six specific types of air pollution. Each of these pollutants causes specific damage both to human health and to the environment.

Ground-Level Ozone

Ozone is a gas that is formed when other common air pollutants mix with sunlight. Most of the pollutants that combine to form ozone are produced by automobiles and factories.

When ozone is located in the upper atmosphere, it helps protect the Earth. At ground level, however, it can cause respiratory problems in humans and animals. It can also damage plants. In fact, ground-level ozone can damage the leaves on many types of plants. This prevents the plants from producing food (photosynthesis). See **Figure 16-2**. Ozone can also weaken a plant's defenses against disease and insects.

Particle Pollution

Particles in the air can also cause air pollution. Some particles are natural, such as dust raised by a high wind. Others are generated by human activities and technology. These include soot, smoke, and chemicals (including acids) emitted from power plants, factories, and automobiles. They mix with liquid droplets and are transported through the air as water vapor or rain.

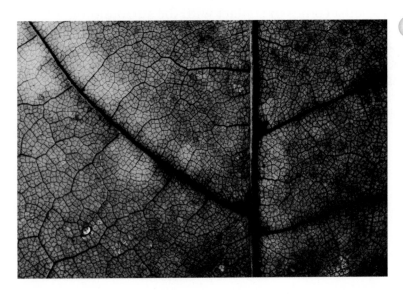

Figure 16-2. Too much ground-level ozone is toxic to plants. It enters through the pores, or *stomata*, of the leaves, causing reddish areas like those shown here.

Particle pollutants are classified as coarse and fine, depending on their size. Fine particles have more potential to harm human health. However, both coarse and fine particle pollutants affect the environment. When particle pollutants mix with rain, they often fall to the ground as *acid rain*. Acid rain can cause lakes and other bodies of water to become acidic, endangering fish and animals.

Carried on the wind, particle pollutants can also change the nutrient balance of an area. This affects the types of plants that can grow in the area. These particles may also reduce crop harvests by slowing plant growth.

Carbon Monoxide

Most vehicles on the road today burn fossil fuel. Because the engines are not 100% efficient, waste gases are produced. One of the most harmful waste gases is *carbon monoxide (CO)*. Carbon monoxide gas is poisonous, but it is also colorless and odorless, so you may not know when it is present. Vehicles account for about 56% of carbon monoxide emissions overall in the United States. See **Figure 16-3**. The percentage is higher in large cities. Other sources of carbon monoxide include industry and forest fires.

Carbon monoxide is hazardous to everyone, even people who are in excellent health. In addition to heart and lung problems, it can cause vision problems and even affect the central nervous system. Environmentally, carbon monoxide is not a direct threat. However, it combines with other gases to form ground-level ozone.

Nitrogen Oxides

Nitrogen dioxide (NO_2) and other nitrogen-containing gases are also a result of engines that use fossil fuels. The EPA groups all of the nitrogen-containing gases into a single category called *nitrogen oxides (NO_x)*. Although these gases are not emitted in vehicle exhaust, they form soon after the exhaust comes in contact with the surrounding air.

Figure 16-3. In spite of strict laws about automobile emissions, vehicles produce more than half of the carbon monoxide emissions in the United States.

Nitrogen oxides can cause both short-term and long-term health issues. Short-term issues include an inflamed throat and difficulty breathing. Long-term issues include asthma and emphysema.

Like carbon monoxide, nitrogen oxides alone do not cause environmental damage. However, they contribute to both ground-level ozone and particle pollution, which can cause significant environmental damage.

Sulfur Dioxide

Most *sulfur dioxide (SO₂)* pollution is generated by power plants and industrial processes. Ocean-going ships and locomotives also contribute to pollution levels. See **Figure 16-4**. High levels of sulfur dioxide can lead to breathing problems, especially in people who have asthma. Sulfur dioxide is also a typical component of acid rain.

Airborne Lead

Until the early 1980s, lead from automobile fumes was a major source of air pollution. Lead has serious effects on our health. It can affect almost every system in the body. It is doubly hazardous because it does not leave the body. It builds up in the bones. It is also an environmental hazard. After being deposited by airborne sources, it can decrease the growth rate of plants and animals in the area. It has also been linked to a loss of biodiversity.

Fortunately, laws regulating the use of lead have resulted in a huge decrease in the level of lead in the air. When the government began requiring the use of lead-free gasoline, lead levels in the air dropped dramatically. Today, lead content in the air is significant only near manufacturers of products containing lead and near some utilities and waste incinerators.

Figure 16-4. Trains provide an economical method of transporting goods across the country, but the exhaust from the locomotive contains sulfur dioxide. How could this problem be prevented?

Science Application

Monitoring Carbon Dioxide Levels

Although people are still arguing about exactly *what* change is taking place, most people agree that Earth's climate is changing. One of the gases believed to be responsible for climate change is carbon dioxide (CO_2). Burning fossil fuels and many other human activities produce large amounts of CO_2. When CO_2 builds up in the Earth's atmosphere, it can cause a "greenhouse effect." In other words, it can trap heat near the Earth's surface, causing the air and water to become warmer. Therefore, scientists have been closely monitoring levels of carbon dioxide gas in the air.

Science Activity

Research online to find CO_2 levels for a period of at least 10 years at a location of your choice. Draw a graph to show the change in CO_2 level over time. On your graph, the horizontal axis should show years and the vertical axis should show the CO_2 levels.

Join the points using a smooth curve. This curve shows the general trend. Assuming that your curved line will continue in the same direction(s), extend your curve to the year 2050. Your extended line represents a prediction. What can you say about your prediction? What factors could cause your prediction to be incorrect?

Land and Soil

Technology and human behavior also affect the land and soil. Some of the effects are good. For example, technology has provided ways to keep soil from eroding on hillside farms. However, some of the effects are harmful. Our activities result in several different kinds of land and soil issues. Many of these issues are related to the careless use of technology.

Technology-Related Waste

One issue around the world is the safe and complete disposal of waste material. Technology is responsible for several different kinds of wastes.

Industry

For example, industry creates waste at every step. Waste is produced when the raw material is prepared for use in the manufacturing process. It is generated during manufacturing of the product and of its packaging materials. When the product is used, the packaging is discarded, creating more waste. The product itself may also be discarded after use. See **Figure 16-5.**

Agriculture

Agricultural technology helps farmers produce a good crop. Machines reduce the time they have to spend in the fields. Chemicals help them produce large, healthy crops with minimal insect damage. Some agricultural waste consists of the fertilizers, pesticides, and other farm chemicals. Agricultural waste also includes manure from cattle and other farm animals, although this waste is not a direct effect of technology.

Some agricultural waste products run off the land and pollute rivers, streams, and lakes. However, many chemicals soak into the soil and affect the plants that grow there. Some chemicals may enter the plants and then be eaten by animals or humans.

Mining

Mining waste includes all of the filtrates and other wastes that result from the removal of raw minerals from the ground. It also includes wastes caused by the initial processing of the minerals at the mine. The specific types of waste depends on the type of mineral being mined. The wastes enter the environment by various means. For example, undetected leaks in underground storage tanks are one cause of land pollution.

A careless approach to surface or open-pit mining can cause other environmental damage. All vegetation is stripped away during the mining process, leaving the land bare and unprotected. See **Figure 16-6.** This can lead to soil erosion and loss of habitat for wildlife. In the United States, surface mining is regulated at the state and federal levels. Mines that are no longer in use are reclaimed and an attempt is made at restoration.

Figure 16-5. Manufactured items generate waste at every stage of their development and use. Most products are eventually discarded in a landfill.

Figure 16-6. Surface mining exposes the bare land, placing it at risk for erosion and other problems. Many countries are now taking steps to reclaim the land to prevent erosion and restore habitat.

Garbage

Technology also contributes to another type of environmental hazard: garbage. Many people use disposable products such as Styrofoam™ cups and disposable diapers without thinking about what will happen to them after they are discarded. Also, with new, improved products being created at a record pace, people tend to throw away their older electronic products. Where do these products go?

Many of them are transported to landfills. A *landfill* is an area that has been prepared to receive garbage. Landfills are lined to prevent leaks into the soil. The garbage is then buried at the prepared landfill site. The EPA has established standards for all landfills in the United States, although the landfills are run by state or local governments.

However, some garbage never makes it to the landfill site. People who are careless about where they discard their trash add to the environmental problems. See **Figure 16-7.** Trash that is discarded by the side of the road cannot be treated to prevent soil contamination. It may also leak hazardous chemicals into the Earth. Discarded automobile batteries, for example, may leak lead into the soil. Discarded plastics can also harm wildlife. See **Figure 16-8.**

Technology-Related Damage

The environment is also affected by practices such as clear-cutting forests. *Clear-cutting* is a method of harvesting lumber in which all of the trees in the selected area are cut down using bulldozers, cables, and other technology. See **Figure 16-9.** Clear-cutting destroys the forest ecosystem. It affects not only animals, but also other plants that depended on the trees for shade. Because the soil is no longer held in place by plant roots, soil erosion also occurs. Nutrients are often washed out of the soil, leaving it barren.

Figure 16-7. Garbage thrown by the side of the road is not just ugly. It can cause damage to the environment.

Figure 16-8. Plastics are a particular threat to wildlife. What will happen if this duck cannot remove the plastic ring from its beak?

Some of the leisure activities people enjoy can also cause harm to the land and soil. See **Figure 16-10.** When used carelessly, off-road vehicles and all-terrain vehicles (ATVs) can damage or destroy vegetation. This, too, can lead to soil erosion.

Water Quality

All living things—humans, animals, and plants—depend on the quality of water. People and most land animals require clean drinking water. Sea life requires clean saltwater.

The same water has been on Earth for centuries. Earth has a natural cycle that cleans the water for reuse. In this cycle, water is filtered through wetlands and natural watersheds. See **Figure 16-11.** *Wetlands* are land areas that are filled with water, such as marshes and swamps. *Watersheds* are areas of land that drain into a lake or river. Between them, wetlands and watersheds remove many kinds of pollutants, including pesticides and metals.

Figure 16-9. Clear-cutting removes all vegetation from the soil.

Figure 16-10. ATVs and off-road vehicles can destroy vegetation and lead to soil erosion if they are used unwisely.

Figure 16-11. Wetlands play an important role in the Earth's natural filtration system.

These natural systems are enough to take care of pollutants that occur naturally, as well as much of the pollution caused by humans. Unfortunately, wetlands and natural watersheds are disappearing. People have filled in many of the swamps and built structures on them.

At the same time, the rate at which pollution enters rivers, lakes, and even the oceans is increasing. The pollutants that affect air quality and soil can affect water quality as well. Mineral runoff from mines and even raw sewage is sometimes pumped directly into them. See **Figure 16-12.** Oil spills and agricultural runoff that includes pesticides and fertilizers are also sources of water pollution. Finally, trash people have thrown out often winds up in the water. See **Figure 16-13.**

Figure 16-13. Sometimes people do not think about how their trash affects the environment.

Figure 16-12. In many areas of the world, raw sewage, runoff from mining operations, and other harmful wastes are piped directly into rivers and lakes.

Acid rain is an indirect source of water pollution. When acid rain falls on land, it affects the land and soil, but when it falls over water, the water becomes more acidic. When this happens, fish and other wildlife that depend on the water begin to die. If the water becomes very acidic, even the plants in and around the water cannot survive.

Hazardous Wastes

Hazardous materials are materials that have been identified by the government to be harmful to people and the environment. When these materials are waste products, they are known as *hazardous waste*. Laws regulate the transport, storage, and disposal of hazardous materials and hazardous waste.

Hazardous waste comes from many sources. Mining and other industrial processes can result in hazardous waste. Many agricultural pesticides also contain chemicals that are considered hazardous. Other sources are less commonly known. For example, medications such as epinephrine and some blood thinners are considered hazardous if their concentration is above a certain level. Metals such as mercury, silver, and lead are also considered hazardous in certain concentrations.

Household Hazardous Waste

In fact, many of the products we use every day are considered hazardous. See **Figure 16-14**. The EPA defines *household hazardous waste* as anything that is:

- Corrosive—causes metals and other materials to break down
- Toxic—poisons people or animals

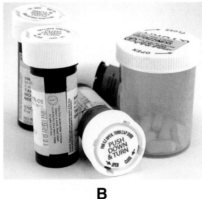

A   **B**   **C**

Figure 16-14. Many of the products we use every day are considered hazardous. Care must be taken when using and discarding these products. A—many household cleaners; B—some medications, including prescription medications; C—pesticides and fertilizers.

- Flammable—catches fire easily
- Reactive—combines with other chemicals to form a dangerous mixture or product

These products include many cleaning products, battery acid, used motor oil, and pesticides, among others.

Radioactive Waste

Radioactive waste is a special type of hazardous waste that gives off energy in the form of radioactive particles. Nuclear power plants, uranium mining, some scientific research, and even some medical treatments produce radioactive waste.

Some forms of radiation are helpful. For example, radioactive particles can be used as "tracers" to help doctors diagnose an illness. Radiation can also be used to treat some types of cancer. However, all types of radiation can damage living organisms. See **Figure 16-15**. Therefore, radioactive materials must be handled carefully and disposed of properly.

The length of time a material remains radioactive varies. Some materials are radioactive for only a few minutes, hours, or days. Others remain radioactive for much longer. For example, uranium can be radioactive for millions or even billions of years.

What should we do with radioactive waste that will remain radioactive for such long periods? How can we protect not only ourselves and today's environment, but future generations also? Government agencies such as the Nuclear Regulatory Commission, the Department of Energy, and the EPA regulate the disposal of radioactive waste. The Department of Transportation regulates how the waste can be transported. In addition, individual states are responsible for laws to protect us from the dangers of radioactive waste now and in the future.

Figure 16-15. Major nuclear accidents like the one at the Chernobyl power plant can kill every living thing within many miles. Because the radioactive material was uranium, the land may remain radioactive for millions of years.

Making a Difference

So far, this chapter has described the harmful effects of technology on our health and on the environment. Can technology be used to fix problems caused by the improper use of technology? Yes! Technology is now being used in several ways to help reduce various types of pollution.

For example, "scrubbers" have been developed for coal-fired electricity plants. Scrubber technology breaks down the waste gases produced by burning the coal. In fact, scrubbers can reduce the levels of sulfur dioxide released into the air by as much as 97%. See **Figure 16-16**.

Farmers are using satellite technology to pinpoint problem areas in their fields. This helps them target specific areas, reducing the amount of pesticide they use. See Chapter 14 for more information about this practice.

New laws and higher public awareness of pollution, technology, and the environment are also helping. As the general public becomes more aware of the issues, more people letting their representatives in Congress know what they think. This helps pass laws restricting the types of technological activities that industries can perform. It also helps convince companies to rethink their use and waste of technological and other resources.

New technology can also be used to fix problems that were not originally caused by the use of technology. For example, technology can be used to repair damage caused by natural disasters such as hurricanes, floods, and tornados. It also provides the means to rescue people and animals who have become stranded by these and other natural disasters.

Figure 16-16. Many coal-burning power plants have begun using scrubber technology to minimize air pollution.

Reduce, Reuse, Recycle

Some people think there is not much they can do as individuals to make a difference. That is not so. Everyone can help. A good rule to follow is "reduce, reuse, recycle."

Reducing Waste

How can we reduce the amount of trash or waste we throw away? We can start by changing our buying habits. For example, we can avoid replacing electronics such as cell phones unnecessarily. Is it really necessary to own the "latest and greatest" if our current phones meet our needs?

When a purchase is necessary, we can avoid buying disposable goods when reusable goods are available. Items such as paper plates and plastic cups are often used once and are then thrown away. See **Figure 16-17**. Instead, we can buy inexpensive dishes that can be washed and used again and again.

We can also choose products that have less packaging. When you buy a new product, as soon as you get home you open the packaging and throw it away. The packaging is instant waste. The EPA estimates that packaging makes up about 30% of the waste we generate every year.

Another important part of reducing waste is reducing the toxicity of the waste we produce. This helps keep hazardous materials to a minimum. For example, read the package before buying disposable batteries to be sure they contain little or no mercury. Use "green" cleaning products that contain no phosphates or other harmful chemicals.

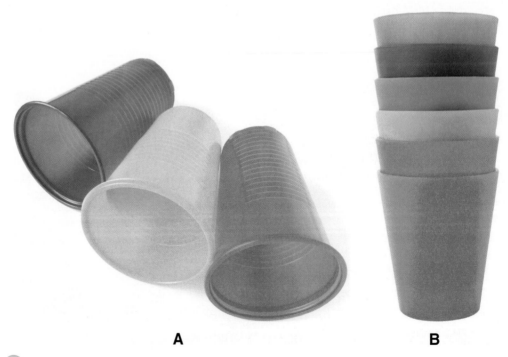

A   **B**

Figure 16-17. A—Many people buy disposable plastic cups for parties because they are colorful and festive. B—Non-disposable plastic cups can be just as colorful and greatly reduce the amount of plastic we throw away.

Technology Application

Finding Substitutes

We use many household products every day without thinking about their effect on the environment. It is easy to use other products that are not as harmful to the environment. The hard part is remembering to think about it! If we make a habit of considering the alternatives on a regular basis, it becomes much easier to do.

Technology Activity

The household items listed below often contain harsh chemicals that are harmful to the environment. Make a table that has two columns. In the first column, list each household item. In the second column, describe alternative methods, products, or technologies that can be used to reduce the harmful effect. Research each item if necessary to find alternatives.

- Drain cleaner
- Flea products for pets
- Toilet bowl cleaner
- Lawn mower gas and oil
- Household pesticides
- Oven cleaner
- Furniture polish

Reusing Items

If we think about it, we can find many ways to reuse products we already own. If an item breaks, can it be repaired? Repairing an item is sometimes less expensive than buying a new one. In these cases, repair can save money as well as reduce the amount of trash we throw away.

Some items can be used for different purposes. For example, if the handle breaks on your favorite mug, you can use the mug as a pencil holder. Your family might use the wood from an old wooden bench to make a coffee table. Before throwing out broken items or items you no longer need, think about other purposes they might serve.

Other ideas for reusing items include:

- Refilling printer cartridges instead of throwing them away
- Holding a garage sale to sell unwanted items to someone who wants them
- Donating unwanted items to charity

All of these ideas can help us reduce the amount of trash we throw away. See **Figure 16-18.**

Figure 16-18. Garage sales, yard sales, and moving sales are good ways to help other people reuse items we no longer want or need.

Recycling

Some items cannot be reused, or cannot be reused safely. *Recycling* is the process of treating materials to make them usable again. Recycling has many advantages. It conserves natural resources such as trees and minerals. At the same time, it reduces the pollution caused by the harvesting of the natural resources. In many cases, recycling products requires less energy than producing new products. Finally, recycling reduces the amount of waste that must be sent to a landfill or incinerated.

Recycling programs exist in almost every community. Some communities recycle only a few types of materials, such as paper and aluminum. Many communities now also accept glass, plastic, and other types of materials for recycling. See **Figure 16-19.** Check with your town or community to find out what types of materials are accepted in your area. If items for recycling are not picked up by the local garbage service, you can generally find a collection center in the community that accepts items for recycling.

Figure 16-19. Even public places such as parks and community centers now provide bins for recycling various items.

Separating items to be recycled from other trash is the first step in recycling. Making sure the items get to a collection center is the second step. Everyone can do these two steps. Processing the collected items and making them into new products is the responsibility of manufacturers. However, everyone can participate in the last step. We can all choose to buy recycled items, when available, instead of newly manufactured items. See **Figure 16-20**.

Other Things We Can Do

We can do many other things to help reduce the harmful effects of technology on the environment. For example, instead of using commercial products for household plants and gardens, we can compost. *Composting* is a method of mixing organic matter such as food waste and lawn clippings and allowing them to decay. This results in nutrient-rich soil that can be used to grow all kinds of plants and vegetables. With today's technology, composting is possible even in apartments and condominiums. Composting reduces the amount of garbage we throw away and at the same time produces rich soil for use with both vegetable plants and decorative plants.

Think **Green**

Composting

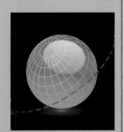

Even if you live in a small apartment, you can set up a working composting system. In fact, you can get started for little or no cost, and in the long run, you can save money or even make money.

You will need a few basic components to get started. First, you need a plastic box or garbage can of a size that is manageable in your home. It does not have to be large. You will also need some soil and grass clippings or dry leaves. If you do not have grass clippings or leaves, you can substitute plain, black-and-white newspaper or cardboard. Finally, you will need a tray to put under the compost box to catch leaks.

To create your composting system, punch holes in the sides and bottom of the compost box. Set the box on the tray. Then add two or three inches of soil to the bottom of the box. Add some grass clippings or dry leaves. Now you are ready to start composting.

The list of items you can compost is long. Examples include pet hair, lint, leaves, grass clippings, paper towels, potato peels, freezer-burned frozen vegetables, coffee grounds, and tea bags. You can search on the Internet for many other items that can be composted.

Be careful not to add things to your compost pile that are harmful to you or the environment. Examples of things *not* to compost include cooking oil, meat products, diseased plants, paint, and motor oil.

Figure 16-20. Items made from recycled materials can be both functional and beautiful. All of the furniture in this photo was created from recycled materials.

People who live in single-family homes can also help the environment and reduce water use by planting trees, shrubs, and grasses that are native to the area. These plants are already adapted for the climate and do not need much, if any, additional watering. By using native plants and composting, people can enjoy a beautiful, low-maintenance yard while helping the environment.

Using less electricity or gas for heating and cooling is another good way to reduce our impact on the environment. In winter, we can put on a sweater and turn the thermostat down a few degrees to use less heat. In summer, we can turn the thermostats up to reduce the air-conditioning. Ceiling fans set on their lowest setting, in combination with minimal air-conditioning, can keep homes comfortable even in hot climates. Also consider opening windows instead of using heating or cooling.

Plastic shopping bags are a serious problem in many communities. People bring home their groceries and then throw the bags away. We throw away millions of bags every day. Because the plastic does not decompose, this generates a huge amount of trash, most of which is not recycled. Some stores accept used plastic bags for reuse. An even better way to fix the problem, however, is to avoid using plastic bags. Instead, we can substitute cloth bags that can be used again and again. Many grocery stores now sell cloth bags at a very low cost. Some stores may even give a small discount when you use their bags. See **Figure 16-21**.

One major way to help reduce air pollution is to use alternative forms of transportation. In some areas, it may be possible to walk or ride a bicycle to school or work. When this is not possible, consider carpooling or using public transportation. Reducing the number of vehicles on the roads is the surest way of reducing the pollutants in vehicle exhaust.

Figure 16-21. We have many options for transporting groceries and other products from a store. Which of the options shown here is best for the environment?

In short, we can try to live a sustainable lifestyle. As you learned in Chapter 2, a sustainable design is a design that has little or no negative impact on the environment. We can extend that concept to include everything we do each day. Living a *sustainable lifestyle* means we do things such as eating, playing, exercising, and going to school or work while causing little or no environmental impact. Scientists and technologists are working to develop more sustainable products and practices. The sustainable design concept has been extended to include new (or reused) homes. Perhaps soon it will be extended successfully to include transportation and other major areas of our lives.

People used technology for hundreds of years without realizing that some of their activities might have a harmful effect on the environment. In earliest times, the effects were not as harmful because technologies such as building a fire to keep warm were used on a much smaller scale.

The introduction of industry in the 1700s and 1800s dramatically increased the harmful effects of technology on air quality. Later inventions of pesticides and other chemicals have had a similar effect on land, soil, and water. We are just now beginning to understand that the careless use of technology can harm not only the environment, but also our health.

Now that we are more aware of the issues, we can begin to use technology and other resources—such as human intelligence—to solve them. We have already begun to make a difference, but much remains to be done.

Summary

- Air quality is affected by pollutants that include toxic gases and airborne particles.
- Land and soil pollution can be caused by waste or other damage from technology, as well as by garbage we throw away.
- Most sources of air and land pollution also affect the quality of our water.
- Hazardous materials and hazardous waste require proper handling, storage, and disposal.
- We can do many things to help reduce harmful effects of technology on the Earth.

Reading Target

Preview and Prediction

Copy the following graphic organizer onto a separate sheet of paper. Do not write in this book. In the left column, record at least six predictions about what you will learn in this chapter. After you have read the chapter, fill in the other two columns of the chart.

What I Predict I Will Learn	What I Actually Learned	How Close Was My Prediction?

Test Your Knowledge

Write your answers to these review questions on a separate sheet of paper.

1. What is smog, and what are its effects on human health?
2. Name the six types of air pollution identified by the Environmental Protection Agency as causing specific damage to the environment.
3. At what step in the manufacture of a product is waste created? Explain.
4. What is clear-cutting, and how does this practice hurt the environment?
5. What is the function of wetlands and marshes in the water cycle?
6. What is hazardous waste?
7. List three examples of household hazardous waste.
8. How long does radioactive waste remain radioactive?
9. What is composting?
10. Name at least four things you can do as an individual to help reduce the effects of technology on the environment.

Critical Thinking

1. Particle pollutants are classified as coarse or fine. The finer the particles are, the more hazardous they are considered to human health. Why do you think this is so?

2. Although lead poisoning has been greatly reduced since the 1980s, some lead poisoning still occurs, even in humans. Why is lead poisoning still a threat?

Apply Your Knowledge

1. Do research to find out more about acid rain. If necessary, refer to the pH chart in Chapter 4 (Figure 4-8). At what level of acidity does lake water become dangerous to wildlife? At what level does it become dangerous to swimmers?

2. Choose one product and list its impacts, both positive and negative, on society and on the environment.

3. Look around your neighborhood. What kinds of trash or garbage do you see? Plan a strategy to convince people in your neighborhood to dispose of their trash properly.

4. Using ATVs and other off-road vehicles in environmentally sensitive areas can destroy habitat for wildlife. Form a group with three or four classmates and brainstorm a plan that would ensure that people who own these vehicles can still enjoy them, while preventing damage to sensitive areas.

5. Make a poster showing how watersheds help remove pollutants from water.

6. Prepare a multimedia presentation to show examples of each category of household hazardous waste listed by the EPA.

7. As a class project, design and implement a school-wide system to compost organic waste produced at your school.

8. It takes about a year for garden waste to change into good compost on a compost heap. Design and make a composting system that speeds up this process. It should be attractive and environmentally sound.

9. Use the Internet to investigate ways in which the following products are recycled: aluminum foil, antifreeze, appliances, asphalt, batteries, bottles, books, boxes, cans, cars, CDs and DVDs, cellular phones, clothes, concrete, furniture, medicine, metals, oil, paint, paper, plastics, and eyeglasses.

10. Do research to find out what plants are native to your area. Design a landscaping plan for your school that includes at least 70% native plants.

Apply Your Knowledge *(Continued)*

11. Research one career related to the information you have studied in this chapter. Create a report that states the following:

- The occupation you selected
- The education requirements to enter this occupation
- The possibilities for promotion to a higher level
- What someone with this career does on a daily basis
- The earning potential for someone with this career

You might find this information on the Internet or in your library. If possible, interview a person who already works in this field to answer the five points. Finally, state why you might or might not be interested in pursuing this occupation when you finish school.

STEM Applications

1. **SCIENCE** Perform an experiment to see what can happen to soil if no plants are growing in it. You will need two equally sized containers of soil: one planted with plants or grass, and one without (bare soil). Design your experiment to see the effect of the following events on the soil in both containers:

   - Windstorm
   - Heavy rain

   You decide how to simulate the weather conditions. Compare the soil in the two pans. What can you conclude? Document your experiment and record your results.

2. **ENGINEERING** Imagine that you are marooned on a desert island in the middle of the Pacific Ocean. The island contains sand, rocks, seashells, and a few coconut palm trees, but nothing else. There is no fresh water on the island. Your first priority must be to create a source of fresh water for drinking. Design and make a device that removes the salt from saltwater, using only materials that are available on the island.

Humans have the ability to think and reason. How can we use our abilities to protect endangered species?

17

Preparing for Your Career

Better by Design

Ariel Shlien created The Mad Science Group

Ariel Shlien is an entrepreneur who, together with his brother Ron, cofounded Mad Science. Their passion for science led them to develop activities that they performed at parties, as well as workshops at local schools and community centers. They provide a unique assortment of hands-on programs, live presentations, theatrical productions, and innovative products. The company is now franchised globally, in all major cities in North America and in 29 countries worldwide. They deliver over 200,000 presentations a year to 4.8 million children and their families.

**The Mad Science Group
makes learning science fun.**

"Being an entrepreneur is all about having a vision, making it happen, and believing in yourself."

Reading Target

Connecting to Prior Knowledge
One good way to prepare yourself to read new material is to think about what you already know about the subject. What do you know about careers in technology? Use the Reading Target at the end of this chapter to record your ideas, even if you are not sure of some of the facts.

Key Terms

aptitude
break-even point
budget
business plan
career
career plan

corporation
customer service
debt
discrimination
distribution system
entrepreneur
entrepreneurship
equity
ethics
free enterprise system
goal
harassment
leadership
liabilities

long-term goals
marketing plan
mentor
overhead expenses
partnership
personal skills
short-term goals
soft skills
sole proprietorship
stock
stockholders
teamwork
work ethics
workplace skills

Objectives

After reading this chapter, you will be able to:
- Identify various career opportunities in technology.
- Develop a career plan.
- List skills needed to be successful in the workplace.
- Identify career opportunities that are available for students.
- Participate in a school-based company.

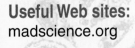

Useful Web sites:
madscience.org

A *career* is an occupation, a way of making a living. Farmers plant, fertilize, cultivate, and harvest crops. Ranchers breed, feed, and care for animals. Horticulturalists grow and maintain plants, shrubs, and trees. Forestry workers cut, transport, and process trees for papermaking, construction, and furniture manufacturing. Geologists locate minerals and fossil fuels. Miners operate the machinery to extract raw materials. Oceanographers study the plant and animal life of the oceans. Environmentalists try to find solutions to problems relating to land use, pollution, conservation of natural resources, and the preservation of wildlife.

Careers in Technology

The variety of careers available in the field of technology is almost endless. As our world becomes more closely tied to technology, even more new careers become available. In this chapter, you will learn about different types of companies. You will learn what is necessary to prepare yourself for a career of your choice. Even if you have not made up your mind about what career you want to pursue, you can take steps now that will help you prepare for any career in the future.

Types of Companies

There are three basic types of business ownership: sole proprietorship, partnership, and corporation. Each type has both advantages and disadvantages.

In a *sole proprietorship*, one person owns the company. See **Figure 17-1**. This person is responsible for all of the company's debts. Sole proprietorships are the easiest type of business to start because one person makes all the decisions.

A *partnership* is a business that consists of two or more people. These people draw up a contract that specifies each person's responsibilities. The contract also specifies how profits will be shared, how disputes will be settled, and how much money each person will contribute to the company. It may also specify how they will close the company and divide the assets, should the need arise.

A *corporation* is a legal entity that is not tied to a specific owner or handful of owners. Instead, it is owned by *stockholders* (people who buy *stock*, or shares, in the company). Large corporations may have thousands of stockholders. The stockholders elect a board of directors to make decisions and manage policies. See **Figure 17-2**. Corporations are different from sole proprietorships and partnerships in that the shareholders are not personally liable for the corporation's debt. If the company fails, their personal assets cannot be used to pay any debt the company has.

Figure 17-1. A sole proprietorship is owned by a single person. In some cases, a small sole proprietorship may require nothing more than a telephone and a computer.

Traditional Career Opportunities

Many technology-related careers are related to manufacturing or construction. The people who work in manufacturing have jobs in factories. Many work on the production lines, but other opportunities are available as well. For example, engineers carry out research to find new technologies. Market research analysts determine the need for new and existing products. Management personnel oversee the production or construction process. They also help ensure that the workplace is safe for workers.

Figure 17-2. A corporation's board of directors is responsible for making major decisions about product lines, investments, and other company matters.

People who work in the construction industry are responsible for planning and building structures. They may work on homes, bridges, dams, hospitals, highways, and shopping centers, among other things. See **Figure 17-3**. Skilled workers specialize in a trade. Examples include carpenters, plumbers, and roofing specialists. Civil engineers and surveyors perform many different tasks. They design and lay out structures, estimate costs, and prepare material specifications. They survey building sites and organize work schedules. Construction managers supervise the building process to ensure that the structure follows approved plans.

A growing number of technology-related careers are service-oriented. People in these careers may work for companies such as computer repair shops, information technology companies. See **Figure 17-4**. They may also work for nontechnical companies such as dry cleaners or a public library. These businesses, too, depend on technology.

Some service-related jobs are offered through the federal, state, or local government. These include public health services and public works facilities. Public works departments take care of many things we take for granted. For example, they keep the streets clean, clear snow from the streets in winter. They also maintain vehicles such as police cars, street sweepers, garbage trucks, and ambulances if these services are offered by the local government.

Entrepreneurship

An alternative to traditional careers in which you work for an established company is to become an entrepreneur. An *entrepreneur* is a person who owns his or her own company. *Entrepreneurship*, or becoming an entrepreneur, takes a lot of planning, determination, and skill. To be successful, an entrepreneur must be willing to work long hours. Success also depends on the person's knowledge of every aspect of the business.

Figure 17-3. Traditional construction workers work for companies that build structures of all kinds. These concrete specialists are pouring the concrete for a new bridge.

Figure 17-4. Computer repair is a technology-based business opportunity that has become necessary only in the last 30 years.

For example, suppose you started a business to sell mugs printed with the mascots of local high schools. You would have to know where to buy the mugs and how to have the mascots or logos printed on them. You would also have to know how many mugs to purchase, how many to print with each mascot or logo, and how much to charge for them to make a profit. You would have to know something about marketing and advertising. You would also have to be able to work with money so you could keep your accounting straight.

Entrepreneurship has many advantages over working for someone else. For example, you can set your own hours. You can choose your own equipment and make your own policies. You do not have to answer to a supervisor.

On the other hand, entrepreneurship also has disadvantages. Even though you can set your own hours, you will find that you need to work many hours a week. In addition, only a percentage of those hours will be spent actually doing the work. The remaining hours will be spent planning, purchasing, paying bills, and doing other accounting tasks. Also, even though you do not have to answer to a supervisor, you *do* have to be responsive to your customers. "People skills," such as getting along with others, are very important. You will be responsible for your own customer service. What if someone buys your product but then does not like it? What if the product breaks? You will need to develop customer service policies to cover these things.

Most entrepreneurs are sole proprietors because sole proprietorships are the easiest type of business to form. This adds an additional level of risk, however. If the business is not successful, the entrepreneur is personally responsible for any debt the business acquires.

Nevertheless, many entrepreneurs are both successful and happy with their career choice. By planning carefully and investing wisely in their company, they overcome any disadvantages and enjoy their independence. See **Figure 17-5**.

Figure 17-5. Even though this entrepreneur spends extra hours each evening working on accounts, he enjoys his work and is glad to be working for himself.

Developing a Career Plan

Whichever type of career you are considering, you can achieve it more easily if you have a plan. A *career plan* is a blueprint that will help you make the right choices to be successful in the career of your choice.

Identifying Interests and Aptitudes

When you were a child, you may have told people, "I want to be a movie star!" or "I want to be a police officer!" These ideas were based on your interests. Interest is an excellent starting point for choosing a career. Most careers, however, require much more than just interest. You must have an *aptitude*, or natural talent or ability, for the work.

Before you can create a career plan, you need to take a good look at your interests and aptitudes. Are you interested in mechanical objects and how they work? Do you like to build things? Do you enjoy solving puzzles? To begin work on your career plan, make lists of the things that interest you and the things you do well.

Identifying Strengths and Weaknesses

In addition to interests and aptitudes, everyone has specific strengths and weaknesses. For example, some people have a knack for working with other people. Someone with this strength might be good at customer service jobs that require calming angry customers. Other people seem to be able to fix just about anything. Someone with this strength would be good at service or repair jobs.

It is important to know your weaknesses as well as your strengths. If you lose your temper easily, for example, customer service is not the career for you. Understanding your strengths and weaknesses can help you choose a career that you enjoy and can do well.

Creating Your Career Plan

You may think it is too soon to begin working on your career plan. What if you change your mind about the career you want? You do not have to choose a specific career now. Keep in mind, though, that many exciting careers require advance preparation. You can do certain things now to make these careers possible later.

A career plan is not a static document. It can grow and change with you. By identifying a general *type* of career that interests you, you can help ensure that when the time comes to choose, you will have everything you need. You will be prepared to pursue any career that interests you.

Your career plan should be suited to your individual needs. It may not look at all like your best friend's career plan. However, all career plans begin with career goals. A *goal* is a statement of what you want to accomplish. Some goals are *short-term goals*—goals you can achieve in a year or two. Others are *long-term goals*. These goals may take many years to accomplish.

An example of a long-term goal is: "Work in the health care industry." This is a very broad goal that includes many different types of careers. Your long-term goal can be flexible. It will change along the way as you develop new interests and skills. You can make it more specific, or even change your mind completely and decide to pursue another goal instead.

After you specify your long-term career goal, think about steps you can take to accomplish it. One of the very first steps is usually to find out more about the career or type of career. What education will you need to work in this area? What classes can you take now or in high school that will make it easier to follow this career? List the steps you will take to begin preparing yourself for this career. **Figure 17-6** is an example of a career plan created by a student who is interested in agricultural technology.

If you do not have any idea of what career you might want, consider making not one, but several career plans. Create one career plan for each type of career you think you might be interested in. Find out more information about all of them. Then you can compare them to see which ones you want to pursue.

Career Plan

Goal: Career in agricultural technology.

  Short-Term Goal: Find out everything I can about agriculture and the people who work in agricultural technology.

  Long-Term Goal: Obtain a fun, satisfying job in agricultural technology that pays well.

Step 1: Research to find all the different types of jobs available in agricultural technology.

Step 2: Find out more about each type of job identified in Step 1, including education and experience needed.

Step 3: Talk to people who currently work in agricultural technology to get first-hand information about their jobs.

Step 4: Identify and take courses that will help me reach my goals.

Step 5: Review my goals to see how much progress I have made.

Step 6: Determine whether my goals are still valid, and revise them if necessary.

Figure 17-6. An example of a career plan for a student who is interested in agricultural technology. Notice that the student has not identified any specific career. She is still exploring options, and her career plan reflects that.

Workplace Skills

To be successful at any job, you will need more than just job-related skills. You will need *workplace skills*, which are sometimes called *soft skills*. These include communication skills and the ability to get along with other people. They also include work ethics, dependability, and punctuality (being on time for work).

Teamwork and Leadership

Teamwork, or working well with others, is one of the most important soft skills. See **Figure 17-7.** Most jobs require you to cooperate with other people on a daily basis. All team members are responsible for the final output or results of the project. However, each member in the team has individual responsibilities. Each member must fulfill these responsibilities if the project is to succeed. Therefore, part of being a good team member is to do the part assigned to you. Taking your share of the responsibility and doing your part will make you a respected member of any team.

All teams need a good leader. *Leadership* is the ability to guide team members and keep the team on track. It involves problem solving, creative thinking, and excellent communication skills. A good team leader respects each team member and inspires team members to contribute. To be a good leader, however, you must first learn to be a good team member.

Work Ethics and Legal Aspects

When you accept a job, you take on certain responsibilities. You are responsible for arriving on time and for doing your work properly. These and similar responsibilities are known as *work ethics*. Your employer relies on you to be dependable and to work within the company rules. In many cases, other people depend on you to do your job so that they can do theirs.

Figure 17-7. Teamwork helps any type of project run smoothly, whether it is related to school, work, or just having fun. These students are working together to finish a school project.

Your employer also expects you to work within the law and to treat everyone equally. *Discrimination* is the unfair treatment of someone based on a personal characteristic such as age or gender. Discrimination is against the law. Employers are not allowed to discriminate when they hire, and employees cannot discriminate on the job.

Harassment is another job-related issue. *Harassment* means tormenting someone by your words or acts. It includes making rude comments about someone or making fun of someone. It can even include teasing if you keep teasing someone even after that person has asked you to stop. In most companies, you can be fired for harassing coworkers.

Personal Skills

Your *personal skills* include communication skills and how well you get along with other people. They also include dressing appropriately for the job and behaving in a businesslike manner. See **Figure 17-8**. Your attitude toward your job, your employer, and your coworkers is another personal characteristic that is often grouped with personal skills. Employers look for workers who have a good attitude. Keeping a good attitude even when things go wrong is a good way to advance in your career.

Figure 17-8. Dressing appropriately for your job and maintaining a neat, clean appearance are important personal skills.

Student Career Opportunities

As a student, you have opportunities to explore many different kinds of careers. Various student organizations help students investigate career opportunities. Joining one or more of these organizations can help you find out more about careers you find interesting.

Technology Student Association

If you have thought of a future as an engineer, scientist, or technologist, you should consider joining the Technology Student Association (TSA). Over two million students have already participated in its activities. TSA provides projects and competitions suited to everyone, including activities related to every area of technology. Examples include agriculture and biotechnology, construction, digital photography, vehicle design, electronic gaming, video production, Web site design, and many others.

The TSA Web site (www.tsaweb.org) provides information about many topics of interest to technology students. State and local chapters allow you to become more fully involved with the organization. You can meet with other technology students, exchange ideas, and work on the TSA's current National Service Project.

Other Student Associations

Depending on your interests, you might want to check other resources. For example, Junior Achievement is a worldwide organization for elementary, middle, and high school students. This organization concentrates on helping students manage the financial aspects of a career, including entrepreneurship. You can go to the organization's Web site (www.ja.org) to find more information about how this organization can help you.

Your school or community might also offer associations that you can join to help you get started in a technology career. Check with your teacher to find out more about these and other opportunities.

School-Based Company

Are you curious about how it might feel to be a part of a company? This chapter provides guidelines for participating in a school-based company. See **Figure 17-9**. You can experience starting and running a company in a risk-free setting. The business you will develop will be one that the entire class can work on under the guidance of your teacher. Your class will select a business that you can set up and run within the school year.

Figure 17-9. Working with your classmates in a school-based business helps you learn about working in a company. It also builds important teamwork and communication skills.

Although the size of your business will not match that of the typical corporation, you will operate in a similar manner. You will become part of our free enterprise system. In a *free enterprise system*, people make most of the decisions about how to use the country's economic resources. The government does not specify who can do what job and for what amount of money. People are free to start a business or work for someone else. Their business can produce anything from athletic shoes to construction cranes. The four parts of the free enterprise system are:

- People who decide how to use resources and purchase goods
- Companies that produce goods and provide services to make a profit
- Markets that buy and sell goods, services, and finances
- Governments that make and enforce rules to protect rights and property

Deciding on a Product

The first place to look for ideas for a school-based business is the school itself. What are the needs in the school? What complaints do students or teachers have? What could be improved? Figure 17-10 contains a few ideas. Use these as a starting point for a class brainstorming session.

Remember that you will not have a lot of time to form your company and make a product. The product you decide to make should be something simple. The equipment to make it should be readily available and easy to use. For example, suppose you decided to recycle paper. You could collect paper that has only been used on one side and then cut it into various sizes. You could then glue or bind the paper together to make notebooks and notepads. The equipment needed for this business would be a mechanical paper cutter, glue, cardboard, and a binding machine.

Product Theme	Possible Products
Sports	Banners, flags, or a large hand that could be carried and displayed by fans
Special Events	Posters, mugs, silk-screened T-shirts
Repairs	School desks, outside benches, tables
Recycling	Wastepaper, leftover food

Figure 17-10. Examples of ideas for products for a school-based business.

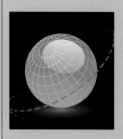

Think Green

Green Product Analysis

Many opportunities exist for providing environmentally friendly products and services. If you select such a product for your class company, you can help your school and community at the same time you practice running a business. After you brainstorm product ideas, test each idea for its "green quotient" by asking the following questions:

- Can the product be made from renewable resources?
- Can the product be made without the use of chemicals or processes that will harm the environment?
- Will the finished product be biodegradable? If not, what plans could be made for reuse or recycling?

Creating Your Business Plan

To start your business, you need a plan. Writing your ideas on paper is the first step in a plan to make your company come alive. A *business plan* serves several purposes. First, it serves as a road map that points the way to success. You may have to change your route if you encounter a roadblock, but the business plan always points to your destination.

A business plan also becomes a standard against which you can compare your success. It provides targets to aim for and completion dates. It lists important details and realistic ways to achieve your business goals.

Initial Plan and Feedback

After you have decided on the product your company will make, create a short version of a business plan. The purpose of this version is to explain your idea to a mentor. A *mentor* is someone who has knowledge about the product the class is proposing and agrees to provide advice. See **Figure 17-11.** This person can be your teacher or someone else in the community. The class should make a short presentation to the mentor using the short business plan. Include the following points:

- State the name of the proposed company and give a brief description of the product to be made

Figure 17-11. Mentors are professionals who have experience with the product or business you want to start. A mentor can help you start out right and stay on the right track as you develop your business.

- Explain why this type of business interests the class
- Explain the type of product and how it will be made
- Describe the advantages of the product over the competition
- Describe the qualifications class members have to operate the business
- State the proposed selling price of the item and the expenses to make it

Schedule 30 minutes after your presentation to listen to the mentor's suggestions. Later, discuss the mentor's ideas and modify the proposal before you begin work on the full business plan.

Avoiding Pitfalls

A detailed, well-thought-out business plan is vital to avoid some of the common reasons businesses fail. In fact, the process of creating a detailed business plan forces you to pay attention to all aspects of the company, including possible pitfalls.

One primary reason businesses fail is lack of research. It is important to learn all about the proposed product or service. Now is the time to think about selling points. What will set your product apart from the competition?

Financial problems are a second reason for failure. They can arise if you do not have enough capital to start. They can also be caused by not having enough cash flow to carry the company until the product begins to make a profit for the company. Poor management, failure to control costs, and failure to collect amounts due are other reasons for financial problems.

If the product does not meet customer needs, it will not sell well. Research into customer wants and needs is essential to avoid this pitfall. Finally, not having enough staff or having staff without the necessary skills can result in business failure.

Detailed Business Plan

Your detailed business plan can help you prevent all of the pitfalls mentioned in the previous section. **Figure 17-12** shows a recipe for a detailed business plan. Use this as a template to create a business plan that is specific to the product your company will make. Many sample business plans are also available on the Internet. You may want to look at some of them for further ideas.

TITLE PAGE

1. TABLE OF CONTENTS

2. EXECUTIVE SUMMARY (one page maximum)
This is the most important part of the business plan. It should include routine information including your company address and a contact phone number. You should also describe:
- What you are planning to do and why
- Who your customers will be
- What facilities you will use
- Your financial needs
- Your estimate of sales and profits

3. DESCRIPTION OF THE COMPANY
This section gives a concise description of the purpose of your company. Include the type of product to be made and its distinctive advantages. Include:
- The competitive advantage of your product
- Partners or contacts
- Clients and suppliers

4. DESCRIPTION OF PERSONNEL AND THEIR SKILLS
Describe the personnel who will work in the company:
- Functions and special skills of each person/group
- Mentors
- Any other support and organizational systems

5. PRODUCT OR SERVICE TO BE PROVIDED
Describe the product in detail:
- Advantages for the user
- Source of materials and supplies
- Amount of sales expected
- Production, delivery, and servicing

6. FEASIBILITY STUDY
Describe the research that has been carried out by the class:
- Market niche identified
- Market projections and trends
- Number of units that could be sold in the local area

7. MARKET ANALYSIS
Describe the clients you are targeting and how you intend to promote sales:
- Description of targeted customers
- Market size

8. COMPETITION
Describe your nearest competitors:
- Strengths and weaknesses of competitors
- Advantages your company can offer

9. FINANCIAL PROJECTIONS
State the actual amounts of money needed:
- To start the company (startup costs)
- To keep it running month by month (operating costs)
- The projected estimate of revenues versus costs
- The expected net income

10. MARKETING STRATEGY
Describe the proposed sales techniques, including methods of advertising, promotion, campaigning, or presentation

11. FACILITY AND EQUIPMENT
Describe the space needed to make the product:
- Why the proposed area meet the needs
- Any necessary modifications
- Tools and equipment to be purchased

Figure 17-12. A template for a detailed business plan.

Obtaining Capital

The amount of startup money your business will need depends on the type of business. Plan for enough operating funds to keep the company going for the school year. You may need to borrow at least a small amount of money. If so, how much is required? When will the money be needed? What is the best source for financing your business?

From a corporate point of view, there are only two ways to raise money. *Debt* occurs when you take a loan from a bank or other financial institution. These institutions do not usually invest directly in a business. They make money from the interest they charge on the money they lend. Your business must repay the money, plus interest, regardless of how well the business is doing. You may hear people refer to debts as *liabilities*.

Equity, on the other hand, does not have to be repaid. Instead, the person who provides the equity invests in the company. He or she owns a part of the business. If the business fails, the investors lose their money. If the business is successful, investors share in the profit. The disadvantage of using equity is that the investors own part of the company and have control in the business. See **Figure 17-13**.

Entrepreneurs often finance their startup costs using personal credit cards or savings. Other alternatives include negotiating supplier credit, obtaining personal loans from banks, or getting loans from family members.

Specifying a Budget

A *budget* is an estimate of costs and expenses that a company does not want to exceed. Planning a budget is an important tool for making a company successful. To create the budget for your company, first determine all of the things you will need. Include the cost of equipment or tools needed, as well as supplies and any *overhead expenses* (utilities, rent, and salaries). The cost of equipment and tools is generally a startup cost. It only happens once. Overhead expenses are ongoing expenses that the company will need to pay on a regular basis.

Figure 17-13. Business owners keep careful track of liabilities and equity to help make sure the debts can be repaid and the investors will make a profit.

Next, list your startup money and expected profits. Compare the expenses and money on hand. Plan to spend only the money that is available at a specific time. The purpose of budgeting is to ensure that you do not spend more money than you make, causing your business to fail. See **Figure 17-14.**

Financial Factors

For a class-based business, you may be able to use school facilities. This will significantly lower your costs. The school may even allow you to use the space and any necessary equipment free of charge. In your business, as in any business, low overhead costs and low spending are better than running up large debts.

To create a realistic budget, you will need to determine the price for which your company will sell your product. Factors that help determine the price include:

- A projection of the number of items that will be sold
- The cost of making the product
- The cost of servicing the product (if you decide to provide service)
- Overhead expenses
- The expected break-even point

The *break-even point* is the point at which the income from items sold equals total expenses. After your company reaches the break-even point, the company begins to make a profit.

Strategies to Minimize Costs

To stay within budget, most companies plan strategies to minimize their costs. For your school business, an example of minimizing costs is to get permission to use school facilities instead of renting space. You can also minimize costs by buying supplies in bulk. However, be careful not to buy more than you need. Plan your manufacturing methods to reduce waste as much as possible. Waste is expensive!

Figure 17-14. Budgeting and then keeping track of your company's expenses helps keep the company profitable.

Math Application

Cost Estimation

Accurate estimation of costs is essential to creating a realistic budget. If your actual costs are greater than you estimated, your company will go over budget.

The list of building materials shown below includes costs for raw materials, labor, and overhead. An example of a cost estimate is shown below.

Estimated Costs

Raw Materials

Glass blocks (24 blocks @ $13.43)	$322.32	
Tile glue	17.67	
Grout	12.96	
Framing lumber	56.11	
Caulk	5.79	
Total materials		$414.85

Labor

12 hours @ $32/hour		384.00

Overhead

Utilities	211.00	
Equipment rental	160.45	
Total overhead		~~371.45~~

Total Cost	**$1170.30**

Math Activity

Estimate the costs for your class business. First, copy the chart below to a separate sheet of paper. Then fill in the chart with your estimates. Be sure to itemize the costs in each category.

Item	Cost
Raw Materials (Itemize)	
Labor (Estimate the number of hours and multiply by the hourly rate you will pay)	
Overhead Costs (Itemize)	

Hiring and Training

The next task is to hire and train the staff. In your school-based company, the staff will consist of members of your class. First, identify the specific tasks that need to be done to run the company and make and sell the product. For each task your class identifies, you will need a task group, or department. The function of each department will vary depending on the product. Figure 17-15 explains the tasks performed by various departments in a typical company.

Team Member Roles

To decide who will play which role in your school-based company, have each class member complete a form. Class members should state which group they would like to work in and why. They should also list their skills and qualifications for the position. As a class, and based on these forms, decide who will belong to each group.

Department	Typical Roles
Purchasing	Determining what needs to be purchased Completing paperwork Ordering items Receiving orders from vendors
Inventory	Maintaining a current count of supplies and products Notifying Purchasing when more supplies are needed
Management	Setting targets and specific goals with deadlines Motivating employees to do their jobs well Solving problems Keeping the business on target and within budget
Financial	Completing daily and monthly cash records Keeping a log of all sales and expenses
Personnel	Ensuring that each member of the team fulfills their work requirements Evaluating work schedules
Information Technology	Maintaining computers and computerized equipment in good working order Installing and maintaining any software used by the business Maintaining the company Web site
Customer Service	Work with customers before, during, and after sales to ensure customer satisfaction

Figure 17-15. Typical roles of departments within a company.

Mentors

One of the first steps in building a successful team is realizing that there are a lot of things you and your classmates do not know. You will probably need the help of experts. You have already identified a mentor to help you with your business plan. As you identify other information you need, you can search for mentors who can supply it. These people may be students, teachers, or other people in the community.

Mentors play a role in many different companies, not just those run by students. See **Figure 17-16**. For school-based businesses, however, experienced business and social leaders often volunteer their help. Smart business owners listen to the advice offered by mentors and other interested people. These people can give you the benefit of long years of experience for free. They can help your team avoid some of the pitfalls of starting a new business.

Mentors are particularly useful for suggesting ways to deal with financial and business management. They can also help you establish public relations. In some cases, they can help you extend product sales beyond the local area. Mentors provide business information, help your team focus, make recommendations, and act as a sounding board for your ideas.

Production and Quality Control

In Chapters 1 through 14, you learned how to design and make products. In Chapter 15, you learned about how products are manufactured and the difference between traditional and lean manufacturing techniques. For best results, you may want to use lean manufacturing methods in your business.

The inspection of products, or quality control, occurs in both types of manufacturing. In traditional manufacturing, products are inspected both during and at the end of the manufacturing process. This system is costly. When defective parts are found, they are usually thrown out.

Figure 17-16. Mentors can help all kinds of companies, especially small businesses, by giving financial advice and business tips.

With lean manufacturing, workers have a responsibility for the quality of the product throughout the manufacturing process. See **Figure 17-17**. When they find a mistake, the assembly is stopped and the situation that led to the problem is fixed. As a result, no products have to be thrown away, which reduces costs. This is why lean manufacturing is the preferred method for most businesses.

Preparing a Marketing Plan

An effective *marketing plan* sets your product apart from your competitors' products. It describes how and to what kind of customers you will market your product. It describes the target market.

You can develop a good marketing plan only after you know the facts about potential customers. A marketing plan identifies the customers, attracts them to see the product, and persuades them to buy it. Who will buy your product? What are the characteristics of people who will buy your product? How can you best appeal to this market?

To start your marketing plan, first describe how your product is different from competing products. Next, think about your customers. You want to know as much about them as possible: their age group, income, education, and many other factors. These factors determine how you will present, or market, your product to the customers.

It is important to know your competition. Your market is influenced by your competitors and the strengths of their products. For example, if one of your competitors is offering a very low price, you may have trouble competing on price. It may be better to focus on their weaknesses and make sure your product excels in those areas. In this example, the competitor may be able to offer low prices because of a high output. If this is so, your company may be able to offer better product customization, with lower output. Customization would become part of your marketing plan.

Figure 17-17. Building quality control and quality assurance into the overall development and production of a product results in less waste. This helps control costs and makes the company more profitable.

Looking for a specific selling point or product need is called *targeting*. If you know your target market, you will know exactly where to focus your attention and resources. You can even design features especially for your target market. Targeting can reduce the competition because it allows you to meet specific needs more effectively than your competitors.

Advertising Your Product

Your advertising campaign is closely tied to your marketing plan. You will want to create advertisements that appeal to people in the target market. See **Figure 17-18**. What type of advertisements or promotions do they use? Do they use flyers, ads in local papers, or discount coupons? Do they buy only from local stores, or do they order from catalogs or the Internet?

To avoid misreading, keep your advertisements simple. Place the key message you want to convey in the headline. Emphasize unique benefits of your product, but do not make false promises. Use color for a greater visual effect and to increase the readability of the text.

Many different advertising opportunities are available to your business. You can use social media to let friends know about your product. You can distribute brochures at school or post a large ad in a strategic place. Contact a local paper and get a reporter interested in your product. You can even find a prominent person in the community who likes and uses your product and ask that person to promote it.

Figure 17-18. You can find many templates for advertising brochures, flyers, and business cards on the Internet. The template shown here might be appropriate if you are targeting customers who care about the environment.

Selling and Distributing Your Product

How and where will you sell your products? Will you need to rent retail (store) space, or will you set up a booth at school? Will your class receive requests by mail and ship directly from your classroom? Will your class sell from a Web site? These decisions depend partly on your methods of advertising and the customers you are targeting.

After you decide on your sales methods, you can determine how to distribute your products. If you rent retail space or set up a booth, you need only consider how to get the products to the store or booth. However, if you set up a Web site to sell your product, or if you accept orders by mail, you will need to set up a distribution system.

A *distribution system* is the method you use to get the product to the customers. For example, you could use the U.S. Postal Service to mail the product to your customers. See **Figure 17-19**. You could make arrangements with another delivery service, such as UPS or Federal Express. Whichever method you use, consider the charges involved. How much are your customers willing to pay for delivery?

Providing Customer Service

Customer service is the process of dealing with customers at all stages of customer contact. It includes helping customers choose the right product for their needs. It also includes making the actual transaction, or sale, easy for customers by being pleasant and helpful. Finally, customer service includes handling any questions or complaints customers have after they have purchased the product. This is also known as *customer support*.

Figure 17-19. Your method of distribution affects the packaging for your product. If you use a delivery service, the packaging must protect the product in transit.

The extent of customer service you provide after the sale depends on the type of product you are providing. At a minimum, your company should provide a way for customers to contact you. Supplying an e-mail address or phone number in your advertising material is one good way to do this.

Customer Service Representatives

Customer comments will vary greatly. Some customers may be delighted with the product. Others may believe it does not perform as advertised. The students who respond to the comments are your company's customer service representatives. They must have good communication skills and good people skills. They must also be familiar with all aspects of the product so they can explain it to people who may be having trouble using it. See **Figure 17-20**.

When a customer contacts your company, listen to the customer's point of view. Reply directly to the points that are mentioned. Pleasing customers is critical to good service. Your customer service representative's response can make the difference between an unhappy customer and a happy customer. Unhappy customers may talk to friends about their bad experience. Happy customers often bring in new clients.

Ethics and Legal Responsibilities

Businesses can be held accountable if they make decisions that are harmful to customers or the general public. In its simplest form, *ethics* is a matter of knowing right from wrong. Creating a reputation for being ethical can help your business become successful and stay successful. A company that makes a reputation for being unethical loses customers quickly.

Figure 17-20. Customers who cannot get a product to work frequently become annoyed well before they call customer service. Your company's representatives need the skills and patience to deal calmly with frustrated customers.

Ethical businesses avoid shipping faulty products. For example, if a product is discovered to be dangerous in some way, the company immediately stops selling it and offers refunds to people who have already purchased it. Although these practices may cost the business money, they can also help keep the company from being sued. They also help your company maintain the confidence of current customers. These customers are then more likely to return to your company and buy future products.

Closing and Evaluating Your Company

Most school-based companies are meant to last only through the school year. Near the end of the year, your class will close the company and liquidate its assets. Liquidation involves selling the company's assets, including leftover supplies, finished products, and any tools and equipment the company has purchased. It also includes paying all outstanding debts.

If money remains after all the debts have been paid, you will need to decide what to do with it. In a real corporation, the money would be divided among the stockholders. In your school-based company, your class may want to consider donating the profits to benefit people in your community. You may choose to donate to an established charity or to address immediate local needs, such as personal hardship caused by a plant shutdown or by a natural disaster. Helping out financially not only benefits the recipients, but also provides publicity for your school, your mentors, and any future school-based companies your school may form.

After liquidating the company, your class can review and discuss the experience of forming a company. What skills have you learned that may help you in later life? If you were to start over now, would you do anything differently? If so, what would you do differently, and why?

As new technologies are developed, we tend to incorporate them into our daily lives. We depend on them. Can you imagine life without a telephone? Prior to Alexander Graham Bell's invention in 1876, people did not have telephones. The same is true of many other technologies we take for granted today. Think about all of the new career opportunities that were created just by the invention of this one device.

New technology almost always results in new career opportunities. In the future, you may have a chance to work in a field that is entirely unknown today. As you consider all of the possibilities now and in the future, do not forget entrepreneurial business opportunities. These can be especially exciting when new technologies are introduced. If a related career does not exist, you can create one!

- Many different types of career opportunities are available in technology, and new opportunities are constantly developing.
- A career plan is a tool that can help keep you focused on your career goals.
- To be successful in a career, you need soft skills as well as job-related skills.
- Students can participate in school-based companies or even form their own entrepreneurships.
- A school-based company provides an opportunity for students to gain experience in a risk-free environment.

Connecting to Prior Knowledge

Copy the following graphic organizer onto a separate sheet of paper. Allow space in each row for one or more sentences. Before you read the chapter, write sentences in the first column to record your current knowledge about careers in general and technology careers in particular. As you read through the chapter, record new concepts you learned by reading the chapter. When you are finished, review the chart to see how much you have learned. If you have questions about topics in the chapter, record your questions at the bottom of the chart. Ask your questions in class or do research on your own to find the answers.

What I Know (Or Think I Know) about Careers	What I Learned in This Chapter
Careers in Technology	
Developing a Career Plan	
Workplace Skills	
Student Career Opportunities	
School-Based Company	
Further Questions:	

Test Your Knowledge

Write your answers to these review questions on a separate sheet of paper.

1. What is the major difference between a corporation and a partnership or sole proprietorship?
2. What is an entrepreneur?
3. What personal traits should you assess before you develop a career plan?
4. Name at least six workplace skills you may need in a typical career.
5. Name at least two student organizations that can help technology students prepare for a career.
6. What is the purpose of a business plan?
7. Why do most school-based companies rely on mentors?
8. What is the difference between debt and equity?
9. What is a break-even point, and why is it important?
10. What skills do successful customer service representatives have?

Critical Thinking

1. Suppose a company has three partners who share equally in the company's debts and profits. After several years, profits are declining, and two of the partners want to dissolve the company. The third partner is convinced the company can be profitable again and does not want to dissolve the company. How could the partners resolve this disagreement fairly?

2. Suppose you own a company that makes toys for small children. Before selling each type of toy, you have it tested professionally to be sure it is safe and that it will not break under normal usage. A customer contacts your customer service department claiming that the toy she bought broke the first time her toddler used it. What should you do in this situation?

Apply Your Knowledge

1. Go to the TSA Web site (tsaweb.org) and find out what is involved in competing in a TSA event. As a class or in small groups, choose a competition in which you are interested. Read the detailed specifications and rules for your chosen event. As a class, hold a similar competition. How might this help prepare you for national competition?

2. Investigate other national competitive opportunities for technology students. Examples include First Vex Challenge, First Lego League, and BEST Robotics.

3. You know that more and more technologies are being developed. Many of these are the result of refining existing products to address special or related needs. Write an essay explaining why it can be said that "specialization of function is at the heart of many technological improvements."

4. Technological activities and advances can be divided into four categories: applying technology, designing technology, producing technology, and assessing technology. Create a chart that shows technological advances in the last 50 years and the category to which each advance belongs.

5. Quality control is a very important impart of any technological process. Prepare a multimedia presentation that shows how quality control can be accomplished in each of the major areas of technology.

6. Think about a business you would like to start. Prepare a short business plan that you could present to a mentor to gain more knowledge about how to start and run the business.

STEM Applications

1. **MATH** Suppose you are operating a store to sell widgets. Your total overhead costs are $10,000 per month. You sell an average of 1,000 widgets every month. Each widget costs $50 to make. If you marked up the price by 20% and sold 1,000 widgets, you would make $10,000, because 20% × $50,000 = $10,000. Since your operating costs are also $10,000, you would not make any profit. Calculate how much you would have to charge to pay yourself a salary of $20,000 per year.

2. **TECHNOLOGY** Throughout this chapter, the point has been made that new technologies provide new career opportunities. However, technology plays a larger role. Investigate how technology has affected the process of searching for, obtaining, and keeping a job. How is job-hunting today different from job-hunting 50 years ago?

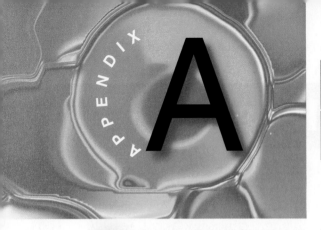

Appendix A

History of Technology

The six simple machines were developed during this period

Period	Food	Shelter	Clothing	Defense	Transportation	Communication	Health
500,000–10,001 BC	-hoe -harpoon -nets for fishing	-cave -painting -hides over frame -oil lamp	-animal skins -needle	-fire -spear -bow and arrow	-dugout	-hieroglyphics	
10,000–1 BC	-Archimedes' water screw -waterwheel -quern to grind corn -trained animals to pull plows -pottery -spoon -fishhook -sickle	-sun-dried mud hut -lock and key -rope and pulley -gear -brick -arch -nail -glass -bath	-vertical loom -cosmetics	-monumental stone buildings -bronze and iron weapons -knives -swords -sling	-sled -wheel -sail -boat -skis -harness (oxen)	-ink and paper -cuneiform writing -Pheonician alphabet -astrolabe -Estruscan alphabet -first coinage -papyrus -maps	-false teeth
AD 1 – 1399	-aqueduct -windmill -horse collar -horseshoe -porcelain drinking vessels	-stained glass -Roman central heating -clock -dome -chimney	-spinning wheel -trousers -felt hat -button -lace	-crossbow -bronze-cast cannon -gunpowder -gun	-magnetic compass -Roman roads -Viking longboat -horse stirrup and saddle -rudder -skates	-movable type -paper from bamboo (Chinese) -stencil -pen -printing	-hospital -spectacles
AD 1400 – 1699	-ice cream -pressure cooker -bottle cork	-wallpaper -watch -theodolite -water closet (toilet) -thermostat -barometer -surveying instrument	-knitting machine -umbrella	-artillery shell -naval mine -hand grenade -rifle -submarine	-diving bell -dredger -telescope -wheelchair	-Gutenberg's press -arithmetic signs (+ – = ×÷) -newspaper -envelope -calculating machine	-toothbrush -artificial limbs -microscope -thermometer -inoculation
AD 1700 – 1849	-canned food -threshing machine -carbonated water -steam tractor -reaper -seed drill -sandwich -fertilizer	-electricity -street lighting (gas) -iron frame building -fire extinguishers -cement -matches -central heating	-spinning jenny -power loom -cotton gin -sewing machine -waterproof coat -dry cleaning	-Winchester rifle -standard parts for guns -machine gun -shrapnel	-pneumatic tire -bicycle -lifeboat -locomotive -hot-air balloon -sextant -roller skates	-metric system -photography -lithography -typewriter -Morse's telegraph -Braille -steel pen -eraser -postage stamp	-bifocals -anesthetics -sedatives -porcelain false teeth -ambulance -vaccination -plastic surgery -stethoscope -blood transfusion

Period	Food	Shelter	Clothing	Defense	Transportation	Communication	Health
1850 – 1899	-barbed wire -refrigeration -condensed milk -milking machine -margarine -cola -breakfast cereal -pasteurization	-high-rise building -plastics -linoleum -electric lighting	-jeans -man-made fibers -zipper -aniline dyes	-dynamite -submarine -automatic machine gun -torpedo	-hang glider -airship -clipper ship -diesel engine -automobile -helicopter -modern bicycle -motorcycle -gasoline engine	-telephone -typewriter -cinematography -wireless telegraph -radio -postcard -fountain pen	-hypodermic syringe -pasteurization -dental drill -antiseptics -incubator -X ray -aspirin
1900 – 1945	-tea bag -frozen food -insecticide (DDT) -combine harvester -supermarket	-prestressed concrete -air conditioning -fluorescent lighting -vacuum cleaner	-electric washing machine -nylon -artificial silk	-poison gas -tank -radar -gas mask -aerial bomb -V2 rocket	-aircraft -tracked vehicles -safety glass -seaplane -traffic lights -helicopter -jet aircraft -subway system	-motion pictures -television -xerox -ballpoint pen	-electro-cardiograph -hearing aid -blood transfusion -chemotherapy -insulin -iron lung -kidney machine
1946 – 1999	-synthetic fertilizers -microwave oven -nonstick pan -domestic deep freezer -foods for use in space -new "miracle" strains of rice and wheat -ultrasonics to detect fish -cloning	-solar panel -geodesic dome -synthetic turf -fiberglass insulation -lightweight modular dwellings -space station	-synthetic fibers -automatic clothes dryer -permanent creases in clothes -metallized fabric	-atomic bomb -ejection seat in aircraft -portable atomic weapon -hydrogen bomb -ICBM	-hovercraft -lunar vehicle -nuclear submarine -monorail train -space shuttle -ultra high-speed train -VTOL aircraft -space station -fuel-injected engines -magnetic levitation trains -space orbiters	-radar -transistor -satellite -laser beam -fiber optics -computer network -silicon chip -videotape -microprocessor -integrated circuits -digitzed typesetting -desktop publishing -compact disc player -laptop computer -space telescope	-artificial voice box -heart-lung machine -artificial heart and other organs -equipment for organ transplants -high-speed dental drill
2000 – present	-soy beans used for oil, flour, cereals, snack foods -genetically modified foods	-self-cleaning windows -robotic vacuum -memory foam mattress	-nanotechnology -wearable fabrics	-unmanned aerial vehicle	-fuel cell bike -hybrid automobiles	-braille glove -mp3 players	-artificial liver

Safety

The workshop in which you design and make products is a very exciting place. It offers you the opportunity to be creative and to explore emerging ideas. It contains many interesting materials and the tools and equipment to shape and form those materials. With care and concentration, you can see the high-quality results of your thinking. The workshop is also a potentially dangerous place, however. The same materials and tools you use to make wonderful objects can also injure you and your classmates. Without constant care and attention to what you and others are doing, accidents will happen.

The safety rules and procedures described in this appendix are designed to minimize the probability that you, your classmates, or your teacher will have an accident in the school workshop. Remember, while you are designing and making products, *you* are the person responsible for your safety. You are also responsible for paying attention to the safety of everyone else in the workshop. At all times, work smart!

Use of Color in the Workshop

Color-coding plays an important role in keeping you safe in the workshop. The examples in **Figure B-1** show how color is used to show safety information.

Symbols and Signs Used in the Workshop

In addition to using colors to convey information, workplaces also use symbols and signs to convey safety information. **Figure B-2** shows some of the symbols you will see posted in various parts of your school workshop.

RED: Danger
Used to show danger and emergency information, such as Stop buttons and fire extinguishers. The red area on the sander shows that using the "up" side of the sanding disc is dangerous. Where else in the workshop is red used to identify potentially dangerous situations?

YELLOW: Hazard, caution
Hazard areas near equipment are often marked off with yellow or yellow-and-black striped tape. Make sure no one else is inside the hazard area before turning on a machine.

ORANGE: Warning
Used alone or with yellow to show areas with high hazards, such as electric shock.

ORANGE-RED: Biohazard
Identifies infectious waste products, such as blood.

PURPLE (MAGENTA) ON YELLOW: Radiation Hazard
Identifies radioactive waste.

GREEN: Safety
Often used for medical supplies. Green is also used for Start buttons on equipment.

BLUE: Information
Provides nonemergency information about machines, protocols, and other topics.

WHITE/BLACK: Directions or boundaries
Used to show pathways, directions, or the location of stairs and other features.

Figure B-1. Color codes used in the workshop.

Fire Extinguisher

Fire Exit

Combustible Materials

High Voltage

Figure B-2. Symbols and signs used in the workshop.

TSA Competition Safety Rules

Are you one of the 150,000 student members of the Technology Student Association (TSA)? Each year, schools compete in TSA competitions at the annual national conference. Both individuals and teams compete to solve a variety of technology-related design problems. The problems are changed slightly every two years. To see this year's list of events, visit the TSA Web site: tsaweb.org.

The TSA competition guidelines include rules to help ensure the safety of everyone. One general rule applies to all competitions. It states that hazardous materials, chemicals, combustibles, wet cell batteries, and other, similar substances are not allowed at the national TSA conference. Other safety rules are specific to the individual competitions. For example, some competitions require students to wear approved safety glasses. Failure to abide by the safety rules can lead to immediate disqualification or a maximum deduction of 20 points from the team's score.

General Workshop Safety Rules

1. Do not wear loose or baggy clothing that can interfere with your work.
2. Tie back long hair and tuck it securely into your sweater or shirt.
3. Remove rings and jewelry.
4. Do not wear open-toe shoes.
5. Wear the workshop apron or coat provided.
6. Use a hand tool, portable electric tool, or fixed machine only after the teacher has taught you to use it correctly and has given you permission to use it. If you do not remember how to use the tool or machine, ask your teacher for help.
7. Use the correct tool for the job.
8. Wear safety goggles or a safety shield when you are using tools and equipment that may create a hazard for the eyes. Examples of hazardous equipment include drills, lathes, milling machines, and grinders. In some cases, you may need ear protection also.
9. Do not run. You might slip and hurt yourself or bump into someone and cause an accident.
10. Pay attention to other people when you move around the workshop.
11. Do not shout. You may distract someone using a tool, causing an accident.
12. Carry sharp tools with the cutting edge pointed toward the floor and the cutting edge turned away from your body.
13. Return all tools to their correct location. A neat workshop is a safe workshop.
14. Return all flammable materials to the storage cupboard immediately after use.
15. Report all accidents. Even a minor cut can become infected if not treated properly.
16. Never use a damaged tool or piece of equipment. Report the damage to the teacher. Never attempt any repair without consulting the teacher.
17. Tell the teacher if you accidentally damage a tool or piece of equipment, so that it can be repaired.
18. Respect all safety zones around machines.
19. Store oily rags or waste in the self-closing metal containers provided.
20. Clean up spilled liquids immediately.

Additional Safety Rules for Hand Tools

The following rules are specific to using hand tools. Follow these rules in addition to the general safety rules stated in the previous section.

1. Wear safety goggles.
2. Do not blow sawdust or metal filings toward other students.
3. Keep your fingers behind a cutting edge.
4. Never cut toward your hand or any other part of your body.
5. Reduce clutter near the work surface by returning tools to their correct place.
6. Never use damaged or broken tools.

Additional Safety Rules for Portable Electric Tools

The following rules are specific to using portable electric tools. Follow these rules in addition to the general safety rules.

1. Wear safety goggles or a full face shield appropriate to the machine.
2. Obtain your teacher's permission before using any portable electric tool.
3. Always disconnect the tool from the electrical outlet before making any adjustments.
4. Disconnect the tool from the electrical outlet immediately after you have finished using the tool.
5. Never use a portable electric tool with a frayed cord.
6. Use an extension cord only when absolutely necessary.
7. Use only an extension cord that is grounded. See **Figure B-3**.

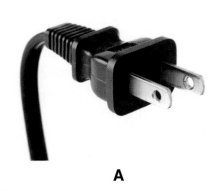

To ground

A   B

Figure B-3. A—Two-prong electrical plugs are not grounded. B—The third prong on a three-prong plug is the grounding prong.

Additional Safety Rules for Fixed Machines

The following rules are specific to using fixed machines. Follow these rules in addition to the general safety rules.

1. Wear safety goggles or a full face shield appropriate to the machine.
2. Obtain your teacher's permission before using any fixed machine.
3. Do not operate a machine unless all guards are in place.
4. When using a machine to cut materials:

   - Be sure to use the correct blade.
   - Use the correct speed for the material being cut.
   - Check the setting of all blade guides and reset them if required.
   - Keep your fingers clear of the blade path.
   - Do not force the workpiece into the blade.
   - When working with a circular saw, use the correct push stick.
   - Keep your fingers a minimum of 4″ (100 mm) away from the blade.

5. Use the correct vice or clamping tools when drilling, milling, or grinding.
6. Keep the floor clear of any debris that may cause you to trip or slip.
7. Always make sure your body is well-balanced when you operate a machine.
8. Use the ventilation system when you are using machines that create dust.
9. Never leave a machine that is switched on unattended.
10. Switch off the machine and wait until moving parts have stopped before moving away.
11. If a piece of material becomes jammed, switch off the machine immediately and inform your teacher. Do not attempt to remove the jammed material yourself.

Good Housekeeping

A cluttered or untidy workstation or workshop can cause accidents and injuries. Follow these good housekeeping safety procedures:

1. Keep the floor area free of shavings, sawdust, and metal cuttings. These materials can cause the floor to become slippery.
2. Keep the benchtop and machine surfaces as clean as possible.
3. Return excess material to its correct storage area.
4. Return tools to their racks when you finish using them.

Fire Safety Rules

Because of the nature of the work done in a school workshop, fire safety rules must be developed and followed by everyone. You should be familiar with your school's emergency exit plan, as well as the procedures for handling small and large fires.

Emergency Exit Plans

If a fire occurs in your school, a warning bell will sound and you will be required to evacuate the building as quickly as possible. To ensure that everyone can leave the building quickly, safely, and without panic, most schools have an emergency exit plan or evacuation plan. This plan includes identification of emergency exits, a floor plan showing routes out of the building, and a safe meeting place away from the building. See **Figure B-4**.

When the emergency bell sounds, you are responsible for the following actions:

- Remain silent so that all students can hear instructions from the teacher.
- Immediately line up in the place designated by the teacher.
- Do not stop to collect your books or personal belongings.
- Move silently and quickly with the class along the evacuation route.
- Do not run; running could cause confusion and panic.
- If you must use a staircase, walk in single file.
- Never use an elevator.
- Stand quietly in the designated meeting place.
- Do not return to the building for any reason until instructed by your teacher.

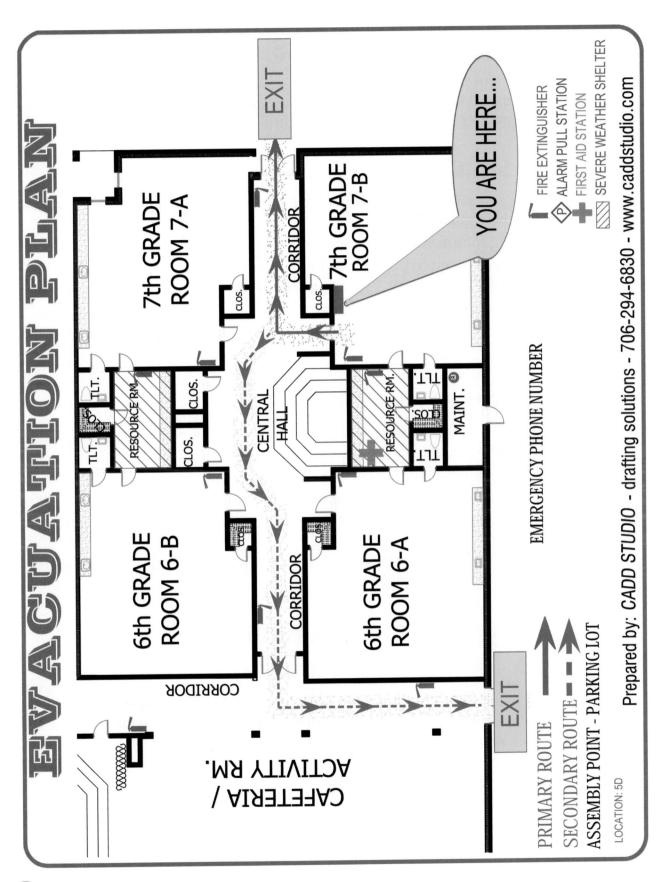

Figure B-4. An example of an evacuation plan.

Fire Procedures

In the unlikely event that a fire breaks out in the workshop, follow the procedures established at your school. Most school workshops have policies for both small and large fires.

Small Fires

If your school does not have a specific policy for small fires, follow these guidelines:

1. Inform your teacher immediately.
2. If requested, sound the fire alarm.
3. If possible, use a fire extinguisher or fire blanket to put out the fire.
4. Remember, human safety is more important than fighting the fire. If you see that you cannot contain the fire, follow your teacher's instructions immediately.

Large Fires

If your school does not have a specific policy for large fires, follow these guidelines:

1. Inform your teacher immediately.
2. If requested, sound the fire alarm.
3. *Do not* attempt to use a fire extinguisher or fire blanket to put out the fire.
4. Follow your school's emergency exit plan or the instructions given by your teacher.

Types of Fire Extinguishers

Fires are classified according to the types of chemicals that can be used to extinguish them. Each class of fire is represented by a different letter and symbol, as shown in **Figure B-5**. Fire extinguishers are rated according to the classes of fires they can extinguish. After use, a fire extinguisher must be refilled and checked for readiness. In many areas, the local fire department checks the fire extinguishers in public buildings on a regular basis.

Fires	Type	Use		Operation
Class A Fires Ordinary Combustibles (Materials such as wood, paper, textiles.) *Requires... cooling-quenching* Old New	**Soda-acid** Bicarbonate of soda solution and sulfuric acid	Okay for use on A		Direct stream at base of flame.
		Not for use on B C D		
Class B Fires Flammable Liquids (Liquids such as grease, gasoline, oils, and paints.) *Requires...blanketing or smothering.* Old New	**Pressurized Water** Water under pressure	Okay for use on A		Direct stream at base of flame.
		Not for use on B C D		
	Carbon Dioxide (CO$_2$) Carbon dioxide (CO$_2$) gas under pressure	Okay for use on B C		Direct discharge as close to fire as possible, first at edge of flames and gradually forward and upward.
		Not for use on A D		
Class C Fires Electrical Equipment (Motors, switches, and so forth.) *Requires... a nonconducting agent.* Old New	**Foam** Solution of aluminum sulfate and bicarbonate of soda	Okay for use on A B		Direct stream into the burning material or liquid. Allow foam to fall lightly on fire.
		Not for use on C D		
	Dry Chemical	Multi-purpose type Okay for A B C	Ordinary BC type Okay for B C	Direct stream at base of flames. Use rapid left-to-right motion toward flames.
		Not okay for D	Not okay for A D	
Class D Fires Combustible Metals (Flammable metals such as magnesium and lithium.) *Requires...blanketing or smothering.* D	**Dry Chemical** Granular type material	Okay for use on D		Smother flames by scooping granular material from bucket onto burning metal.
		Not for use on A B C		

Figure B-5. Classes of fires and the types of fire extinguishers that can be used on each.

Hazardous Materials and Hazardous Waste

As you may recall from Chapter 16, a *hazardous material* is any material that negatively affects the health of humans or animals or that damages the natural environment. Hazards posed by these materials include the possibility of explosion, fire, poisoning, or damage to the skin, eyes, or other organs.

Hazardous materials in your workshop include:

- Fumes from soldering, welding, and diesel engines
- Dust from sanding wood or grinding metal
- Aerosol mists from spray paints and cutting oils
- Vapor from paint, solvents, or photographic chemicals
- Carbon monoxide from internal combustion engines

Exposure to these materials can cause a number of health problems, including skin irritation, eye irritation, and in extreme cases, loss of eyesight.

A *hazardous waste* is any waste material that can cause damage to your health or to the health of the environment. Some hazardous wastes are highly combustible (catch fire easily). These include gasoline and some solvents. Others are corrosive—they eat away any material they contact. Most acids are corrosive. Other hazardous wastes are toxic. These wastes are poisonous if consumed or absorbed by the body.

When handling hazardous materials or hazardous wastes, always follow these safety precautions:

- Wear protective gloves
- Wear eye protection
- Wear the appropriate breathing apparatus, if required

Occupational Safety

Keeping people safe is not only a concern in the school workshop. Safety in the workplace affects millions of people every day. In the United States, two organizations are responsible for regulating and enforcing safety standards.

Occupational Safety and Health Administration

The *Occupational Safety and Health Administration (OSHA)* is a department of the United States Department of Labor. It regulates and enforces health and safety regulations for business and industry.

One very important role for OSHA is to provide education and training for both employers and employees. With the proper training, everyone can contribute to recognizing and preventing unsafe or unhealthy working conditions. OSHA workers also meet with employers and employee groups to discuss ways to minimize the chance of occupational injury or illness.

OSHA sets standards that employers must follow to protect workers from hazards. For example, Standard Number 1926.100 states that all construction workers must be provided with a hardhat (protective helmet) whenever there is a possible danger of head injury from falling or flying objects. OSHA also sets standards to reduce the risk of injury from falls, to ensure that workers are not exposed to harmful materials, such as asbestos, and to ensure that machines are equipped with appropriate guards.

OSHA also requires employers to display *material safety data sheets (MSDS)* for every potentially harmful substance available in the workplace. The MSDS are often collected in a binder and placed in a prominent location so they are easy to find. The MSDS provides guidelines for safe handling of the substance, as well as information about what to do if a spill or accident occurs.

You can learn more about the role OSHA plays in workplace safety by visiting the Web site www.osha.gov/. This site contains information and resources for workers, including the location of regional offices, information about personal protective equipment (PPE) and other safety topics, and how to contact OSHA for help.

National Institute for Occupational Safety and Health

The National Institute for Occupational Safety and Health (NIOSH) is part of the Centers for Disease Control and Prevention (CDC) in the Department of Health and Human Services. NIOSH conducts research and makes recommendations to help prevent work-related injury and illness.

When an occupational hazard is identified, NIOSH designs a research study, collects and analyzes data, and recommends improvements. After employers and employees take the suggested actions, NIOSH evaluates the results. For example, NIOSH research showed that miners using drills were exposed to noise levels that could lead to hearing loss. Working with designers, engineers, and manufacturers, NIOSH developed an enclosure that trapped much of the noise. This reduced the risk of injury to the miners. You can learn more about NIOSH by visiting www.cdc.gov/niosh/.

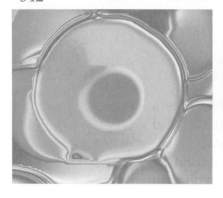

Glossary

Each term in this glossary is followed by the number of the chapter in which it is described. Items shown in **blue** are academic terms.

3D printer: Any machine that can process a CAD file to create a 3D part for prototyping or production purposes. (Ch. 3)

A

abutments: The supports where a bridge arch meets the ground. Abutments resist the outward thrust (push) and keep the bridge up. (Ch. 6)

acid: A substance that has a sour taste. (Ch. 4)

acid rain: Rain that mixes with acidic gases and particle pollutants before it falls to the ground. (Ch. 16)

additive fabrication: The process of creating a "3D print" of an object from a CAD file by depositing layer after layer of liquid, powder, or sheet material according to computer instructions and fusing the layers together. (Ch. 3)

adhesion: A chemical joining process that relies on a natural attraction between molecules. (Ch. 5)

adobe: A mixture of clay and straw used as a structural material in the southwestern United States. (Ch. 7)

alloy: A metal formed by a mixture of metals. (Ch. 4)

alphabet of lines: The standard line types and widths used on technical drawings. (Ch. 3)

alternating current (AC): An electric current whose strength and direction vary on a regular basis. (Ch. 11)

amperage: A measure of the amount of current in a circuit. (Ch. 11)

ampere: A unit of measurement of electric current. (Ch. 12)

analog signal: An electronic signal that changes continuously in a smooth manner. Analog signals appear as sine curves on an oscilloscope. (Ch. 13)

AND gate: An electronic gate or switch that has two or more inputs and produces a 1 if all of the inputs are 1. (Ch. 12)

anode: The positive electrode in a dry cell. (Ch. 11)

antibodies: Special proteins developed by the body's immune system to fight various diseases. (Ch. 14)

aptitude: A natural talent or ability. (Ch. 17)

aquaculture: A type of farming in which farmers raise fish or shellfish in sheltered areas for commercial purposes. (Ch. 14)

arch bridge: A type of bridge in which the compressive stress created by the load is spread over the arch as a whole. (Ch. 6)

artisan: An individual responsible for every step in producing a finished product; also, someone skilled in a trade; a craftsperson (for example, a woodworker). (Ch. 15)

assembly line: An arrangement of workers, machines, and equipment used to quickly produce a large number of identical products by assembling parts in a planned sequence. Also called a *production line*. (Ch. 15)

automated guided vehicle (AGV): A mobile robot used in industrial applications to move materials from point to point following a set path. (Ch. 9)

automation: The use of computers or automatic machines to control machine operations and make a product. (Ch. 15)

B

balance: In design, the arrangement of mass over the space used. The three types of balance are symmetrical, asymmetrical, and radial. (Ch. 2)

bamboo: A woody member of the grass family that can be processed and used for everything from flooring and furniture to clothing. (Ch. 4)

base: A substance that tastes bitter and feels slippery or soapy to the touch. (Ch. 4)

batter boards: Boards that support the lines set up to locate the building on a site so excavating can begin for the foundation. (Ch. 7)

battery: A single package containing several cells connected together to produce a DC current. (Ch. 11)

beam: A horizontal structural member, usually used to support floor or roof joists. (Ch. 7)

bending: A method of shaping sheet material by folding it like a sheet of paper. (Ch. 5)

binary digital code: A code used by digital electronics to convey information; based on the binary number system. (Ch. 13)

biofuels: Fuels created from plant or biomass sources. (Ch. 10)

biotechnology: The use of living organisms to make goods or provide services. (Ch. 14)

bit: An on or off signal, known as a binary digit, in the binary code used by computers. (Ch. 13)

Bluetooth® technology: A short-range technology that connects electronic devices wirelessly and has a maximum range of about 100 feet. (Ch. 13)

brainstorming: A group problem-solving method of generating new ideas in which everyone's ideas are welcome and no idea is too crazy. (Ch. 2)

biomass energy: Energy produced from plants and other organic matter. (Ch. 10)

biomaterial (biocompatible material): A special material that is compatible with living tissue and is able to function with living tissue. (Ch. 4)

bioplastics: Plastics made from renewable, biodegradable sources such as corn, rice, sugar cane, soybeans, or other plant material. (Ch. 4)

break-even point: The point at which the income from items sold in a business equals total expenses. (Ch. 17)

budget: An estimate of operating costs and expenses that a company does not want to exceed. (Ch. 17)

business plan: A detailed plan of how the business will operate and what goals it will achieve, using specified resources. (Ch. 17)

byte: A group of eight binary bits that are read together as a single "word." (Ch. 13)

C

cantilever bridge: A type of bridge in which a beam is capable of supporting a specified load at one end when the opposite end is anchored or fixed. (Ch. 6)

capacitor: A device designed to store an electrical charge. It consists of two metal plates (conductors) separated by an insulator. (Ch. 12)

carbon monoxide (CO): A colorless, odorless, and poisonous gas produced by burning fossil fuels. (Ch. 16)

career: An occupation or way of making a living. (Ch. 17)

career plan: A plan that helps you make the right choices to be successful in the career of your choice. (Ch. 17)

cathode: The negative electrode in a dry cell. (Ch. 11)

casting: A method of shaping parts or products by pouring liquid material into a mold. (Ch. 5)

cell: A single DC power source consisting of a positive electrode, a negative electrode, and an electrolyte. Also called a *dry cell*. (Ch. 11)

central processing unit (CPU): The part of a computer that accepts inputs, performs calculations and other processing tasks, and sends the results to output devices. (Ch. 13)

ceramic: Any of the various hard, brittle, heat-resistant, and corrosion-resistant materials made by shaping and then firing a nonmetallic mineral, such as clay, at a high temperature; used for making items such as pottery and bricks. (Ch. 4)

chemical energy: Energy locked away in different kinds of substances; often released by burning. (Ch. 10)

chemical joining: A method of fastening joints by using chemicals such as glues, adhesives, solvents, or cements. (Ch. 5)

chiseling: A technique used to shape material by cutting away the excess with a chisel and mallet. (Ch. 5)

circuit: A closed path in which current can travel; three basic types of circuits are series, parallel, and series-parallel. (Ch. 11)

circuit breaker: A device made of a bimetallic strip that heats up and bends if too much current is flowing through the circuit, stopping current flow. (Ch. 12)

circumference: The distance around the border of a circle. (Ch. 3)

clear-cutting: The practice of harvesting lumber by cutting down all of the trees in the selected area using bulldozers, cables, and other technology. (Ch. 16)

cloning: The process of creating an exact copy of any kind of biological material. See also *reproductive cloning*. (Ch. 14)

closed-loop system: A system that includes a feedback device to provide control. (Ch. 7)

cloud computing: A form of distributed computing that uses the Internet to provide access to computer applications and other files. (Ch. 13)

coatings: Materials applied to a surface. (Ch. 5)

cohesion: A chemical joining process in which the materials to be joined melt and mix together to form a strong bond. (Ch. 5)

communication: The process of exchanging information or ideas. (Ch. 3, 13)

communication system: A means of transmitting and receiving information between two or more points. (Ch. 7)

communication technology: The equipment and technology used to transmit and receive information. (Ch. 3, 13)

commutator: The split ring in a DC generator that sends current through the circuit in one direction only. (Ch. 11)

complementary colors: Contrasting colors; colors found on opposite sides of the color wheel. (Ch. 2)

composite: A material that is made from a combination of different substances. (Ch. 4)

composting: A method of mixing organic matter such as food waste and lawn clippings and allowing them to decay to create nutrient rich soil. (Ch. 16)

compound: A substance made up of more than one element (type of atom). (Ch. 4)

compression: A squeezing force. (Ch. 6)

computer-aided design (CAD): A method of making drawings using a computer and drafting software. (Ch. 3)

computer numerical control (CNC) machine tool: A tool in which movements and operations are controlled by a computer program. (Ch. 15)

concurrent engineering: A system of designing, developing, and manufacturing a product that involves every department, and sometimes the customer, from the beginning of the design stage. (Ch. 3)

conductivity: A material's ability to conduct electric current. (Ch. 12)

conductor: A material through which electric current flows easily. (Ch. 4, 11, 12)

construction lines: Light, thin lines made to start a sketch or drawing. (Ch. 3)

continuous improvement: A theory that all products and processes can and should be improved on a continual basis. (Ch. 15)

contrast: A clearly evident difference between two things. (Ch. 2)

convection currents: Air currents caused by the expansion and rising of air as it warms and the movement of cooler air to replace it. (Ch. 10)

corporation: A type of company that is a legal entity owned by stockholders who buy stock in the company. (Ch. 17)

customer service: The process of dealing with customers at all stages of customer contact: before, during, and after a product or service is sold. (Ch. 17)

cycle: In alternating current, a period during which electricity flows or pulses in one direction, then in the other direction. (Ch. 11)

cyclic loading: The repetitive movement that occurs in the normal use of a product. (Ch. 4)

D

data: A set of facts. (Ch. 13)

debt: Money borrowed from a bank or other institution that must be paid back, usually with interest. (Ch. 17)

degree of freedom: The term used to describe each joint or direction of movement in a robotic arm. (Ch. 15)

deoxyribonucleic acid (DNA): A long chain of molecules in the nucleus of every living cell in an organism. DNA contains the genetic information needed to reproduce the organism. (Ch. 14)

design brief: A statement that clearly describes the design problem to be solved. (Ch. 2)

designer: A person who creates and carries out plans for new products and structures. (Ch. 2)

designing: Generating and developing ideas for new and improved products and services that satisfy people's needs. (Ch. 1)

design problem: A situation or condition that can be solved or improved through the application of technology. (Ch. 2)

design process: An orderly set of steps or skills used to generate and develop design ideas. The actual process varies depending on the design problem. (Ch. 2)

development: A pattern that shows how sheet metal must be cut and folded to create an object. (Ch. 5)

diameter: The distance from one side of a circle to the other through its center. (Ch. 3)

diesel engine: An engine in which air is squeezed to very high pressure and very high temperature inside the cylinder. Diesel fuel is then injected, and spontaneous ignition occurs. (Ch. 9)

digital signal: An electronic signal that has only two values: on or off. Digital signals appear as square waves on an oscilloscope. (Ch. 13)

diode: A device that allows current to flow in one direction only. Diodes are most commonly used as rectifiers to change alternating current to direct current. (Ch. 12)

direct current: Current that maintains a constant flow in one direction. (Ch. 11)

discrimination: The unfair treatment of someone based on a personal characteristic such as age or gender. (Ch. 17)

distributed computing: A computer system that consists of many individual computers working together. Problems and tasks are distributed among the individual computers, and the results are gathered by the controlling computer. (Ch. 13)

distribution lines: Lines that transport lower-voltage electricity from a step-down transformer at an electrical substation to individual buildings such as homes, schools, and offices. See also *transmission lines*. (Ch. 11)

distribution system: The method a business uses to get its products to the customers. (Ch. 17)

division of labor: A system in which each person is assigned one specific task in the making of a product. (Ch. 15)

drafting: The process of creating drawings to specify the exact size, shape, and features of a design idea. (Ch. 3)

drilling: A process used to make holes in wood, plastic, metal, and other materials. (Ch. 5)

dry cell: A single DC power source consisting of a positive electrode, a negative electrode, and an electrolyte. Also called a *cell*. (Ch. 11)

durable product: One that is designed to last for a long time. (Ch. 15)

dynamic load: A load on a structure that is always changing. (Ch. 6)

E

efficiency: Making good use of energy, time, and materials. (Ch. 15)

elastic materials: Materials that obey Hooke's Law, which states that the amount a spring extends is proportional to the force with which you pull it. (Ch. 10)

electrical energy: The movement of electrons from atom to atom. (Ch. 10)

electrical system: The circuits that carry electricity in a product or that provide electricity for lights and appliances throughout a home. (Ch. 7)

electric current: The flow of electrons in a circuit. (Ch. 11)

electric motor: A machine that changes electrical energy into mechanical energy. (Ch. 9)

electric vehicle: A vehicle that uses an electric motor. (Ch. 9)

electrode: A conductor, usually metal, that transfers electrical energy into and out of a dry cell or battery. (Ch. 11)

electrolyte: A solution in a cell that reacts with the electrodes in a chemical reaction that creates electric current. (Ch. 11)

electromagnetism: Magnetism created by passing an electric current through a wire to create a magnetic field around the wire. (Ch. 11)

electromotive force (EMF): The force that pushes electrons through a circuit. (Ch. 11)

electronics: The use of electrically controlled parts to automatically control or change current in a circuit. (Ch. 12)

electron theory: Theory of electricity in which electrons jump from one atom to the next along a conductor to create an electric current. (Ch. 11)

element: The simplest form of matter, made up of only one kind of atom. (Ch. 4)

elements of design: The things you see when you look at an object, including line, shape and form, texture, and color. (Ch. 2)

energy: The capacity to do work. (Ch. 10)

engine: A machine that converts energy into mechanical force or motion. (Ch. 9)

engineer: Someone who applies scientific and design principles to create practical solutions for problems. (Ch. 1)

engineered wood: Wood that is created from natural wood and/or other components to meet a specific need or to solve a problem associated with natural wood. (Ch. 4)

engineering: Applying scientific and design principles to create something that solves a specific problem. (Ch. 1)

engineering design process: A design process used to design an item that is scientific or technical in nature. (Ch. 2)

entrepreneur: A person who owns his or her own company. (Ch. 17)

entrepreneurship: Becoming an entrepreneur; the process of owning your own business. (Ch. 17)

equilateral triangle: A triangle in which all the sides are the same length and all the angles are equal. (Ch. 6)

equity: Money or other resources contributed by investors, who in turn receive partial control of the company. Equity does not have to be repaid. (Ch. 17)

ergonomics: The study of how a person, the products used, and the environment (our surroundings) can best be fitted together. (Ch. 2)

ethics: Knowing right from wrong. (Ch. 17)

experimentation: Trying different ideas to solve a problem to see which one works best. (Ch. 2)

external combustion engine: An engine that uses fuel burned outside of the engine. (Ch. 9)

F

factory: A building in which products are manufactured. (Ch. 15)

farad: A unit used to measure the storage potential of a capacitor. (Ch. 12)

fatigue resistance: A material's ability to resist constant flexing or bending. (Ch. 4)

feedback: In communication, a receiver's response to a source's question or statement. (Ch. 13)

ferrous: Any metal or metal alloy that contains iron. (Ch. 4)

filing: A process used to smooth a material (usually metal) by removing small amounts from the surface with a toothed tool called a *file*. (Ch. 5)

finishing: A process that changes the surface of a product by treating it or placing a coating on it. (Ch. 5)

flexible manufacturing system: A manufacturing system in which machine tools can be changed quickly and easily to produce several different kinds of parts. (Ch. 15)

floor plan: A scale diagram of a room or building drawn as if seen from above. (Ch. 7)

footing: The lowest portion of a building's foundation; the foundation wall rests on top of the footing. (Ch. 7)

form: A three-dimensional representation of an object. (Ch. 2)

forming: Changing the shape of sheet material, often through the use of a mold. (Ch. 5)

foundation: The footing and foundation wall that support a building and spread its load over a large ground area. (Ch. 7)

free enterprise system: A system in which people make most of the decisions about how to use a country's economic resources. (Ch. 17)

freehand sketching: Sketching by hand without the aid of mechanical instruments such as rulers and templates. (Ch. 3)

frequency: The number of cycles of a wave pattern that occur in one second. (Ch. 10) In alternating current, the number of complete cycles per second. (Ch. 11)

friction: A force that acts like a brake on moving objects. (Ch. 8)

fuel cell: A device that allows hydrogen and oxygen to combine, without combustion, to generate electricity. (Ch. 10)

fuel cell vehicle: A type of vehicle that mixes hydrogen from an on-board tank with oxygen from the air to power electric motors for turning the wheels. (Ch. 9)

function: What an object does or how it works; a functional object or product solves the problem described in the design brief. (Ch. 2)

fuse: A current-limiting device consisting of a thin filament that melts when too much current reaches it. This stops current flow in the circuit. (Ch. 12)

G

gasoline engine: Engine in which a mixture of air and gas is ignited by an electric spark, pushing down pistons to drive a crankshaft in a rotary motion. (Ch. 9)

gear: A rotating wheel-like object with teeth around its rim used to transmit force to other gears with matching teeth. (Ch. 8)

gene: A sequence of DNA that defines a specific trait, such as hair color or eye color. (Ch. 14)

generating station: A facility that uses an energy source to operate turbines and produce electricity. (Ch. 11)

generator: A machine that produces electricity when turned by an outside force, such as a turbine. (Ch. 11)

genetic code: The set of genes that makes up a specific living organism. (Ch. 14)

genetic engineering: The process of recombining genes or adding genes from another species to change the traits or characteristics of a living organism. (Ch. 14)

geothermal energy: Energy from the heat at the Earth's core. (Ch. 10)

goal: A statement of what you want to accomplish. (Ch. 17)

gravitational energy: The energy associated with the Earth's gravitational field; the natural force of attraction created by an object that draws other objects toward itself. (Ch. 10)

greenhouse: A building made mostly of glass or transparent plastic that helps control conditions such as humidity and temperature for growing plants. (Ch. 14)

H

haptic device: A device that allows people to touch and move objects in a virtual reality system. (Ch. 13)

harassment: Tormenting someone by words or acts. Harassment includes making rude comments, making fun of someone, or even teasing if you persist even after that person has asked you to stop. (Ch. 17)

hardware: The physical input, processing, and output components of a computer, such as a standard keyboard, mouse, CPU, and monitor. (Ch. 13)

hardwood: The wood from broad-leaf, deciduous trees. (Ch. 4)

harmony: A condition in which chosen colors or shapes in designs naturally go together. (Ch. 2)

hazardous materials: Materials that have been identified by the government to be harmful to people and the environment. (Ch. 16)

hazardous waste: Hazardous materials that are discarded as waste. (Ch. 16)

heating, ventilation, and air-conditioning system (HVAC): The furnace and/or air-conditioner and associated ducts or pipes used to distribute heated or cooled air or water throughout a house. (Ch. 7)

heat joining: A process that melts the material itself or a bonding agent (such as solder or another filler metal) to secure a joint. (Ch. 5)

household hazardous waste: Hazardous waste produced in most households in the United States, including products that are corrosive, toxic, flammable, or reactive. Examples include many cleaning products and battery acid. (Ch. 16)

human-powered vehicle (HPV): Any vehicle that is powered solely by one or more humans. (Ch. 9)

hydraulics: The study and technology of the characteristics of liquids at rest and in motion. (Ch. 8)

hydroelectricity: Electricity generated from moving water over turbines connected to electric generators. (Ch. 10)

hydroponics: A type of agriculture in which plants are grown in a water-based system, without soil. Liquid nutrients are added to the water that surrounds the roots of the plant. (Ch. 14)

hypertext transfer protocol (HTTP): The protocol, or language, that allows people to access the Internet through the World Wide Web. (Ch. 13)

hypothesis: A guess or suggestion based on scientific data that attempts to explain the data. (Ch. 1)

I

implants: Mechanical body replacement parts. (Ch. 14)

improper fraction: A fraction greater than 1 in which the part that is greater than 1 is converted to fractional units and added to the fractional part. For example, 1½ can be written as the improper fraction 3/2. (Ch. 5)

inclined plane: A simple machine in the form of a sloping surface or ramp, used to move a load from one level to another. (Ch. 8)

information: Data that has been arranged or organized in a meaningful way. (Ch. 13)

information technology: The equipment and systems used in storing, processing, and extending the human ability to communicate information. (Ch. 13)

innovation: The process of modifying an existing product or system to improve it. (Ch. 2)

insulation: Material used in the walls and ceiling of a building to help prevent heat loss in winter and heat gain in summer. (Ch. 7)

insulator: A material through which electric current does not flow. (Ch. 4, 12)

integrated circuit: A single electronic component that replaces a whole group of separate components. One integrated circuit may contain the equivalent of about 1,000,000 separate components. (Ch. 12)

interchangeable parts: Parts that are the same shape and size. (Ch. 15)

intermodal transport: Transportation involving more than one type of vehicle or transportation system. (Ch. 9)

internal combustion engine: An engine that uses fuel burned inside the engine. (Ch. 9)

invention: The process of turning ideas and creativity into new devices and systems. (Ch. 2)

irradiation: A safe and effective method of exposing foods to a low level of radiation to kill bacteria and other parasites. (Ch. 14)

isometric axes: Three intersecting lines that form the basis of an isometric drawing. One of the lines is vertical. The other two are at 30° from the horizontal, running from the vertical line in opposite directions. (Ch. 3)

isometric sketching: A type of sketching that shows three sides of an object by placing the picture axes at 120° from each other, or at 30° from the horizontal. (Ch. 3)

J

jet engine: An engine that sucks in air at the front, squeezes it, mixes it with fuel, and ignites it. This creates a strong blast of hot gases that rush out of the back of the engine at great speed, propelling the engine forward. (Ch. 9)

jig: A tool that is used to hold the workpiece and guide the tool or machining process. (Ch. 5)

joist: A horizontal member of a house framework that supports the floor. (Ch. 7)

just-in-time delivery: A system in which only the parts needed for the current day's production are made. (Ch. 15)

K

kinetic energy: Energy an object has because it is moving. (Ch. 10)

L

laminating: A process that involves gluing together several veneers (thin sheets of wood) to form a strong part. (Ch. 5)

landfill: An area that has been prepared to receive garbage and has been lined to prevent leaks into the soil. (Ch. 16)

landscaping: Designing the exterior space that surrounds a home. (Ch. 7)

laparoscopic surgery: Surgery performed through small incisions in the abdomen; also called *minimally invasive surgery*. (Ch. 14)

laser: A form of radiation that has been boosted to a high level of energy and produces a strong, narrow beam of light. "Laser" is an acronym for Light Amplification by Stimulated Emission of Radiation. (Ch. 12)

leadership: the ability to guide team members and keep the team on track. (Ch. 17)

lean manufacturing: A system of manufacturing that focuses on reducing waste, improving quality, and increasing efficiency. (Ch. 15)

lever: A simple machine that consists of a bar and fulcrum (pivot point). Levers are used to increase force or decrease the effort needed to move a load. (Ch. 8)

liabilities: Debts incurred by a business. (Ch. 17)

line drawings: Objects and ideas represented by lines and shapes. (Ch. 3)

lines: Design elements that describe the edges or contours (outlines) of shapes; they show how an object will look when it has been made. (Ch. 2)

linkage: A system of levers used to transmit motion. (Ch. 8)

load: The weight, mass, or force placed on a structure (Ch. 6); in an electrical circuit, any current-using device. (Ch. 12)

logic gates: Electronic switches used to control and modify current in an electronic circuit. (Ch. 12)

long-term goals: Goals that may take many years to accomplish. (Ch. 17)

lubrication: The application of a smooth or slippery substance between two objects to reduce friction. (Ch. 8)

M

machine: A device that does some kind of work by changing or transmitting energy. (Ch. 8)

magnetic: Attracted by or to a magnet. (Ch. 4)

magnetism: The ability of a material to attract pieces of iron or steel. (Ch. 11)

manufacturing: The process of using raw materials to create products. (Ch. 15)

marketing plan: A plan that describes the target market for a business. (Ch. 17)

marking out: Measuring and marking material to the dimensions shown on a drawing. (Ch. 5)

mass production: The process of making products in large quantities to reduce the unit cost. (Ch. 15)

mass transit: A method of moving many people at once. (Ch. 9)

material fatigue: The wearing out or failure of a material due to constant use. (Ch. 4)

mechanical advantage: In a simple machine, the ability to move a large resistance by applying a small effort. (Ch. 8)

mechanical energy: The energy of motion. Mechanical energy is often, but not always, associated with machines. (Ch. 10)

mechanical joining: The use of physical means, such as a bolt and nut, to assemble parts. (Ch. 5)

mechanism: A way of changing one kind of effort into another kind of effort. (Ch. 8)

media: Sources of communication and information, such as newspapers, radio and television broadcasts, and the Internet. (Ch. 13)

mentor: Someone who has knowledge about the product a company is proposing and agrees to provide advice. (Ch. 17)

microelectronics: Traditional electronic devices, such as transistors, diodes, and resistors, that have been made very small. (Ch. 13)

minimally invasive surgery: Surgery performed through small incisions in the abdomen; also called *laparoscopic surgery*. (Ch. 14)

mixed fraction: A number that includes both a whole number and a fraction, such as $1\frac{1}{2}$. (Ch. 5)

model: A three-dimensional likeness of an object that is used to communicate design information and to evaluate a design. (Ch. 2)

modular construction: A building system that involves basic units of different sizes and shapes that can be combined on site. (Ch. 7)

modulate: Change the frequency, amplitude, or other characteristic of an electronic communication signal. Modulation allows the signal to carry information over long distances. (Ch. 13)

molding: A method of making shapes by forcing liquid material into a shaped cavity. (Ch. 5)

moment: The turning force acting on a lever; effort times the distance of the effort from the fulcrum. (Ch. 8)

monoculture: The spreading or planting of a single plant species over a large area. (Ch. 14)

motherboard: The main circuit board in a computer. (Ch. 13)

N

NAND gate: An electronic gate or switch that has two or more inputs. It produces a 1 only if all inputs are 0; produces a 0 if one or more inputs are 1. (Ch. 12)

nanomaterial: A material that is so small it must be measured in billionths of a meter, or nanometers. (Ch. 4)

nanotube: A carbon molecule that has a cylindrical or donut shape without gaps between its atoms. This structure makes it very strong. (Ch. 4)

nitrogen oxides (NO_x): Harmful nitrogen-containing gases that are produced by burning fossil fuels. (Ch. 16)

nondurable product: A product that is meant to be used once and then discarded. (Ch. 15)

nonferrous: Any metal or metal alloy that does not contain iron. (Ch. 4)

nonrenewable energy: Energy from sources that will eventually be used up and cannot be replaced. (Ch. 10)

NOR gate: An electronic gate or switch that has two or more inputs. It produces a 0 only if all inputs are 1. It produces a 1 if any or all of the inputs are 0. (Ch. 12)

NOT gate (inverter): An electronic gate or switch with only one input. If the input is 1, it produces a 0. If the input is 0, it produces a 1. (Ch. 12)

nuclear energy: Energy that occurs when the atoms of certain elements are split (nuclear fission) or when atoms are fused together (fusion). (Ch. 10)

nuclear fission: The splitting apart of atoms to produce nuclear energy. (Ch. 10)

nuclear fusion: The joining together of atoms to produce nuclear energy. (Ch. 10)

O

off-grid: A building in which electricity is generated at the site, with no connection to an electric utility. Electricity is typically generated using solar, wind, hydro, or a fuel-powered generator. Batteries are used to store energy to provide a continuous supply day and night. (Ch. 7)

ohm: The unit of measurement for electrical resistance. (Ch. 12)

Ohm's law: The law of electricity stating that voltage equals the current times the resistance. (Ch. 12)

on-site transportation: Types of vehicles that transport people and materials from one place to another within a defined location, such as a building or mine. (Ch. 9)

opaque: An optical characteristic of a material that allows no light to pass through it. (Ch. 4)

open-loop system: A system that is not controlled through the use of a feedback device. (Ch. 7)

OR gate: An electronic gate or switch that has two or more inputs. It produces a 1 if any one or more of the inputs is 1. (Ch. 12)

orthographic projection: A kind of drawing that shows each surface of the object from a view at right angles to the surface. Orthographic views can be measured and used for manufacturing. (Ch. 3)

overhead expenses: Business expenses such as rent, utilities, and salaries. (Ch. 17)

overload: A condition that occurs when devices on an electrical circuit demand more current than the circuit can safely carry; a cause for a fuse or circuit breaker to interrupt current flow. (Ch. 12)

ozone: A gas that is formed at ground level when other common air pollutants mix with sunlight. (Ch. 16)

P

parallel circuit: A circuit that provides more than one path for electron flow. (Ch. 12)

partnership: A business that consists of two or more people who share the financial responsibilities and profits of the business. (Ch. 17)

pattern: An element or shape repeated many times in a design. (Ch. 2)

perspective sketching: Sketching method that provides the most realistic picture of objects. (Ch. 3)

photosynthesis: The process used by trees and plants to change water and carbon dioxide into carbohydrates using light as an energy source. (Ch. 4)

photovoltaic cells: Solar cells that convert the sun's energy directly into electrical energy. (Ch. 10)

photovoltaic panels: Solar panels; panels of solar cells that convert energy from the sun into electric energy. (Ch. 7)

pH scale: A scale that specifies how acidic or basic a substance is. (Ch. 4)

pier: A structure that extends from a bridge to the ground below to form a firm foundation to support the bridge. Also a platform extending from shore over water and supported by piles or pillars, used to secure, protect, and provide access to ships or boats. (Ch. 6)

pitch: A quality of sound that can be arranged or sequenced from low to high. (Ch. 10)

planing: A process for smoothing wood that uses a sharp blade to remove very thin shavings. (Ch. 5)

plasticizers: Various chemicals that act as internal lubricants for thermoplastics. (Ch. 4)

plumbing system: The system used to bring in clean water and dispose of wastewater in a home. (Ch. 7)

pneumatics: The study and technology of the characteristics of gases. (Ch. 8)

polymer: A chainlike molecule made up of smaller molecular units; the scientific name for plastic. (Ch. 4)

post and lintel: The simplest form of a framed structure, with horizontal framing members (lintels or beams) supported by vertical members (posts). (Ch. 7)

potential energy: Stored energy; energy that is derived from an object's position or condition. See also *kinetic energy*. (Ch. 10)

potentiometer: A type of variable resistor that raises and lowers voltage in a circuit. (Ch. 12)

power: The rate at which work is done or the rate at which energy is converted from one form to another or transferred from one place to another. (Ch. 8)

precision farming: A farming method in which farmers gather data from satellites to locate problems in their fields. They can then target that specific area for repairs or pest control. (Ch. 14)

prefabrication: A system of building in which components such as wall framing or roof trusses are built in a factory, rather than on the job site. (Ch. 7)

pressure: The effort applied to a given area; effort divided by area. (Ch. 8)

primary cell: A device that chemically stores electricity; its electrode is gradually used up during normal use, and it cannot be recharged. (Ch. 11)

primary colors: The three colors of the spectrum (red, blue, and yellow) that cannot be created by mixing other colors. (Ch. 2)

primary materials: Natural materials; materials that exist in nature. (Ch. 4)

principles of design: The guidelines for combining the elements of design: balance, proportion, harmony and contrast, pattern, movement and rhythm, and unity and style. (Ch. 2)

production technology: The systems and processes that are used to manufacture products. (Ch. 15)

proportion: The relationship between the sizes of two things. (Ch. 2)

prosthetic devices: Replacement body parts that mimic the original human body part. (Ch. 14)

prototype: The first working version of the designer's solution to a problem. (Ch. 2)

pulley: A simple machine in the form of a wheel with a groove around its rim to accept a rope, chain, or belt; it is used to lift heavy objects. (Ch. 8)

R

radiation: Particles released by unstable nuclei when an atom tries to become stable. (Ch. 10)

radioactive waste: A special type of hazardous waste that gives off energy in the form of radioactive particles. (Ch. 16)

radio frequency identification (RFID): A system that uses a microchip equipped with a radio transmitter. The microchip contains information such as expiration dates, medical records, and pet ownership details. (Ch. 13)

rapid manufacturing system: A 3D printer that is capable of creating real parts for use in actual products. Rapid manufacturing systems make on-demand manufacturing possible. (Ch. 3)

rapid prototyping (RP): A system that creates a prototype part quickly from a CAD file. Because the part can be created quickly, RP is often used to experiment with ideas during the design stage. (Ch. 3)

rectangle: A four-sided shape in which the opposite sides are parallel and the adjacent sides are at 90 degrees to each other. (Ch. 15)

rectifier: An electronic component that converts alternating current to direct current by allowing current flow in only one direction. (Ch. 12)

recumbent: Lying on one's back or stomach. (Ch. 9)

recycling: The process of treating materials to make them usable again. (Ch. 16)

refractory material: A material that retains its strength at high temperatures. (Ch. 4)

reinforced concrete: Concrete in which steel or plastic rods have been embedded to increase the concrete's resistance to tension. (Ch. 6)

renewable energy: Energy from sources that will always be available. (Ch. 10)

reproductive cloning: The process of creating a genetic "twin" of an existing animal or microorganism. See also *cloning*. (Ch. 14)

resistance: Opposition to the flow of electricity. (Ch. 12)

retrieval: The process of accessing information that has been stored in a storage device. (Ch. 13)

rheostat: A type of variable resistor that raises and lowers electrical current. (Ch. 12)

rhythm: A quality or feeling of movement, provided by repeating patterns. (Ch. 2)

robotics: The development of technology and applications for robots. (Ch. 13)

roof truss: A structure that forms a framework to support the roof and any loads applied to it. (Ch. 7)

rotor: In a generator, electromagnets mounted around a rotating shaft that are mounted inside the stator. (Ch. 11)

router: An electronic device that allows two or more different networks to communicate. (Ch. 13)

S

sawing: The process of cutting material with a tool that has a row of teeth on its edge. (Ch. 5)

scale drawing: A drawing in which all of the components are enlarged or reduced by a fixed ratio. (Ch. 3)

schematics: Diagrams of electrical or electronic circuits that use symbols to show the location of circuit components. (Ch. 12)

science: The field of study that is concerned with the laws of nature and the natural world. (Ch. 1)

scientific method: A method of inquiry developed by Sir Francis Bacon in 1620 in which scientists collect data, analyze facts, state a hypothesis, and then try to prove or disprove the hypothesis with further testing. (Ch. 1)

scientists: People whose field of study is science. (Ch. 1)

screw: A simple machine that is an inclined plane wrapped in the form of a cylinder. (Ch. 8)

secondary cell: A device for storing electrical energy as a chemical; it can be charged, discharged, and recharged. (Ch. 11)

secondary color: A color obtained by mixing equal parts of two primary colors. (Ch. 2)

semiconductor: Material that allows electron flow only under certain conditions. Semiconductors have some characteristics of conductors and some characteristics of insulators. (Ch. 4, 12)

series circuit: A circuit that provides only one path for electron flow. (Ch. 12)

series-parallel circuit: A circuit in which some loads are wired in series and others are wired in parallel. (Ch. 12)

server: A central, powerful computer that "serves," or provides storage and services, to other computers on a network. (Ch. 13)

shape: A two-dimensional representation of an object. (Ch. 2)

shear: A multidirectional sliding and separating force. (Ch. 6)

shearing: A process used to cut thin material with a scissors-like tool, called *shears* or *snips*. (Ch. 5)

short-term goals: Goals you can achieve in a short period of time, usually a year or two. (Ch. 17)

sintered: Formed by heat. (Ch. 4)

site: The land on which a building is to be built. (Ch. 7)

smart material: A material that changes in response to its surroundings. (Ch. 4)

smog: A combination of smoke and fog. (Ch. 16)

social media: Any digital information tool or service that uses the Internet. (Ch. 13)

soft skills: Personal skills such as communication skills and the ability to get along with other people. Also known as *workplace skills*. (Ch. 17)

software: The operating system and applications that allow us to perform meaningful tasks using a computer. (Ch. 13)

softwoods: Wood from coniferous trees that retain their needlelike leaves and are commonly called *evergreen trees*. (Ch. 4)

solar energy: Energy that travels in waves from the sun. Light energy moves at approximately 186,000 miles (300,000 km) per second. Also called *radiant energy*. (Ch. 10)

sole proprietorship: A business that is owned by one person who makes all the decisions and is responsible for all business debts. (Ch. 17)

sound energy: The kinetic energy of sound waves. Sound energy moves at about 1100′ (331 m) per second. (Ch. 10)

static load: A load that is unchanging or that changes slowly. (Ch. 6)

stator: Loops of wire, or coils, mounted around the inside of a generator housing. (Ch. 11)

stays: Cables that support a bridge deck from above. (Ch. 6)

steam turbine: A heat engine powered by steam. (Ch. 9)

step-down transformer: A transformer that reduces the voltage and increases amperage in an electrical wire. (Ch. 11)

step-up transformer: A transformer that increases the voltage and decreases amperage in an electrical wire. (Ch. 11)

stock: A "piece" or share of a company that is purchased by stockholders. (Ch. 17)

stockholders: People who own stock in a corporation. (Ch. 17)

storage: In communication systems, an electronic device that retains information until it is needed. (Ch. 13)

storyboard: A group of illustrations or photos that help a design team communicate design ideas or the proposed functions and purpose of a specific design idea. (Ch. 3)

strain energy: The energy contained in an elastic material that has been stretched or compressed from its original shape. (Ch. 10)

structure: Something that encloses and defines a space; also, an assembly of separate parts that is capable of supporting a load. (Ch. 6)

strut: A rigid structural member that is in compression. (Ch. 6)

style: A feature or quality that is typical of designs created by a specific person or during a specific time period. (Ch. 2)

subassemblies: Components of an assembly that consist of several individual parts that are preassembled before the final product is assembled. (Ch. 15)

subfloor: A covering over joists that supports other floor coverings. (Ch. 7)

subsystem: A smaller system that operates as part of a larger system. (Ch. 7)

sulfur dioxide (SO$_2$): A pollutant that is generated by power plants, industrial processes, and large diesel engines such as those used in locomotives and ocean-going ships. (Ch. 16)

superconductor: A quality of a material that allows current flow without resistance at low temperatures. (Ch. 12)

surface: Any figure that has length and width but no thickness. (Ch. 7)

suspension bridge: A bridge in which the deck is suspended (hung) from hangers attached to a continuous cable, which passes over towers and is anchored to the ground at each end. (Ch. 6)

sustainability: Using resources to meet current human needs, while preserving the environment so that these needs can continue to be met in the future. Sustainability considers the potential for long-term maintenance of human well-being. This, in turn, depends on the well-being of the natural world and the responsible use of natural resources. (Ch. 7)

sustainable lifestyle: Doing things such as eating, playing, exercising, going to school or work, and performing other daily activities while causing little or no harm to the environment. (Ch. 16)

sustainable manufacturing: The use of processes designed not only to reduce waste, but also to conserve resources such as electrical energy and water. Products designed and produced sustainably also consider what happens to the product after it is no longer useful. (Ch. 15)

switch: A device that allows a circuit to be turned on and off. (Ch. 12)

symbol: A simple picture or shape used as a means of communicating without using words. (Ch. 3)

synthetic: A human-made material; often something that was originally natural and has been treated and changed chemically to become a synthetic material. (Ch. 4)

system: A series of parts or objects connected together for a particular purpose. (Ch. 7)

T

teamwork: Working well with others to accomplish a common goal or task. (Ch. 17)

technologist: A person who uses the laws of nature to solve problems by designing and making products or structures. (Ch. 1)

technology: The knowledge and process of using the laws of nature to solve problems by designing and making products or structures. (Ch. 1)

telecommunication technology: Technology developed to send and receive information over a distance. (Ch. 13)

telemedicine: The practice of using the Internet and other communication devices to transmit medical data between patients and medical organizations. (Ch. 14)

tension: A pulling force. (Ch. 6)

tertiary color: A color obtained by mixing a primary color with a secondary color. (Ch. 2)

texture: A design element that determines the way a surface feels or looks. (Ch. 2)

thermal conductivity: A measure of how easily heat flows through a material. (Ch. 4)

thermal energy: Energy that occurs as a material heats up and its atoms become more active. Also called *heat energy*. (Ch. 10)

thermal expansion: Expansion of matter caused by heat. (Ch. 4)

thermoplastics: Plastics that can be repeatedly softened by heating and hardened by cooling. (Ch. 4)

thermoset plastics: Materials that assume a permanent shape once heated. (Ch. 4)

thrust: Pushing power; based on the principle that for every action there is an equal and opposite reaction. (Ch. 9)

tie: A rigid structural member that is in tension. (Ch. 6)

tolerance: The amount by which a part can vary in size or shape and still be interchangeable with similar parts. (Ch. 15)

torque: A measure of turning effort. (Ch. 8)

transistor: An electronic device with three terminals: a base, emitter, and collector. Voltage applied to the base allows transistors to act as electronic switches. (Ch. 12)

translucent: An optical characteristic of a material that allows some light to pass through it. (Ch. 4)

transmission lines: Lines that transport high-voltage electricity from the generating station to a step-down transformer at an electrical substation. See also *distribution lines*. (Ch. 11)

transparent: An optical characteristic of a material that allows all light to pass through it. (Ch. 4)

transplants: Living organs, usually from a human donor, that replace faulty organs. (Ch. 14)

transportation: The process of moving people or material from one place to another. (Ch. 9)

transportation system: An organized set of coordinated modes of travel within a set area that are usually run by the local government; also refers to an internationally coordinated collaboration of businesses utilizing land, air, and water transportation vehicles to deliver passengers, foods, and other goods. (Ch. 9)

triangulation: The process of sending a signal to three different satellites. The time it takes the signal to bounce off each satellite and return to the sending device is then computed to identify the exact location of the sending device. (Ch. 13)

troubleshoot: Systematically eliminating possible causes to a problem to isolate the actual cause. (Ch. 2)

truss: A structural element made up of a series of triangular frames. (Ch. 6)

turbine: An energy converter that includes a large wheel with blades turned by water, a jet of steam, or hot gases. (Ch. 9)

turbofan: A type of jet engine in which the gas stream drives a large fan located at the front of the engine. Thrust is as great as that in a simple jet, but the engine is quieter. (Ch. 9)

turboshaft: An engine that uses a stream of gases to drive turbine blades connected to a shaft which, in turn, is connected to rotors or propellers. (Ch. 9)

U

unity: A sense of belonging or similarity among the parts so that they fit well together visually. (Ch. 2)

V

vaccines: Weakened or dead pathogens that are introduced to the human body to stimulate an immune response. The body creates antibodies to the pathogens to prevent the person from infection by the pathogen. (Ch. 14)

velocity: The speed of an object; also the rate of change in an object's motion. (Ch. 8)

views: In orthographic projection, a drawing of the front, top, or right side of an object taken at right angles to the surface. (Ch. 3)

virtual reality: An artificial environment provided by a computer that creates sights, sounds, and sometimes a sense of touch through the use of a head-mounted display or glasses, speakers, and gloves or other sensory devices. (Ch. 3, 13)

viscosity: The thickness of a liquid, caused by the friction of its molecules rubbing together. (Ch. 8)

Voice over Internet Protocol (VoIP): An Internet protocol that converts voice signals into a digital format that can be sent over the Internet, allowing people to use their computers and the Internet to place telephone calls. (Ch. 13)

volt: A unit of measurement for electrical pressure. (Ch. 12)

voltage: A measure of electrical pressure. (Ch. 11, 12)

voltaic cell: A simple cell consisting of copper and zinc rods immersed in a weak sulfuric acid solution. (Ch. 11)

W

wall studs: Vertical framing members to which gypsum board, paneling, or other wall coverings are attached. (Ch. 7)

watersheds: Areas of land that drain into a lake or river; watersheds filter many kinds of pollutants, including pesticides and metals, out of water. (Ch. 16)

watt: A unit of measurement of the work performed by an electric current. (Ch. 12)

Watt's law: The law of electricity stating that power equals voltage times current or current squared times the resistance. (Ch. 12)

wavelength: The length of one wave cycle. (Ch. 10)

wedge: A simple machine that consists of two inclined planes placed back-to-back. (Ch. 8)

wetlands: Land areas that are filled with water, such as marshes and swamps. These areas filter many kinds of pollutants, including pesticides and metals, out of water. (Ch. 16)

wheel and axle: A simple machine that is a special kind of lever; effort applied to the outer edge of the wheel is transmitted through the axle. (Ch. 8)

Wi-Fi™ technology: A wireless technology often used to provide Internet access in public places such as restaurants and airports. (Ch. 13)

work: A measurement of the amount of effort needed to change one kind of energy into another. (Ch. 8)

work cell: All of the equipment and processes used to make a specific part in lean manufacturing systems. (Ch. 15)

work ethics: Responsibilities you take on when you accept a job, such as getting to work on time and being dependable. (Ch. 17)

workplace skills: Personal skills such as communication skills and the ability to get along with other people. Also known as *soft skills*. (Ch. 17)

World Wide Web: The processes that use HTTP to transmit information over the Internet. (Ch. 13)

Z

zero defects: A philosophy that no defects are acceptable; parts are to be made right the first time. (Ch. 15)

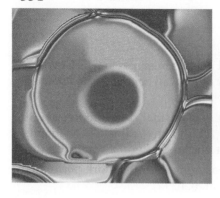

Illustration Credits

Chapter 1
Technology and You—An Introduction
Animas Canada, Figure 1-7
Better Energy Systems Ltd., Better by Design (bottom right)
Ecritek, Figures 1-4, 1-5, 1-6
Fishbol, Better by Design (top left)
Jonano.com, Better by Design (top right)
Karim Rashid, Inc., Better by Design (bottom left)
www.uisreno.com/~photography/, Figure 1-8

Chapter 2
Generating and Developing Design Ideas
Christopharo, Figures 2-20B, 2-23A, 2-26A, 2-28
Ecritek, Figures 2-1, 2-3, 2-11, 2-16, 2-19B, 2-20A, 2-21A, 2-23B, 2-29, 2-32
Fishbol, Figure 2-39
IDEO, Better by Design
TEC, Figures 2-17A, 2-19A, 2-22B, 2-26B

Chapter 3
Communicating Design Ideas
Chris Hardwicke, Figure 3-6
Ecritek, Figures 3-1, 3-3, 3-36
IDEO, Better by Design
Nadia Graphics, Figures 3-4, 3-23, 3-27, 3-28, 3-29, 3-30
Philippe Soucie: Riverboard, Figure 3-9

Chapter 4
Materials
Alcan, Figure 4-28
Canstar, Figure 4-43
Christopharo, Figure 4-22
Ecritek, Figure 4-4
Karim Rashid, Inc., Better by Design (top, bottom left), Figure 4-1
Milovan Knezevic, Better by Design (bottom right)
Mirel™ Bioplastics, courtesy of Telles, a Metabolix and Archer Daniels Midland Company joint venture, Figure 4-36
Nadia Graphics, Figures 4-2, 4-11, 4-14, 4-15, 4-16, 4-17, 4-18, 4-19, 4-20, 4-29, 4-32, 4-39, 4-40, 4-41
NFB, Figure 4-9
PBHS.com, Figure 4-46
Pellerin Studios, Figures 4-5, 4-6, 4-8

Chapter 5
Processing Materials
Ecritek, Figures 5-19, 5-26,
Jonano.com, Better by Design
Nadia Graphics, Figures 5-1, 5-2, 5-3, 5-4, 5-5, 5-6, 5-8, 5-9, 5-10, 5-11, 5-12, 5-13, 5-17, 5-18, 5-20, 5-22, 5-23, 5-24, 5-25, 5-28, 5-29, 5-30, 5-31, 5-32 (left), 5-33, 5-34, 5-35, 5-36, 5-37, 5-39, 5-40, 5-41, 5-43, 5-44
TEC, Figures 5-21, 5-32 (right)

Chapter 6
Structures
Cervelo, Figure 6-5
Christopharo, Figures 6-1, 6-7
Earl Carter, Better by Design
Ecritek, Figures 6-2, 6-4, 6-25B, 6-27
Nadia Graphics, Figures 6-8, 6-9, 6-10, 6-11, 6-12, 6-14, 6-15, 6-16, 6-17, 6-18, 6-19, 6-20, 6-21, 6-22, 6-23, 6-25A, 6-26, 6-28
Pellerin Studios, Figure 6-29
TEC, Figures 6-3, 6-13

Chapter 7
Construction
Ecritek, Figures 7-3B, 7-7B, 7-8B
Green Planet Homes, Better by Design (bottom left)
Nadia Graphics, Figures 7-2, 7-3A, 7-4, 7-5, 7-6, 7-7A, 7-8A, 7-9A, 7-10, 7-11, 7-12, 7-15, 7-16, 7-17, 7-20, 7-21, 7-23, 7-25B, 7-27, 7-32, 7-33
Rena Upitis, Better by Design (bottom right)
Tom Severson, Figure 7-1
Wintergreen Studios, Better by Design (top)

Chapter 8
Machines
Charles Harrison, Better by Design
Ecritek, Figures 8-35, 8-48B, 8-58A
Hewitt, Figure 8-30
Jack Klasey, Figure 8-58B, C, D
Nadia Graphics, Figures 8-1, 8-2, 8-4, 8-5, 8-6, 8-7, 8-8, 8-9, 8-10, 8-11, 8-12, 8-13, 8-14, 8-15, 8-18, 8-20, 8-21, 8-22, 8-23, 8-24, 8-28, 8-31, 8-33B, 8-34, 8-37, 8-38, 8-39, 8-46, 8-49B, 8-50, 8-52A, 8-53, 8-54, 8-55, 8-56, 8-59, 8-60A, 8-64
NFB, Figure 8-62
TEC, Figures 8-41, 8-42, 8-43, 8-44

Chapter 9
Transportation
Alta Velocidad Española, Dewet, Figure 9-5
Bell Helicopter, Figure 9-29
DG Flugzeugbau GmbH, Figure 9-12
Easy Racer, Figure 9-18
Eclipse solarcar team, École de technologie
supérieure (ÉTS), Montreal, Figure 9-35
Ecritek, Figures 9-1, 9-3, 9-11
Ford Motor Company, Figure 9-22A
Herman van Ommen, Better by Design
Jack Dysart, Figure 9-9
McWethy, Figure 9-4
Nadia Graphics, Figures 9-28A, 9-30A
NASA, Figures 9-38, 9-39, 9-40
Polaris, Figure 9-22B
Pratt and Whitney, Figure 9-25
Tesla Motors, Inc., Figure 9-33
Toyota, Figure 9-32

Chapter 10
Energy
Ecritek, Figure 10-25
Nadia Graphics, Figures 10-1, 10-4B, 10-7, 10-13,
10-17, 10-20, 10-21, 10-22, 10-26, 10-27, 10-28, 10-29,
10-30, 10-32, 10-33, 10-35, 10-36
NASA, Figure 10-23
Pellerin Studios, Figures 10-11, 10-19, 10-37
Tom Rielly, Better by Design

Chapter 11
Electricity and Magnetism
AEC, Figures 11-4, 11-5
Better Energy Systems Ltd., Better by Design
Ecritek, Figure 11-34
Nadia Graphics, Figures 11-5, 11-6, 11-7, 11-8, 11-13,
11-14, 11-15, 11-16, 11-17, 11-20, 11-21, 11-22, 11-24,
11-25, 11-26, 11-27, 11-28, 11-29, 11-30, 11-31, 11-33,
11-35, 11-36, 11-38, 11-39, 11-41, 11-42
TEC, Figures 11-37, 11-40
Transalta, Figure 11-32

Chapter 12
Using Electricity and Electronics
Leah Buechley, Better by Design
Nadia Graphics, Figures 12-1, 12-2, 12-3, 12-4, 12-5,
12-6, 12-8, 12-9, 12-10, 12-11, 12-12, 12-13, 12-14,
12-15, 12-16, Science Application, 12-21, 12-22,
12-23, 12-24, 12-26, 12-27, 12-29, 12-32, 12-33, 12-35,
12-38B, 12-40, 12-41
TEC, Figure 12-25
Tom Severson, Figure 12-37

Chapter 13
Information and
Communication Technology
Celluon, Inc., Figure 13-20
John W. Rauchly Papers, Rare Book & Manuscript
Library, University of Pennsylvania, Figure 13-16
Nadia Graphics, Figure 13-1
One Laptop per Child, Better by Design
University of Michigan Virtual Reality Laboratory,
Figures 13-25, 13-26

Chapter 14
Agricultural, Medical, and Biotechnology
Armstrong Healthcare Limited, Figure 14-8
Blake Kurasek, Better by Design (top, bottom right)
Dickson Despommier, Better by Design (bottom left)

Chapter 15
Manufacturing and Production
Ecritek, Figure 15-8
Ford Motor Company, Figures 15-6, 15-7
Nadia Graphics, Figures 15-1, 15-4
Pellerin Studios, Figure 15-21
Tony Cenicola, Better by Design

Chapter 16
Technology and the Environment
Vestergaard-Frandsen, Better by Design

Chapter 17
Preparing for Your Career
Ariel Shlien/The Mad Science Group, Better by Design

Appendix B
Safety
CADD Studio, Figure B-4

Index